PEBBLES ON THE COSMIC SEASHORE

A LIFETIME OF ASTRONOMICAL DISCOVERIES

Charles M. Safford

Splash Books
Reno, Nevada

Pebbles on the Cosmic Seashore: A Lifetime of Astronomical Discoveries

For information or additional copies contact author: csafford@pershing.k12.nv.us

Library of Congress Control Number: 2019919265

ISBN: 978-1-938814-71-6

To my wife and best friend, who supports me in all my undertakings and whose kind forbearance during this project was key to its completion; and to my children and grandchildren, who may one day journey among the celestial wonders described herein.

Charles M. Safford

Table of Contents

Acknowledgments

This book's chapters were planned as a series of standalone essays that were originally intended to form the basis of an astronomy class at Coal Canyon High School in Lovelock, Nevada. In the course of this three-year writing project, the manuscript, as they say, took on a life of its own, and it became clear to the author that the essays could come together to form a book-length work.

Numerous individuals contributed to the successful completion of this book. Specific mention is due of the generous and skillful feedback offered to the author by Dr. Elizabeth A. Harvey, Ms. Shari Shirley, and Sandra M. Safford-Olsen.

A number of volunteer proofreaders among the students at Coal Canyon High School examined the chapters as they were produced and made invaluable suggestions. Many were members of a writing class at Lovelock Correctional Center, in Lovelock, Nevada. Students in the Coal Canyon High School art class supplied a number of beautiful illustrations and computer graphics. Their willingness to help with the project despite their own daunting daily challenges was a continual source of wonder to me.

Mistakes that remain after the help of the many reviewers and proofreaders are solely the responsibility of the author.

Special thanks are due to the National Aeronautics and Space Administration, for the creation of their truly wonderful library of images.

The author wishes to express thanks and admiration for the numerous ingenious men and women over the course of the centuries who have dedicated themselves to the challenging and noble enterprise of communicating scientific knowledge to the general populace. Among the luminaries, living and dead, whose works are exemplars of scientific exposition for popular audiences, and to whom I wish humbly to express particular gratitude—not presuming to number myself among them—are: Carl Sagan, Neil deGrasse Tyson, Michio Kaku, Jacob Bronowski, and Brian Cox.

The author is deeply indebted to Christopher Myers, whose brilliance and countless contributions transformed *Pebbles* into a book possessing power, eloquence, and artistry.

Thank you all.

Charles M. Safford

Lovelock, Nevada, 2019

Introduction

Pebbles on the Cosmic Seashore: A Lifetime of Astronomical Discoveries is a beautifully written introductory work on the science of astronomy. A compilation of original essays intended for the enlightenment of a general audience, Pebbles on the Cosmic Seashore is divided into two parts. In Part I, the author transports the reader on a temporal, and at times poetic, voyage of discovery. Hypothesizing that the science of astronomy emerged out of the sense of awe with which paleolithic peoples regarded the heavens, he traces its advance into the realms of mythopoetic thought, in which the motions of the heavenly bodies were attributed to gods and enshrined in myth. Charting this transformation through such intellectual milestones as Stonehenge and the pyramids, monumental forms of architecture that helped our ancestors to map and at least partially understand the motions of the planets and the stars, the author continues the story of astronomy into Greco-Roman times. Moving deftly from the achievements of the Classical Age to the astronomical discoveries of the Scientific Revolution, and then on to astronomy's breathtaking achievements in our own times, the author completes Part I with a discussion of the physical world's underlying reality and the beauty of the cosmos. Part II is a magisterial depiction of the wonders of the Solar System and beyond. Opening with a description of the startling beauty of the Earth, "our home," Part II then introduces its readers to the earth's closest neighbors within the solar system. It then moves on to a discussion of the constellations of stars that lay beyond our own region of the heavens, and, finally, suggests that someday humankind will explore this greater world so seemingly at present beyond our reach. Humanity, the author believes, is perched upon the cusp of a great decision: Should we explore our universe with sentient beings (ourselves), or continually hurl finely-crafted, heroic probes into the heavens, on one-way trips to reconnoiter the Stygian depths of surrounding space. No matter which choice humanity selects, our future will be intertwined, as our fate was in the past, with the investigation of the heavens. The magnificence and grandeur of this quest, detailed in the pages of Pebbles on the Cosmic Seashore, will entrance and captivate the readers of this text.

Elizabeth Harvey, Ph.D.
Professor of History, Merced College

Preface

Whirlpool Galaxy

Astronomy compels the soul to look upwards
and leads us from this world to another.
Plato

Studies serve for delight, for ornament, and
for ability.
Francis Bacon

Equipped with his five senses, man explores
the universe around him and calls the
adventure Science.
Edwin Hubble

The book you are reading is the culmination of a passionate three-year pursuit of excellence and fulfillment by the author, standing deeply humbled before his subject matter, to assemble and present a distillation of facts, ideas, anecdotes, and tales relating to astronomy. By entertaining, educating, and enlightening, the author hopes to contribute to the dissemination of knowledge and traditions about astronomy while promoting a better future for humanity.

When we incline our eyes to gaze upon the night sky, we behold something genuinely amazing—figures etched in light parading across the cold, black depths of space. Like a young child exploring its ever-expanding universe, we are all explorers on an unending voyage of discovery. But it was not always so. Our ancient predecessors were tiny, inconsequential creatures that cringed at loud noises, cowered in shallow burrows to hide from rampaging thunder-lizards, and timidly crept through the underbrush only under cover of darkness. Today, we have become bold seekers of light as we confidently roam the entire Earth. We hike the open plains, frolic in vast oceans, and climb the tallest mountains simply because they are there. We have developed into remarkable beings capable of flying over the loftiest of forest canopies, soaring far above the tallest peaks, and even leaving our atmosphere behind as we sail through the void of space towards new horizons. We observe the mighty Sun's majestic prominences, its cold spots, its twisting magnetic fields, and its startling eclipses. We have learned that the electromagnetic spectrum is a resplendent repository packed with information, and visible light, which constitutes only a small portion of the spectrum, is but one volume in this expansive library of light. Therefore, we build magnificent telescopes that enable us to capture data about all the varieties of radiation that impinge upon us from space. From that accumulated data we are able to deduce truths about the composition, power, and nature of the universe. Ambitious dreamers, we unabashedly thirst for knowledge, delight in discovery, and dare to envision voyaging among the stars to distant abodes.

We have an intimate, profound connection with the stars. As Carl Sagan often said, "We are made of starstuff." Thus, the roots of our family tree stretch back to the origin of the universe. In that distant epoch, near the beginnings of time and space, the lightest atoms and the most powerful forces of the cosmos burst into existence. Relentlessly, inexorably, those forces seized control and created the first stars. At the core of those massive celestial bodies, lighter atoms were compressed into heavier ones in a process called heliosynthesis. A byproduct of this process was the production of heat and light, which began to course through the universe. Eventually those first stars ran out of fuel. At the end of their death throes, they exploded, hurling newly constructed, more-complex matter far out into space, where it slowly reaggregated to become a more fertile seedbed—a seedbed enriched with carbon, nitrogen, oxygen, and other vital elements. These newly formed elements combined and recombined to produce a warm, hospitable, and auspicious environment containing soil, water, air, and organic compounds—the primordial ingredients from which life arose.

The simplest life forms emerged from this primitive broth, grew, replicated, combined, and mutated into more complex ones. Life in its varied, versatile, and robust manifestations learned to swim, fly, run, hunt, graze, and mate. But life never stays the same—life evolves. Today, it has evolved to the point where it can not only contemplate its own existence; it can also laugh, cry, and experience the sublime beauty of itself and its surroundings. We humans have now become much more than the sum of our elemental components. At this stage of our journey, we have developed the ability to appreciate physical grace, potent intuition, and deductive thought.

We utilize those innate abilities in our ongoing, quintessentially human quest for self-fulfillment, and a crucial component of fulfillment is the search for knowledge and understanding of our surroundings. As we examine the universe through enhanced senses, our keen and probing minds seek to grasp all its seemingly mechanistic and chaotic components. Eager to penetrate the mist-shrouded inner workings of the cosmos, our creative minds generate leaps of imagination as we search for clarity. Our centers of logic, honed by years of daily practice and experience, perceive the subtlest nuances and formulate increasingly clever, useful, and rational explanations for the phenomena that envelop us in their exquisite celestial beauty.

While humans pursue all types of knowledge, astronomy in particular holds a powerful, almost mystical appeal. Long before we ever wiggled our way out of a salty ocean, our atoms were synthesized inside stars, and those original stars had collected their substance from the disparate particles and streaming energies of the infant universe. Thus we can directly trace our ancestral lineage to the first moments of the cosmos. Astronomy's charm stems, at least in part, from this awareness of our ultimate connectedness with the stars. From the innermost depths of our being, we wish to increase our knowledge of our cradle, our mother, our home: the cosmos. The placid lights that adorn the night sky are beacons emphatically blazing forth the truth of our origins. Their mere presence urges us to seek out our origin and discover our future among the stars.

Like mathematics and physics, which sprang from it, astronomy stirs our spirit, intensifies our curiosity, and provides a grand opportunity to rove physically and mentally among undiscovered realms. Furthermore, astronomy has the power to make us happy. Contemplating the awesome majesty of Saturn, the overwhelming grandeur of the Pillars of Creation, and the silver-tinged iridescent glow of the Milky Way, we are filled with joy as we stare like children beholding our first Christmas tree all lit up. Likewise the elegance, symmetry, and logic of Newton's laws of motion, Einstein's special and general theories of relativity, and Hubble's Law of universal expansion—to name just a few of the most impressive and successful astrophysical models—move us in mysterious and ineffable ways. They appeal to the innermost sense of beauty and order within the mind. A concise mathematical expression, derived from the simplest of axioms, can describe diverse interactions of the elements of nature with precision, consistency, and cohesiveness, and this creates a sense of satisfaction and enjoyment that can approach euphoria.

Astronomy is beauty, logic, and science, but it is much more. The glittering ornamentation of the night sky; the smoothly flowing logic of a mathematical demonstration; the accouterments of natural philosophy; all of these are key aspects of this uniquely human endeavor which has transformed campfire fables into scholarly dissertations. Through rigorous examination of the matter and light of the universe, astronomy becomes a treatise to be perused, scrutinized, and comprehended. It typifies and embodies the collective striving for knowledge that unites and distinguishes us as human beings. We are explorers, and astronomy satisfies our deepest yearnings for exploration as we become enthusiastic pathfinders and patient sojourners on our way towards greater understanding of the stars, galaxies, space, and, ultimately, ourselves.

Part I
Quintessentially Human
(Chapters 1 - 9)

1
Why Study Astronomy?

M81

I would rather discover one cause than be king of Persia.
Democritus

To give man ever more knowledge of the universe and 'to help him to learn humility and to know exaltation,' that is the mission of astronomy.
Robert Aitken

My goal is simple. It is a complete understanding of the universe.
Stephen Hawking

There is no need to try to encourage anyone to study astronomy. Every human being naturally and automatically studies what is in the sky. Children's natural curiosity tugs their eyes upward, and the sight of a full Moon in the daytime produces such delight that a three-year-old will exuberantly point and proclaim, "There's the Moon!" Even very young humans are studying astronomy as they behold the Moon arcing across the sky, the Sun as it illuminates the dancing dust motes through the schoolroom window, and the pretty stars that paint portraits across the heavens after bedtime. A child seeing these things is studying astronomy.

It is a rare person indeed who can stand beneath the stars without experiencing awe and joyful amazement. Virtually everyone adores those picturesque formations and admires the milky, swirling phenomena that transpire just within our range of vision. Humanity earnestly seeks understanding, revels in the knowledge gained with each new discovery, and yearns to listen to the "music of the spheres." We long to become one with the nearby stars, the distant nebulae, and the far-flung archipelagoes that were once known as "island universes" but are now called galaxies. In essence, we wish to comprehend the entire cosmos.

The study of astronomy is an expression of one of the most fundamental, innate, and universal traits of the human species: the urge to know. This urge has led humankind repeatedly throughout its history to redefine how it perceives the universe and the place that human beings occupy within it. The seeker who approaches the banquet table of astronomical delights is presented with a scene that tickles the taste buds of the imagination while fueling the mind. By careful observation of the universe around them, and by using logic and reason to draw conclusions from what they see, humans have enhanced all aspects of their lives. We no longer struggle merely to survive. Employing science,

Jet Plane and Moon

which has been inspired in large part by the study of astronomy, we have extended our lifespans, enriched our lives, and expanded our potential.

From the initial trickles of astronomical data accumulating drip by drip through the generations, the original stream has become a raging torrent as facts accrue at an accelerating rate. Today, data about the universe around us are pouring in from huge telescopes, speedy space probes, and sophisticated instrument-platforms orbiting the Earth. It is as if human culture and civilization are being carried on a spherical vessel in the center of an Amazon-like river of stellar information. From the middle, the opposite banks are too distant to be seen, yet the mighty current pushes us with irresistible force toward the vast ocean of cosmic knowledge yet to be discovered. Knowledge comes to our bobbing vessel encoded in far-traveling waves of light and gravity, some so faint that they can be detected only by exquisitely clever and sensitive instruments. Therefore astronomers strive constantly to devise better tools and techniques for seeing farther out into space and farther back in time so that they may occasionally find a "smoother pebble" like the ones Newton so eagerly sought beside his sea of undiscovered truth.

Crab Nebula

Advances in astronomy enrich human society by augmenting our understanding of the universe. When Copernicus put forward the idea that the Earth was not the center of the universe, he revolutionized humanity's view of the Solar System. Although some thought at the time that this diminished the importance of humankind, in fact, Copernicus's paradigm-shifting vision was a step on the path to greater comprehension of humanity's place in the cosmos. By looking to the heavens, Copernicus was able to upend millennia of dogma, and he was not the only one. Such intellectual revolutions rarely happen, but when they do, they are transformational moments of unsurpassed importance in history. A similar revolution occurred in ancient Greece when the philosophers first sought to explain the universe without recourse to the actions of the gods. Likewise, Einstein in the twentieth century melded time and space into a new universe, a previously undreamed-of non-Euclidean fabric that stretches and deforms under the influence of the stars and galaxies spread throughout its expanse.

When Newton grafted a falling apple into a unique and ultimately brilliant new vision of how planets and stars interact with each other across the vast gulfs of space, he was bringing astronomy down to the orchard and teaching a lesson that would be reiterated throughout future centuries. Every time an airplane flew, a rocket launched, a space probe rode the gravitational corridors to a distant ringed planet, or a slice of jelly-covered toast spun down onto a carpeted floor, they were moving in accordance with the laws of motion that Newton devised.

Humanity's survival through its primitive stage was at least partially due to knowledge of astronomy. For the procurement of food, the seasons marked by the motions of the heavenly lights had to be strictly observed. As the ages rolled on, humankind steadily improved its agricultural production, its skills at navigation, and its ability to communicate between regions, peoples, and nations. All of these advances were ultimately linked to advances in astronomy.

With each advance, humanity's lifestyle grew more productive and comfortable. The following are just a few of the technological innovations designed for astronomy that have been adapted for alternative uses to positively impact the lives of human beings.

Pillars of Creation

Much of our current imaging technology was initially developed for astronomical purposes. The ubiquitous digital camera seemingly in every pocket and smartphone was invented for use in advanced space-based telescopes. The technology on which the digital camera is based, called the charge coupled device, was first employed in astronomy in 1976. In the area of medical treatment, many of the advanced imaging machines that enable doctors to non-invasively view internal organs arose through the process of astronomical research. Astronomers and medical practitioners face a similar challenge: how to observe the small, the dim, and the hidden—and to do so unobtrusively. The astronomer's quarry is the unusual or thought-provoking stellar object, often hidden in dust or so far away as to be frustratingly faint, while the targets of the physician's attention are the disease-causing abnormalities concealed deep inside the human body. In either case, the task is to use technology to make visible the invisible. In one of the most impressive examples of success in this activity, radio astronomers developed a technique for combining multiple inputs from more than one receiver to compile a single higher-resolution image. CAT scanners and MRI machines now routinely use the same technique.

Today, we are accustomed to plotting our way to a distant location or knowing our exact position to the nearest foot by consulting our electronic maps and automobile navigation systems. Stopping at a service station to ask for directions has now become as much of an anachronism as the dial telephone or the rawhide buggy whip. This is possible because a fleet of specialized satellites encircles the Earth at all times, linking huge networks of GPS devices and supplying the world with a steady stream of up-to-the-second data.

In the security realm, chemical detectors invented for interplanetary exploration are sniffing out specific molecules in airports to detect explosives or other harmful substances. In the area of energy, solar panels and specialized batteries developed for space applications have proved immensely useful for meeting energy needs on Earth and will surely contribute even more in the future.

Long-distance communication once consisted of whistles piercing the air, drumbeats reverberating through the jungle, or smoke signals rising above the desert. Now when we pick up a phone or watch a broadcast from the other side of the world, we are effortlessly and routinely bouncing signals off a satellite orbiting 22,236 miles above the Earth's Equator. We can even communicate beyond the boundaries of the Solar System, across billions of miles to the Voyager probes as they careen toward other star systems.

Astronomical knowledge will assuredly prove of vital importance for the long-term survival of human life on Earth. The region of space near our beloved home planet is peppered with a large number of speeding, spinning, jagged, stony, dense missiles heading in our general direction. These gyrating chunks of destructive doom, otherwise known as asteroids, have bombarded the Earth in the past, and it is only a matter of time before the next cataclysmic impact occurs. Astronomy has revealed this dire circumstance, and much more research and effort will be necessary before we will be able to claim that we have defused it. If there ever was a reason to study astronomy, this life-and-death issue is it. The asteroid that struck the Yucatán Peninsula 65 million years ago and exterminated the dinosaurs was just one of many that have hammered our planet in the past. If an asteroid of that size hit today, it would end human civilization. There are steps which must be taken. We need to improve detection so that we can see the dark, nearly invisible asteroid menace as it approaches. For this purpose, among others, we study the sky with radio waves, which fly out into space at the speed of light and are subsequently reflected back to us by approaching asteroids. However, relatively few of Earth's radio telescopes are dedicated to asteroid detection. In addition to detection, we need to evaluate the many interdiction tactics that could be tried when the inevitable asteroid emergency arises. Gravity tractor? Laser beam nudges? Nuclear missiles? The survival of humanity could depend on it. In the words of Neil deGrasse Tyson director of the Hayden Planetarium and host of *Star Talk*, "Asteroids are nature's way of saying, 'How is that space program coming along?'" Imminent danger will galvanize the nations of the Earth, spurring them on to ever-greater exertions in research and innovation.

NGC 4302 and NGC4298

Galaxy NGC 1365

Space will become a rich source of materials and energy as soon as human knowledge and capability rise to the level where we can exploit it. The resulting surpluses could contribute to peace on Earth, since competition for scarce natural resources would no longer provide impetus for warfare. Scientists envision sailing to nearby asteroids, mining them, and returning to Earth laden with cargoes of the newly acquired minerals and metals, which will surely invigorate economies and contribute to the improvement of the quality of life for people of all nations in a new era of prosperity and cooperation.

The advantages of studying astronomy are manifest, but encompass more than the merely practical. For millennia, the stars have tugged on the human mind, pulling it out of the mundane and drawing it upward toward the wonders of the heavens. The presence of those sublime, mysterious, luminous entities is a continuous reminder that we occupy this cosmos in company with things much greater than ourselves. We are made of a minimal amount of matter in a humdrum neighborhood on an out-of-the-way spur of one spiral arm of a rather normal, if lovely, galaxy that twirls through space not giving a twinkle for our lives and all our grand achievements. The stars teach us humility. Yet as they gleam above, they also beckon us by their very loftiness toward exaltation. This combination of effects is unique in all the fields of human activity and produces feelings and thoughts akin to the spiritual.

Soaring above our elevated cathedral mountains to the edge of the universe, we may find ourselves traversing forsaken voids on the way to explore the tenuous filaments of galaxies. From the phoneless solitude of the country roads of West Virginia to the Great Sloan Wall, from Big Sky Country to the Coal Sack, beyond those distant hinterlands lies the unknown. We tread between starlight shadows paler than an astronomer's dreams and see fire raining down on airless worlds so hot that rocks melt like wax while our intrepid probes skirt the volcanic eruptions of icy moons and dart nimbly between the steamy plumes that fall flash-frozen in icy shards shinier than Murano glass. We record the pulsing vibrancy of wildly spinning gravity wells, powerhouses of energy firing colossal fountains of matter at a significant fraction of the speed of light across vast reaches of space. Eventually these tremendous jets ripple through clouds of dust and gas, energizing them and causing them to glow and shimmer in greens, pinks, and other colors according to the wavelengths absorbed and retransmitted by their constituent molecules. Over billions of years, this process stimulates the production of embryonic solar systems that may one day resemble our own. Who would not long to behold such vistas? These and other even more extravagant, enticing, and exhilarating panoramas will surely open up to our view, but with our present level of technology, we cannot journey there to see them up close, except in our imagination.

The more we explore the universe, the farther back in time we peer. What we see today when we look at a distant galaxy is what it was like 13 billion years ago, because it has taken that long for the light from that galaxy to reach us. As we keep pushing the boundary, going further and further back in time, we cannot know what we will find until we actually glimpse it, nor can we predict what new wonders will arise. To find out, we must continue to seek, continue to learn. This is the essential impulse that drives us to study astronomy.

Hubble Space Telescope image of the Carina Nebula.

2
Threads of Culture

A people without the knowledge of their past history, origin, and culture is like a tree without roots.
Marcus Garvey

Nature uses only the longest threads to weave her patterns, so that each small piece of her fabric reveals the organization of the entire tapestry.
Richard Feynman

Knowledge, customs, and traditions endure through time and can be likened to threads. So, too, can living species. The varied threads of culture and nature interweave to form the vast and diverse tapestry that is humanity. Life's thread is long—it first began to be spun on nature's spinning wheel 3.5 billion years ago, forming intricate, delicate, and often deeply hidden patterns that humankind is still endeavoring to disentangle. Human culture came into being when life, after billions of years, began to express itself in human form; and the warp of human culture helps bind the weft of life.

Astronomy is but one thread of culture. This thread runs forward and back through time. It proceeds through Newton, Galileo, Omar Khayyam, Claudius Ptolemy, Aristarchus, and Democritus, to long lost and forgotten scribes and acolytes who recorded the signs from the heavens, and to storytellers, troubadours, and bards who passed on myths and legends. It also travels forward to the present through Annie Jump Cannon, Albert Einstein, Cecilia Payne, Edwin Hubble, Carl Sagan, Neil deGrasse Tyson, and a myriad of others. Finally, it concludes with you, a child of the future, who may be inspired by these essays to pursue greater knowledge and add to humankind's burgeoning comprehension of the cosmos, to utilize every scrap of knowledge for the betterment of all, and, ultimately, to contribute to the creation of our species' ultimate future on the loom of destiny.

Culture is the thread that runs through and binds one generation to the next. Out of the foibles of humanity, our ancient ancestors wove fanciful tales that enthralled children and adults alike. These same stories passed down values, beliefs, and attitudes to following generations. Thus, although each generation weaves new patterns and stretches the fabric differently, the thread of culture remains unbroken.

Each succeeding generation is heir to the legacy of wisdom bequeathed by its predecessors. Armed and fortified with values, knowledge, and shared tales, human society has expanded, thrived, and succeeded against all obstacles. These diverse elements are like disparate fibers that become stronger when spun together.

Today, this synergistic thread is kept intact through a variety of instructional methods ranging from rigorous classes taught at universities to felicitous stories of the past told by revered elders. Storytelling is an ancient and deeply embedded human impulse knitting together generations while exploring every aspect of human life. Today, our ancient astronomical culture is kept alive by seekers, keepers, and sharers of knowledge who pursue and disseminate it for the betterment of humankind. All these individuals stretch the limits of human capability through word, mind, and deed, and yet they invariably maintain their connection with their illustrious forebears. The work of each succeeding generation keeps alive the memory of the preceding generations. Treasuring one's cultural legacy is a way to pay homage to and show respect for one's esteemed ancestors.

Culture is inherited from the past, preserved and enriched in the present, and passed on to future generations. The intangible attributes of culture, including social mores, traditions, customs, aesthetics, spiritual beliefs, artistic expression, and astronomical knowledge, are resources of immense worth and their preservation is a moral obligation. A civilization that does not maintain its historical culture is like a tree that has lost a section of its roots. Such a tree will struggle in its uptake of nutrients and will become diminished. Its dying leaves may fall upon a barren field and never be replaced. So, too, a civilization that loses its intangible culture will become diminished and will find it difficult to transmit its values to its offspring.

Ancient cultures passed down their knowledge to us, bestowing upon us the spirits of their philosophies. It is our duty to retain the finer aspects of their traditions, which include the appellations they assigned to astronomical objects. In ancient times, the names given to planets were the names of gods. Undoubtedly, some among the ancients who assigned those names thought that the wandering lights whose movements through the obsidian sky they reverently studied from night to night were the very gods they worshipped in marble temples on Earth. Astronomers of later centuries preserved this thread by continuing to name planets in this manner as they were discovered. The planet Neptune, for example, was never seen by the ancients, but when it was discovered in the nineteenth century, it was named after an ancient god. The tradition continued with the naming of Pluto in the twentieth century. When the United States of America dedicated itself to the manned exploration of the Moon, phases of the plan were named for ancient gods: Mercury, Gemini, and Apollo.

Each of the three names is apt: speedy Mercury was the fastest vehicle ever flown by Americans up to that time. Gemini held two astronauts like twins. Apollo traversed the heavens while seeking wisdom. Credit for naming Mercury and Apollo has been ascribed to Abe Silverstein, the Director of Space Flight Programs at NASA at the time the programs began. In a published statement, he recounted his dawning realization, after reading a book on mythology, that the image of "Apollo riding his chariot across the sun was appropriate to the grand scale of the proposed program." According to NASA lore, the name "Gemini" was the suggestion of an attendee at a lecture given by Robert C Seamans, Jr., NASA Associate Administrator, and was almost simultaneously suggested by Alex P Nagy in NASA's Office of Manned Space flight.

The constellations, too, bear the names of heroes and other characters from mythological stories. In this way, the stars themselves serve as a visual repository of cultural knowledge. As the ancient stories continue to be passed down, we can contemplate the sky and be reminded of the millennia-old adventures and fantastic deeds of legendary heroes like Hercules, Perseus, and Orion, whose exploits appear to be woven into the sparkling blanket that covers the Earth at night.

Today we remember the constellations according to their Greek origins, but the Greeks were strongly influenced by the Babylonians who preceded them, and after the Greeks fell from prominence on the world stage, the Arabs preserved their legacy. This is why, although we remember the constellations according to the Greek stories they are associated with, many stars today bear Arabic names.

It is likely that the personages, narratives, and morals passed down to us from thousands of years ago will remain in use far into the future. If so, the dread shears of Fate will not sever our cultural connection with the past. Human beings who gaze into the sky a millennium from now will very likely find themselves reminded, just as we are today, of cultures long past, who, despite the decay and disappearance of their homes, temples, literature, and religions, preserve cultural memories arrayed in the heavens.

Our cultural identity is bound up with the lights and pictures in the sky. As we cast out various lines of discovery and exploration into the vast cosmic ocean of space all around us, it is fortunate and wise that we, as a species, preserve a strong cultural thread of our shared images of heroes and legends of yesterday. We can appreciate that this thread readily twines with the sturdy strands of mathematics and science and is solidly anchored in the bedrock of our character and the time-tested values of our revered ancestors.

Sometimes cultural memories are literally preserved as patterns in thread. A millennium ago, needle workers recorded a pivotal moment in English history. In February of 1066, the comet that would later be known as Halley's Comet soared over Europe, at a time when comets were viewed as "the terror of kings." To later chroniclers, its arrival seemed to foretell the defeat and death of King Harold II of England at the Battle of Hastings the following October. These events were memorialized on an artwork known as the Bayeux Tapestry (right), a 231-foot-long strip of woven linen containing 72 scenes embroidered in woolen yarn.

Each epoch encompasses but one embroidered scene within history's multilayered tapestry that can potentially be unwound as years pass. Thus, it is incumbent upon each generation to protect and strengthen the thread. Our children and our children's children will weave the thread they take from us into new patterns and images, but all will remain part of the same grand tapestry, built upon the past and foreshadowing the future.

3
Dogma versus Science

Crystal Spheres

In questions of science the authority of a thousand is not worth the humble reasoning of a single individual.
Galileo

Unthinking respect for authority is the greatest enemy of truth.
Albert Einstein

Science is the poetry of reality.
Richard Dawkins

As human beings, we naturally and unavoidably observe the world around us. Not only do we observe; we also seek explanations for what we observe. When we encounter the unknown, we search relentlessly until we find a satisfactory answer to our mystery. Imagine, for instance, that you are a prehistoric human hunting in the woods, alone but for your faithful canine companion, when a sudden thunderstorm develops and it starts to rain. You seek cover. While sheltering under the bough of a tree, you watch the sky darken with swollen obsidian clouds, you see far-off flashes of light, and eventually you hear the roar of distant thunder. Suddenly a nearby tree is struck with a mighty bolt of lightning, causing its limbs to shatter and its trunk to burst into flames. You hear a deafening roar and feel a powerful rush of air as you are flung through the air and knocked flat to the ground. Shaken, you huddle upon the damp needles, and your thoughts race to find the reason for these dreadful occurrences. Without an explanation, you will probably be as frightened as the dog with the tucked tail cowering and whimpering beside you.

How can you find the explanations you need for such events? There are two methods. The first and easiest is to ask someone. In our thunderstorm example, having survived the lightning strike you might seek out a respected member of your tribe and ask him for his thoughts on the topic. This person could be a shaman, chieftain, soothsayer, parent, or grandparent, but almost certainly someone of venerable age. He would most likely tell you that thunder and lightning result from the actions of the gods. He might smile gently and offer illumination by first thoughtfully and reverently intoning, "It was Thor, the god of thunder," and then declaiming with an exuberant motion, "He was pounding a frost giant with his hammer, Mjöllnir!" Some anthropologists suggest that it was in this manner, nourished by innate curiosity, that the variety of mythologies originally sprouted from the fertile imaginations of humanity.

However, there is a second way to find explanations. It is not easy, and if you embark upon it, you are certain to encounter many obstacles and difficulties. The second way is the scientific method.

In the scientific method, you use careful observation, logical reasoning, and experimentation to arrive at useful information about the world around you. In our thunderstorm example, you might observe other electrical phenomena like static electricity. You might note that the damage from a lightning strike is located farther away from you as the difference in time between the lightning flash and the thunderclap increases. At some point, you might even fly a kite weighted with an iron key during a lightning storm to see what happens. As you progress along this path, you will likely cease seeking supernatural explanations for violent weather. The conclusions you arrive at through this kind of observation and experimentation are likely to be more accurate and useful, both now and in the future, than the "truth" some well-meaning elder espouses, no matter how highly respected he is. Science becomes a way to expose the ignorance or outright deceit of those who promote falsehood as truth. Science can thus serve as a bulwark against the threats posed by delusion, nonsense, fantasy, and flimflam.

Throughout history, it has been common for the most highly respected authorities to believe in and promote erroneous ideas. The ancient author Plutarch, in the first century CE, helped disseminate the notion that there were bodies of water on the Moon. Aristotle, revered for centuries as the ultimate authority on everything, to the point that even the Catholic Church accepted his writings as perfectly reliable, believed that the Earth was the center of the universe. (See illustration at the beginning of this chapter.) He also surmised that if two weights were dropped from an equal height, the heavier weight would fall faster. Such beliefs seemed quite reasonable at the time, but we now know that they are wrong. Aristotle's mistake was that he often eschewed experimental verification of his conclusions. He could easily have dropped two unequal weights simultaneously from a tall tower and observed the results—they fall at the same speed—but he did not feel it necessary, and that experiment was not carried out until almost two millennia later by Galileo. A similar experiment was performed on the Moon in July of 1971 by Apollo 15 mission commander David Scott. He simultaneously released a one-ounce falcon feather and a three-pound geologist's hammer from a height of about five feet. Moments later, accelerated by the Moon's gravity and unimpeded by air resistance, both objects struck the Moon's dusty surface at the same instant.

Many in the past who believed in correct ideas, ideas that we now take for granted, were ridiculed, reviled, or, if they were lucky, ignored. The Greek philosopher Democritus over 2,400 years ago (even before the time of Aristotle) articulated the view that matter is composed of atoms. He also wrote about planets that were destroyed by fiery cataclysm, the presence of innumerable worlds of different sizes populating our cosmos, planets without either sun or moon, and planets with more than one sun. Today we know that all of these things exist, but in his own time, Democritus received little recognition or acceptance of his ideas. His intellectual achievements are astonishing since he arrived at his conclusions without the benefit of the advanced instrumentation and huge amounts of data available today, nor did he use the fully developed scientific method of today. Instead, he relied on pure reasoning and fortunate guesswork.

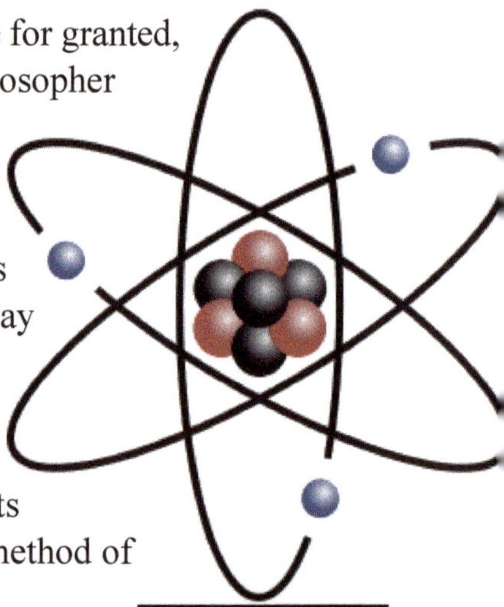

Lithium Atom

Giordano Bruno, in the late 1500s, wrote that the stars were distant suns surrounded by their own planets, which is something that today we know to be true. He also taught that the universe was infinite and therefore the Earth could not be at its center. This, as well as his denial of other accepted doctrines, brought him into sharp conflict with the Catholic Church. He was found guilty of heresy by the Inquisition and died a fiery death at the stake in Rome's Campo de' Fiori. After his death Giordano Bruno gained considerable fame, being particularly celebrated by nineteenth- and early twentieth-century sympathizers who considered him a courageous champion of free speech and unfettered scientific inquiry.

Then there was Albert Einstein. When he first put forth his surprising and counterintuitive theory of relativity, it was slow to be accepted. A book published in 1931 contained the authoritative opinions of no less than one hundred highly esteemed contributors, every single one of whom was convinced that Einstein was wrong. Einstein's response to the book was that if he was wrong, then one author would have been sufficient. Today, of course, relativity is one of the most thoroughly tested and proven ideas in all of science.

Scientific knowledge today advances as a worldwide collaborative effort of researchers and theorists based on the proven mode of thinking encapsulated by the scientific method. Any scientific idea is subject to eventual disproval if information is found that contradicts it. With unrelenting skepticism, kernels of truth are winnowed from the chaff of nonsense. This self-correcting aspect of science and the scientific approach to knowledge stands in sharp contrast to the opposite approach, which is dogmatism.

Dogmatically held beliefs are by definition not disprovable. A person who accepts a dogma accepts it without regard to any evidence either supporting or contradicting it. In fact, such a person will hold to his beliefs despite the clearest disproof. Dogmas are accepted as incontrovertible truths despite their lack of foundation. This is why today's scholars are trained not to allow themselves to be restricted by habitual modes of thought.

Dogma stifles creativity, dampens the pursuit of knowledge, and, worst of all, prevents the transmission of superior concepts and more accurate information to our children, whom we are counting upon to continue the pursuit of knowledge where we leave off. Science, in contrast to dogma, promotes creativity, enhances the pursuit of knowledge, and encourages communication between the generations for the betterment of humankind. If we as a species are to continue bettering the world for each succeeding generation, then open, honest, and vigorous sharing of knowledge will be required. In other words, science must be embraced in preference to dogma.

Unfortunately, the distinction between dogma and science is not always perfectly clear. Sometimes scientists hold on to old beliefs too long, especially if the old beliefs are associated with a towering figure of the past. For example, few scientists of the early twentieth century were ready to reject Isaac Newton's idea of gravity just because Albert Einstein proposed a better idea. The German physicist Max Planck once said that getting a new scientific idea accepted is a two-step process: Step 1: Present compelling evidence that supports your new theory conclusively. Step 2: Wait for all the old scientists to die. Planck was jesting, of course, but his point is well taken: Even scientists, with their desire for objectivity and willingness to accept new ideas based solely on evidence, often fall into the trap of dogmatism without meaning to.

Therefore, how should one proceed? Tirelessly search for the truth, record the facts, and accept the results whether positive or negative, for even a negative result can be profound. One famous example of a negative result leading to a great discovery was the Michelson-Morley experiment performed in 1887 to detect the presence of ether (or aether). Ether was the conjectured medium through which light waves were thought to travel. Water waves travel through the medium of water, and sound waves travel through the medium of air, so it was natural for scientists to believe that light waves must travel through a medium. This hypothetical medium was referred to as the luminiferous ether.

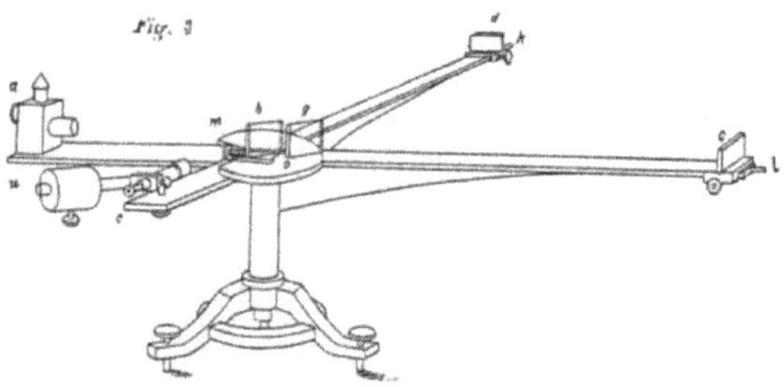

The Michelson-Morley experiment was a clever attempt to detect the velocity of the Earth with respect to this ether. The investigation relied upon a specially designed device called an interferometer (left), which was capable of measuring minute differences in the time required by light beams to traverse mutually perpendicular paths. The scientists reasoned that a light beam traveling in the same direction as the Earth's motion should move at a different speed than if the light beam were traveling perpendicularly to the Earth's motion. However, the two researchers found no difference in the speed of light, regardless of the direction in which the light was traveling. How to explain this null result? No one could. No experiment in the history of physics has roused such energetic efforts to explain its result.

The problem was not satisfactorily solved until 1905, when Albert Einstein created his theory of special relativity. We now know that there is no ether. Instead, at velocities that approach the speed of light, the fabric of space and time alters itself so that the velocity of light always remains constant. This profound truth about the invariability of the speed of light has been consistently proven correct for over 100 years, regardless of the type of experiment performed. In fact, when one travels by car, the car's GPS system is constantly using Einstein's theory as it dependably determines its location to an accuracy of less than three feet.

The constancy of the speed of light is now accepted as fact, but it is not dogma. It could be overturned or improved upon at any time. If that were to happen, even Einstein himself, from beyond the grave, would be the first to voice support for the new theory, as long as it was supported by the evidence.

When Einstein was publishing some of his most famous discoveries, he believed that the universe was static, which is, neither increasing nor decreasing in size. However, when presented with the evidence discovered by Edwin Hubble that the Universe was expanding, Einstein called his ideas about a static universe his "greatest blunder." When presented with compelling evidence, Einstein was willing to modify his perception of reality.

As we prepare ourselves to set sail for the stars, our boundaries of knowledge will inevitably expand and require that our understanding of the universe adapt. During this process, we ought to emulate Einstein and not allow ourselves to become wedded to any dogma, whether a scientific concept or a "revealed" truth like Thor's hammer. The future of humanity could depend on it. The problem is that many people, even scientists who should know better, think that changing their opinion is an admission of error. In fact, rejecting an antiquated and inaccurate idea is the exact opposite—it is the acceptance of truth. Refining our perception of the universe and changing our understanding does nothing to impugn our character or integrity. It simply means that new evidence has emerged and the old concepts must be reconsidered. In actuality, the discarding of a mistaken or inaccurate scientific idea is an occasion for celebration. It means that our understanding of the universe is improving, which we should appreciate. Every time an outmoded paradigm is discarded and replaced by a better one, humanity has achieved a gain in fundamental knowledge, and this, ultimately, leads to improvement in the quality of life for all.

In the meantime, we ought to cultivate sincerity, honesty, and candor regarding uncertainty. For example, astrophysicists today, contemplating observed data about how galaxies rotate or how the universe is expanding, readily admit that they cannot integrate recently discovered data into existing theories. Humbled before the indisputable observations, they are developing theories concerning dark matter and dark energy. The future of these constructs is uncertain, and competent scientists readily admit this. Scientists assiduously examine their preconceptions, refine existing theories, and propose bold, new ideas in the hope and expectation of improving their physical understanding and mathematical depiction of nature.

Like those scientists, we must not allow our versions of reality to become enmeshed in the cement of dogma. We must remain open to new ideas no matter how bizarre and counterintuitive. We must be willing to ask courageous questions and diligently search for meaningful answers. It is better to understand the universe as it really is and accept the truth—even if it is something as unexpected and incomprehensible as dark matter or dark energy—rather than to believe in dogmatic nonsense.

Our civilization has made an arduous, protracted, and treacherous journey. We are no longer the superstitious cave-dwellers of millennia past, grasping at fictitious, mythological explanations of the world around us. Nor do we inhabit a dark and ignorant era such as the Middle Ages, when blind dogmas, which ruled the political, religious, and mental realms, stifled human potential, hindered technological progress, and strangled intellectual freedom. We experienced a glorious Renaissance, when a new thirst for knowledge arose and the modern scientific method, beginning with the work of Roger Bacon in the thirteenth century, began to take shape, later blossoming under the guidance of Francis Bacon and Galileo in the seventeenth century and culminating in nearly universal acceptance sometime around the nineteenth century.

Today, despite our unparalleled scientific and technological achievements, our species continually finds itself teetering on the brink of calamities perpetrated against us by those who long to return to—or never left—a world bound by ancient dogmas. Humankind must staunchly, resolutely, even fiercely stand on the side of truth—rational truth, supported by evidence, bolstered by skepticism, and bestowing upon those who embrace it the power to protect life, to enhance the quality of human existence, and to change our world for the better.

4
What is a Theory?

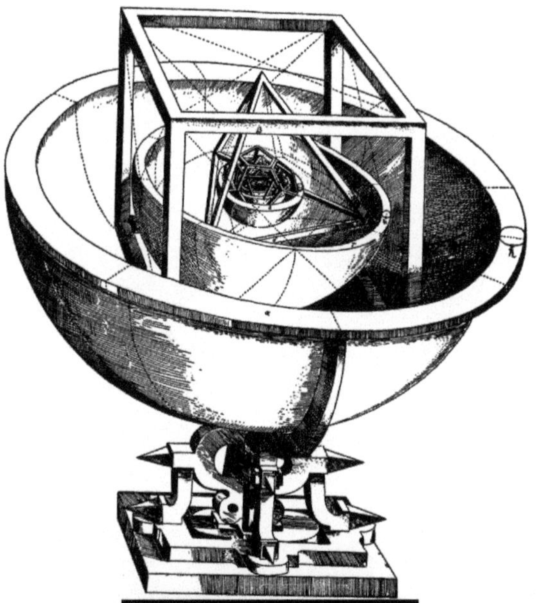
Nested Platonic Solids

The history of science shows that theories are perishable. With every new truth that is revealed we get a better understanding of Nature and our conceptions and views are modified.
Nikola Tesla

The facts will eventually test all our theories, and they form, after all, the only impartial jury to which we can appeal.
Louis Agassiz

In the world of science, the word "theory" conveys a special meaning. Quite a bit of controversy and confusion have arisen about this word, and a clearer understanding of its meaning will go a long way toward resolving that controversy. Before we can understand what a scientific theory is, though, it is necessary to know what science is. Some people believe that science is a collection of facts. Others say that science is an attitude or a worldview. These definitions are at best incomplete. Science is, in fact, simply the pursuit of knowledge by employing the scientific method.

The scientific method is as complex and varied as its practitioners, but it always relies on observation of the world around us and the drawing of logical conclusions from those observations. Scientists strive to perform very carefully controlled observations, called experiments, which ideally produce results that are repeatable and indisputable. A scientist begins an experiment with an idea of what the outcome will be, and this preliminary understanding, or best guess, is referred to as a hypothesis. The measured results of scientific experiments are called "data." Data are the evidence that either supports or disproves the hypothesis. But no result, no matter how convincing, can ever prove a hypothesis. This is because it is impossible to know in advance whether evidence from a future experiment will someday disprove the hypothesis. No matter how many data are gathered that corroborate it, there may still be unexpected data waiting to be collected that will show we do not have a complete understanding of a particular phenomenon.

This brings us to the idea of a theory. A theory is an explanation for a perceived phenomenon. It can bring together a disparate body of related ideas and serve as a guide for scientists in their ongoing quest for new knowledge. Theories are often expressed in mathematical language, although this is not necessary. The ideas that comprise a theory are ones that have been supported by numerous experiments, but there will always remain some gaps in the structure where knowledge is incomplete. Pondering these gaps, scientists can formulate hypotheses and plan experiments to gather new data that will continue to expand human knowledge. When evidence is discovered that does not fit into an existing theory, then true scientific progress can ensue. A new theory must now supplant the old one, or the old theory must be refined so that it will account for the new data and all the preceding data.

The fact that a scientific theory is always subject to being modified or discarded if contradictory evidence is discovered might suggest that science is somehow unsure or unreliable. This is not the case at all. Scientists must remain skeptical, or their enduring quest for knowledge could not continue. All scientists know that our current understanding is never the last word on any scientific subject. You will never hear a scientist claim to have proved anything. The best science can do is to provide so much evidence for a theory that, in the words of Steven Jay Gould, "it would be perverse to withhold provisional assent."

Sometimes the most useful outcome of an experiment is what is called a null result. In this type of result, no data in support of the hypothesis are observed. An example is the famous Michelson-Morley experiment described in the previous chapter. The null result of this experiment, effectively disproving the ether hypothesis, was one of the most important results in the history of science and proved to be one of the necessary pillars upon which Einstein erected his famous theory of relativity.

Some theories have lifetimes measured in thousands of years, only to be ultimately found incorrect. A prime example is the geocentric theory of the universe. For millennia, humans have observed the planets in the night sky. These observations were the beginning of scientific inquiry into planetary motion. Science always starts in this way, from observation. Ancient scientists formulated many theories to explain the movements of the planets. In the geocentric theory (right), the Earth was the center of the universe and the other planets orbited around the Earth. Scientists, with this theory in mind, carried out careful and accurate observations in attempts to support it. After many centuries of intense scrutiny, however, it was finally realized that the observational data did not support the geocentric theory, which was therefore discarded. According to the theory that replaced it, called the heliocentric theory, the Earth and the other planets orbit the Sun.

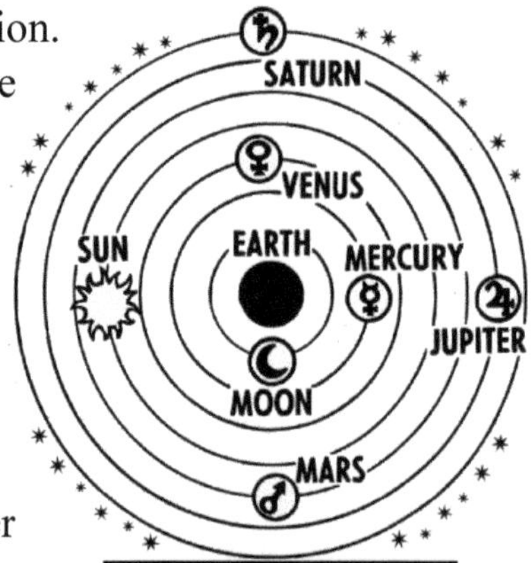

Geocentric Theory

When the heliocentric theory was first proposed, some people thought it was morally wrong because it contradicted the teachings of scripture. However, no theory can be morally right or wrong. A theory is just a tentative explanation subject to continual testing and evaluation. The very concept of moral judgment does not apply to scientific theories. However, it might be argued that vociferous proponents of demonstrably false theories are behaving in an immoral way. By promoting falsehood, these doctrinaire fanatics are causing harm. However, that is not their theory's fault. Theories are utterly dispassionate sets of ideas and are incapable of making moral claims. Only people make moral claims, so only people are capable of attaching a moral evaluation to a theory.

As the evidence in favor of the heliocentric theory accumulated over centuries, moral objections to it diminished and it became almost universally accepted. Johannes Kepler refined the theory by showing that the planets' orbits are ellipses. (Until his time, it was thought that orbits were circles.) Isaac Newton added to the theory by supplying an elegant mathematical description of the force that holds planets in their orbits. As is typical with science, however, Newton's beautiful formula for the gravitational force was later shown to be slightly inaccurate in some cases. It took Einstein to add the next refinement, his famous theory of relativity. In a sense, theories are self-correcting, because any theory that does not fit the data will inevitably be refined, modified, or discarded and replaced with one that better matches the data.

Thus the collective knowledge of humankind advances through the scientific method. Observation, hypothesis, experiment, and data are funneled into ever-growing and changing theories that describe what we know while leading us toward discovery of what we do not know. Of all human constructs, the scientific method is the soundest procedure known for querying nature to arrive at knowledge of objective truth.

Love of learning drives the human spirit. Fathoming the seemingly unfathomable is a joy that transcends the mundane mediocrity commonly called existence. In the pursuit of knowledge, humankind has arrived at an abyss that forms a boundary between what can be learned by large-scale methods and what cannot. Today we can achieve uncanny accuracy in our measurements, down to the millionths of an inch, by bouncing massless photons off a macroscopic object like a basketball or a peach. However, those same photons, when used to measure the subatomic world of the electron, are unwieldy and so energetic that they alter the state of what we seek to measure. As we patrol this boundary between the knowable macroscopic and the presently unknowable submicroscopic, humanity seeks a bridge: a bridge that will unite the Einsteinian gravitational world of the macro with the quantum world of the electron where action at a distance really happens. After making his great discovery of the General Theory of Relativity, Einstein spent the rest of his life seeking such an equational bridge, which has lately been referred to as the Theory of Everything (TOE). Humanity wants to explore the vast reaches of space but is thwarted by space's immense distances. Perhaps a TOE will one day alter our ability to seek knowledge of this gargantuan emptiness that we call the universe and reveal the hidden pebbles of wisdom that will enable humanity to explore the cosmos, colonize distant worlds, develop a complete understanding of nature, and achieve its highest aims. Great minds are seeking this equation, our astronomers are using revolutionary telescopes to catch the light of inhabitable planets light years away, and NASA is experimenting with prototypes of exploratory vessels that will carry the seeds of humanity or its progeny to these distant, future abodes.

Throughout this book, and, indeed, throughout any and every book that contains scientific information, you will encounter numerous theories. Theories are, after all, the crux of science. Although the theories you will see here are well supported and widely accepted, you should remain skeptical, as all good scientists do. No reputable scientist will ever claim to have the last word on any scientific subject. Certainty prevents learning.

The idea of theory, then, is central to all of science. Scientists create theories in order to enhance understanding and better explain the natural world. In astronomy, theories are born, tested, and continually refined to better explain the cosmos. Astronomers and authors work to cast light on the hidden depths of space, to illuminate the universe, and to promote a better world for humankind.

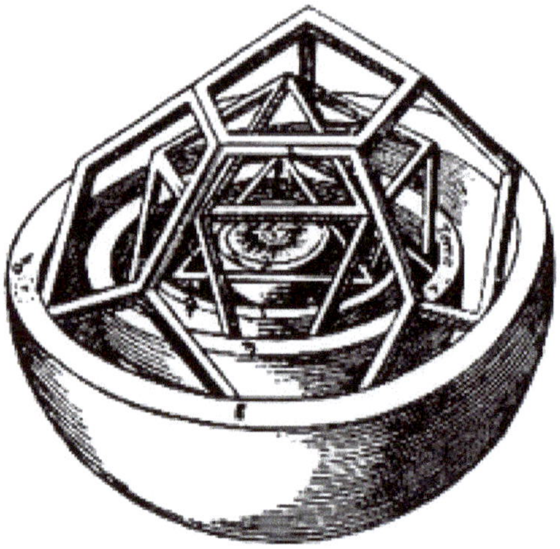

Illustration drawn by Johannes Kepler for his book *Mysterium Cosmographicum,* published in 1596. Kepler hypothesized that the orbits of the known planets corresponded to the five Platonic solids as shown here and at the beginning of this chapter. The hypothesis has since been discarded. This picture is an expanded view of the un-viewable compact center of the chapter's opening figure.

Saint Mark's geocentric clock on the Piazza San Marco in Venice, Italy.

5
Einstein and Astronomy

Matter tells space how to curve; space tells matter how to move.
John Archibald Wheeler

Reaching for worlds beyond our grasp is an essential driver of progress and necessary sustenance of the human spirit.
Brian Cox

Albert Einstein, whose name is universally recognized as synonymous with genius, is probably the most famous scientist of all time. He is best known for his special theory of relativity as represented by the iconic equation $E = mc^2$. Utilizing this equation and his other discoveries, he sought to express the workings of the universe mathematically. With $E = mc^2$, he compressed the beauty, majesty, and power of the universe into a single equation that has become a pillar of modern science.

The story of Einstein's great discovery, known as the general theory of relativity, contains many fascinating twists and turns. Today we know that in order to succeed, Einstein had to recognize and solve two apparently unrelated problems. First, he had to imagine a version of reality that no one before him had envisioned or considered. This required the ability to think visually, four-dimensionally, and creatively. Second, he had to conquer advanced mathematics. This two-prong challenge took Einstein about ten years to overcome.

Starting with Einstein, scientists could no longer depend entirely upon their senses. Light, time, space, and matter work together to create a reality that lies beyond the natural range of human perception, beyond our awareness. Perhaps Einstein's greatest talent was his ability to perform mind experiments. Unlike other physicists, who conducted experiments and gathered data in laboratories, he had the ability to conduct experiments entirely within his own imagination. Einstein coined the word *Gedankenexperiment* to refer to these mental investigations. He was able to discover truths about the universe simply by sitting in his study and thinking through what would happen on a moving train or inside a windowless elevator accelerating through space. These musings allowed Einstein to use the power of pure thought to develop new theories without being restricted by the limitations inherent in laboratory experimentation. Relying on nothing more than his ability to think logically and visualize alternatives, he discovered profound truths about the cosmos. Einstein referred to the fruits of his mental experiments as his "happiest thoughts." In the same vein, he once said, "Imagination is more important than knowledge. Knowledge is limited. Imagination encircles the world."

In addition to the challenge of creating purely mental investigations, Einstein faced a second problem. In order to formulate his theory, he had to master two exceedingly challenging varieties of mathematics: non-Euclidean geometry and tensor analysis. A charming story is often told of a student who complained to Einstein of the trouble that she was having mastering mathematics. Einstein commiserated, "Do not worry about your difficulties in mathematics; I can assure you that mine are much greater."

For his theory, Einstein needed non-Euclidean geometry, the mathematics of curved spaces. This field of mathematics was relatively new, only about 50 years old. Fortunately, mathematicians had made enough progress during those 50 years for their work to aid Einstein in his quest. The mathematicians who developed this abstruse theory probably did not suspect that their work could be applicable in the real world. However, Einstein showed that the curved space of his thought experiments was reflected perfectly in the theorems of non-Euclidean geometry.

The exceptional complexity and unfamiliarity of the mathematics, as well as the near-impossibility of conducting experiments entirely inside the scientist's mind, have caused many to regard the formulation of the general theory of relativity as one of humankind's most impressive mental achievements. Einstein's theory of relativity was so novel and recondite, in fact, that very few contemporary scientists, even among the world's greatest, fully understood it. A scientist once asked a physicist of the era, "I have heard that only three men in the world understand Einstein's theory. Is that true?" The physicist did not answer but sat staring off into space. "Is something wrong?" the scientist asked. "No," said the physicist. "I'm just trying to figure out who the third one might be." The physicist was Sir Arthur Eddington.

Einstein's extraordinary ten-year quest to solve physics' most vexing puzzle would eventually end in a gloriously successful, world-altering conclusion. One person working almost entirely alone fundamentally changed virtually everything we know about gravity. But to solve the puzzle required a modicum of help from astronomy.

Our story begins with Einstein's graduation from the University of Zürich in 1905 with a doctorate in physics. Despite having earned the most prestigious of academic degrees, Einstein could not find a job at a university. His professors mistakenly thought he was lazy and unmotivated. He would often miss lectures, and he did not fit into the rigid authoritarian structure that characterized

European universities at that time. As a result, his professors did not reward him with glowing recommendations.

Later, one of those professors, the renowned, Russian-born, German-trained Herman Minkowski, coined the phrase "spacetime continuum." In 1908 he said, "Henceforth space by itself, and time by itself, are doomed to fade away into mere shadows, and only a kind of union of the two will preserve an independent reality." The importance of Minkowski's work to the development of Einstein's theory cannot be overstated. At first, Einstein viewed Minkowski's treatment as a mere mathematical trick. Einstein eventually realized, however, that a geometrical view of spacetime would be necessary to complete his own work in general relativity. Unfortunately, Minkowski died of appendicitis in 1909, and he lamented on his deathbed that he would not be able to participate in the revolution Einstein had initiated.

Eventually, the father of one of Einstein's friends secured a job for him as a third-level patent officer. It was 1902, Einstein was 23, and the friend in question was Marcel Grossman, who had taken notes for Einstein in geometry so that Einstein would not have to attend the tedious lectures. Grossman was talented at geometry and later became a professor of geometry at the Polytechnic Institute in Zurich. It was he who suggested to Einstein that non-Euclidean geometry might be useful for describing a curved and warped four-dimensional spacetime.

Becoming a patent clerk was the first of two particularly fortuitous events in Einstein life. As a patent clerk, he did not have to spend time preparing lectures and teaching students. The patent-office job did not tax his abilities or claim an inordinate amount of his time. Thus, he could spend most of his time thinking about physics. Einstein had the ability to sit and ponder a single problem for long periods of time, even for years. This prodigious ability is called *sitzfleisch*, a force of will to continually contemplate a single problem, even when it is not certain that a solution will ever be found.

During the first half of 1905, Einstein published four groundbreaking papers in quick succession that ultimately revolutionized physics. The year 1905, therefore, has become known as Einstein's miracle year or *annus mirabilis.* It has parallels with Newton's most productive years when he was forced to remain at home during the Great Plague of 1665-1667.

The first paper, published in June of 1905 explained the photoelectric effect. This phenomenon, in which light shining on metal causes an electric current, had never been satisfactorily explained. For centuries, scientists had felt they needed to determine whether light was a particle or a wave, but Einstein succeeded in resolving the difficulties by treating the photon, the smallest unit of light, as a discrete packet, or quantum, of energy in which the amount of energy in the quantum is determined by its frequency. This revolutionary idea would eventually win Einstein the Nobel Prize. Einstein's work helped lead to the current understanding of light as having characteristics of a particle, a wave, or both.

In the second paper, published in July of his *annus mirabilis,* Einstein described the behavior of atoms and calculated their size. This paper helped promote belief in atoms, which at that time was far from universal. It also explained Brownian motion, which is the constant erratic movement of tiny particles suspended in a fluid or gas. This phenomenon is easily observed any time dust motes are visible when a single ray of sunlight shines through a windowpane into a dusty room of a big old house. Up to that time, the mysterious jostling and jiggling of the dust motes had not been properly explained.

The third paper, published in September of 1905, explained his special theory of relativity, which allows the calculation of changes in time, length, and mass as one approaches the speed of light. (Length decreases and mass increases, while time slows down. The slowing down of time is called time dilation.) The discoveries in this paper relied on only two basic principles, which Einstein discovered: First, the speed of light in a vacuum is constant, no matter who is measuring it or how fast the light source is moving. Second, the laws of physics remain the same, whether one is moving at a constant speed or not moving at all. The first finding resolved once and for all the luminescent-ether conundrum. Today, we know that this paper laid the foundation for general relativity. It would take Einstein another decade (1907-1916) to solve this problem.

The fourth paper, published in November of 1905, capped off his third paper. Here he derived his famous equation $E = mc^2$, explaining the previously unknown mechanism by which stars produce their immense power. The equation, so simple and elegant yet so profound in its implications, states that matter and energy are one and the same. It is an equation that schoolchildren today know by heart, but at the time it was an astonishing revelation.

Any one of these papers should have elevated its author to instant academic success, but Einstein heard nothing from academia for two more years. At that time, Max Planck, a future Nobel laureate and possibly the most respected physicist in the world, read the papers. Planck immediately recognized their importance and the importance of this young physicist. Partly due to Planck's acceptance, Einstein was eventually awarded the position of *Privatdozent* at the University of Bern. This allowed him more time to think and develop his theories.

By 1907, Einstein realized that he needed to generalize his special theory and include time and gravity in his equations. The impetus for this realization came from the Nobel Prize-winning physicist Johannes Stark, who asked Einstein to publish a review of his work up to that time. In writing the review, Einstein was led to organize and generalize his thoughts. He understood that the universe is awash with various forces that cause objects to change their directions in space and alter their velocities. Isaac Newton in his day had superbly explained this concept by mathematically stating that an object, or mass, experiences accelerations when subjected to forces. For about 240 years, Newton's mathematical description of the universe had dominated the science of physics, but a few cracks were beginning to appear in this venerable structure. For instance, the orbit of Mercury minutely deviated from Newton's theory. Newton himself knew that his theory was not complete when he published his elegant mathematical law of gravity, but he could find no way to make further progress. Later in life, Newton wrote, "...we have explained the phenomena of the heavens...by the power of gravity, but have not assigned the cause of this power." He made his exasperation plain in a letter to a friend, in which he wrote that the idea of gravity acting at a distance without a medium "is to me so great an absurdity, that I believe no man who has in philosophical [scientific] matters a competent faculty of thinking, can ever fall into it."

Around 1907, Einstein was being mentored by the great German physicist Max Planck, who was considered the dean of German science. The two men shared an interest in music. Einstein played the violin, and Planck was an organist and an accomplished singer. With Planck's son, the trio would sometimes entertain guests at Planck's home. When Einstein asked Planck about the problem of general relativity, Planck counseled him not to waste his energy trying to solve this particular problem since it would be too difficult. Planck said, "As an older friend, I must advise against it....You will not succeed, and even if you succeed, no one will believe you." Luckily, Einstein chose not to heed the elder physicist's advice and pressed on for nearly a decade. This was a characteristic of Einstein's personality—he did not blindly accept the pronouncements of authority, no matter how well respected.

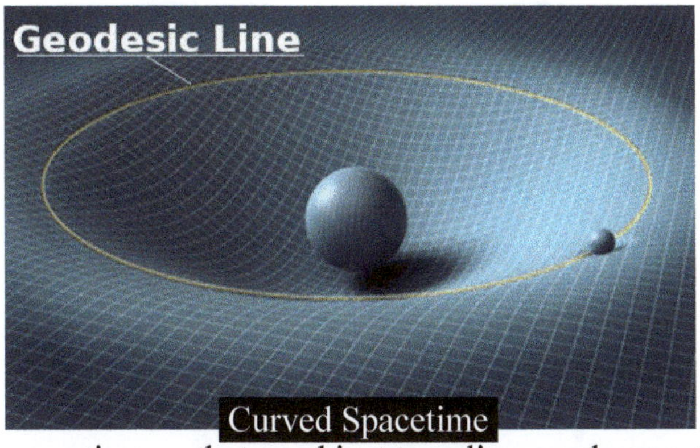

Curved Spacetime

The essence of Einstein's discovery is that spacetime is curved by mass. This is a counter-intuitive idea, to say the least. Rejecting Newton's notion of gravity as a mysterious force acting at a distance, Einstein realized that it could be described as curvature of the four-dimensional coordinate system that includes both space (three dimensions) and time.

The presence of an object with mass bends spacetime so that an object traveling nearby must follow a rounded trajectory that moves it closer to the mass. Thus, when we think we are seeing an object under the influence of gravity moving along a curved path—for example, the Earth's orbit around the Sun—what we are really seeing is the body traveling along the straightest possible path, or "geodesic line," through curved spacetime (above). Space without any matter in it would be "flat." Einstein showed us that gravity is a manifestation of the geometry of the universe as shaped by mass.

Einstein knew all this before he became a full professor in 1911. He needed, however, a testable, measurable hypothesis. Without that, his theory would be nothing more than unverifiable speculation. Casting about in our own Solar System, he naturally came to the most massive object in our vicinity: the Sun. If matter causes the curvature of space, as he was convinced, then surely this massive object should be able to create a distortion of spacetime that would be observable and would lend credence to his ideas.

Gravitational Lensing

His calculations verified his intuition. If he was right, the Sun had enough matter to curve space so that rays of light passing near it would be deflected enough to be detectable by humans. Light that would have streaked past the Sun and missed the Earth by a wide margin would be bent by the curvature of space in the vicinity of the Sun and redirected to our observing eyes on Earth. Thus, a star on the opposite side of the

Sun from Earth would not be where it appeared to us to be, since its light would have been bent before reaching us. In effect, we would be able to see a star that was actually behind the Sun. This effect, the bending of light around any massive object, is called gravitational lensing. (See illustration on previous page.)

Einstein had calculated this curvature and found that light passing close to the Sun would be bent by a tiny but theoretically measurable distance. He could not test his calculation, however, since ordinarily the enormous amount of light emitted by the Sun overwhelms the light of any stars behind it and blinds observers to the curvature. The solution was obvious—wait for an eclipse. He came to the realization that only during a total solar eclipse would it be possible to take photographs of stars whose light passed near the Sun, and then the apparent positions of these stars (little more than pinpricks of light on the photographic plates) could be measured to either support or disprove his theory.

Einstein, a theoretical physicist, was not an astronomer. He was not going to mount an expedition and dash off to some distant country to peer through a telescope for only about five minutes. Luckily, there was no shortage of astronomers who loved doing that sort of thing. The next total eclipse was to occur in the Crimea in southern Russia in 1914. Einstein publicized his wish for astronomers to test his theory in this way. Two teams of astronomers, one from Germany and one from the United States, decided to take up the challenge and set out for the Crimea. The German team was led by an astronomer named Erwin Finlay Freundlich, and the American team was led by the director of the prestigious Lick Observatory, William Wallace Campbell.

Now circumstances beyond Einstein's control took command of his fate, and the second fortuitous event in this narrative occurred (fortuitous from the point of view of Einstein and his theory, although disastrous for the peace of the world). The Austrian Archduke Franz Ferdinand and his wife were assassinated by a Serbian nationalist, and after a series of tragic and seemingly unavoidable diplomatic dominoes fell, all of Europe found itself embroiled in World War I. The German Kaiser declared war on Russia in June 1914, and soon thereafter, Russian soldiers marched into Freundlich's camp.

Freundlich had been eagerly awaiting the upcoming eclipse and did not even realize that war had been declared. As soon as the soldiers saw that he had cameras and telescopes, they assumed he was a spy and arrested him. However, the Americans in Campbell's camp were recognized as neutral—the United States would not enter the war for another three years—so they were allowed to keep their telescopes, cameras, and other equipment. Unfortunately, clouds totally obscured the eclipse when it occurred in August of 1914, and Campbell was not able to get any pictures that provided meaningful data.

When the war broke out, Einstein was living in Berlin and working at the Prussian Academy of Sciences and the University of Berlin. A group of 93 famous German scientists, scholars, and artists issued the "Manifesto of the 93," which declared that the war was justified. Einstein, who was a pacifist, wrote his own manifesto condemning the war, which was signed by only four men and was never published. He fell into disfavor. It did not help that he was Jewish. Jews in Germany at that time were routinely discriminated against. Therefore, he retreated into intensive study of physics and reexamined the details of his theory. As he reviewed his theory, he made a serendipitous discovery. He realized that he had made a mistake in his math and needed to adjust his theory. As he worked on it, he would often think he had succeeded, only to realize later that it was still incorrect. With hindsight, we know that he had hit upon the correct path in 1912 but had rejected it as he tried various ways to manipulate the relevant equations with his mind experiments.

At the time, Einstein believed that the failure of the astronomers to obtain measurable photographs of the eclipse was a stroke of bad luck, but in fact, it was a fortunate turn of events for him because he had made a mistake in his calculations. He expected the observed stars near the Sun's position in the sky to be out of place by 0.84 seconds of arc due to the bending of their light by the Sun's gravity, but this was an underestimate. Successful observation by Freundlich or Campbell of the eclipse of 1914 would have led to findings that contradicted Einstein's inaccurate calculations.

Einstein was already in disfavor for his views about the war. If his theory's specific and exact predictions had been shown to be false, the theory would have been discredited, and his reputation would have been tarnished. Most likely, given the circumstances of the war and the additional stain on his character due to his pacifism, his life in academia would have been terminated and his career, at least temporarily, derailed. He would have had to get some other job and would not have had the time he needed to contemplate and work out the details of his theories. In essence, he would not have been allowed to daydream.

During this time in world history, Jews were not the only ones who were discriminated against. Being female was also a serious disadvantage in many fields, and in particular, it was almost impossible for a woman to be a mathematician or scientist. Einstein's friend, the famous mathematician Emmy Noether, was a great theorist whose expertise in abstract algebra would prove to be pivotal in the field of relativity. Despite her recognized abilities and unique talent, as a Jewish woman, Noether encountered great difficulties. At the University of Göttingen, she was at first denied a job, even after David Hilbert, (incidentally the finest mathematician in the world at that time and one of the greatest mathematicians of all time) pointed out during the acrimonious debate over hiring her that "the faculty senate is not a bathing establishment!" Later, she was allowed to "assist" Hilbert but was still not paid until 1923, when she received a low-paying job as a lecturer.

During 1915, Einstein was able to collaborate with Doctor Noether and made substantial progress on his theory. Einstein wrote, "Yesterday I had an interesting conversation with Fräulein Doctor Noether concerning her work with invariants. It struck me that you could look at this thing [relativity] from several different angles...It seems to me that [she] understands her work well." Tying differential invariance to linear algebra, Noether was able to provide a vital part of the mathematical foundation for relativity, a feat that no one, including Einstein, could have completed without her abstract algebra. As a physicist, Einstein knew that a mathematical proof was necessary, but he was not able to generate it himself. (Whether he knew it or not, he was living up to Hilbert's famous tongue-in-cheek comment to the effect that physics is much too difficult for physicists to do themselves.) Noether's revolutionary techniques allowed her to approach Einstein's theory in a new and effective way and to do the calculations at the only possible level—that of abstract algebra. Noether had found a connection between symmetry—that is, group theory, an aspect of abstract algebra—and the conservation laws of physics. In the summer of 1915, after eight years of strenuous mental exertion, the math and the physics were beginning to come together.

However, 1915 was a tumultuous year, both for Einstein and for the world. Einstein's personal life was in turmoil, with his marriage failing and his academic career in jeopardy due to his pacifism. Meanwhile, the countries of the world had chosen sides and were in the midst of fighting the First World War, one of the deadliest conflicts of all time. At this chaotic juncture in his life and the world at large, he was asked to present his general theory of relativity to the prestigious Prussian Academy of Sciences. However, he still had two problems: one, his theory was still unproven; and two, he knew that his math was incorrect and that his theory was still not finished. As a prelude to his speech before the Prussian Academy, he was to give a series of four lectures at Göttingen University. Göttingen was then the world's preeminent center of mathematics, and foremost among the geniuses there was David Hilbert, to whom Einstein was particularly eager—perhaps too eager, it would turn out—to explain all the intricacies of relativity. After hearing Einstein lecture, Hilbert became enchanted with Einstein and his theory and decided to try to solve Einstein's math problem himself. In effect, he started a race with Einstein to see who could solve the problem first.

Einstein now faced two ticking clocks: he knew that Hilbert was closing in on the correct equations, and he had agreed to give four weekly lectures on his theory at the Prussian Academy. Einstein was working furiously in the throes of one of the most concentrated frenzies of scientific creativity in history. As he worked, he noticed that several of the theory's key indicators were pointing to ideas he had discarded in 1912. This led him to the solution.

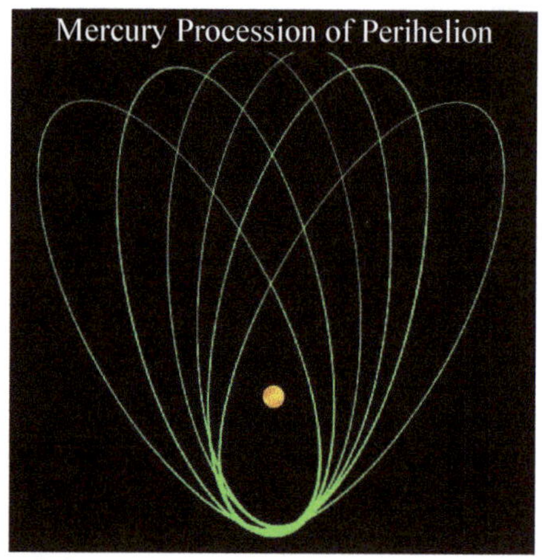

Newton's laws of motion predict an elliptical orbit for Mercury, yet Mercury deviates from this expected orbit and traces a flower petal of orbits around the Sun (right). All planets do this, but Mercury's precession is more pronounced because it is closer to the Sun. The anomaly is referred to as perihelion precession, and the difference between the observed position of Mercury and the position predicted by Newton's laws was approximately 43 arc seconds per century (a truly minuscule amount). Mercury's orbital oddity was first noted in 1855 by the French mathematician and astronomer Le Verrier, who had made similar calculations when he deduced the existence of the planet Neptune from anomalies in Uranus' orbit.

Einstein painstakingly calculated the orbit of Mercury based on his general theory of relativity. When he had finished, he found an almost perfect match of 42.98 arc seconds per century. Clearly, Einstein's general theory of relativity accurately accounted for the discrepancy. This was the moment when Einstein knew his theory was correct. When he later recalled this time, he described his feeling as "ecstatic" and said he had heart palpitations. After a decade of difficult mental mathematical gymnastics, he had reached his goal, only to find that Hilbert had also reached the same result, producing equations almost identical to Einstein's. A minor dispute arose over who had been first. Hilbert, who was already world famous, was gracious and said, "The theory is Einstein's." He was right: All the theoretical underpinnings of the equations were due to Einstein's deep and creative thought, whether or not Hilbert formalized the equations first.

After Einstein officially presented the general theory of relativity to the Prussian Academy, he published the full paper on December 2, 1915, with the title "The Field Equations of Gravitation." Arthur Eddington of England received a translation of the paper and recognized its importance. Eddington, like Einstein, was a fervent pacifist. He hoped that scientists from warring countries could work together to promote peace.

The American eclipse-hunter, Campbell, had had to leave all of his equipment behind in Russia because of the war. However, on June 18, 1918, there was going to be an eclipse in Goldendale, Washington, and Campbell would have the field to himself. Campbell scrambled to find replacement equipment. On the day of the eclipse, the clouds moved in, and Campbell thought he was about to relive his disappointment in the Crimea, but at the last minute the skies cleared and Campbell took excellent pictures of the eclipse.

Part of the problem with viewing light is that it jiggles, or is continually refracted, by molecules of air as it passes through the atmosphere. (This is why stars appear to twinkle—a lovely phenomenon, but one that astronomers could do without.) With uncertainty caused by atmospheric conditions, and the unprecedented accuracy necessary for the detection of such a tiny difference in the position of the stars, it was far from certain that Campbell would be able to succeed in either supporting or contradicting Einstein's prediction. After many months of measuring and reviewing the pictures, and with the rest of his team in agreement, he announced to the world that they had seen no deviation and that Einstein was wrong.

Eddington went to Principe, a small island off the west coast of Africa, for the total solar eclipse on May 29, 1919. Eddington took a large number of photographs, but most were ruined by clouds, and others were so blurry that the stars looked like smudges. Luckily, a few of the last pictures came out well. Back in England, Eddington meticulously examined the photographic plates. He had to use a micrometer to measure the deflection because it was so minute. Although his data seemed to support Einstein's theory, Eddington was in no hurry to publish his results.

The world waited for Eddington's results. On November 6, 1919, he addressed the Royal Society. Gesturing toward the stately portrait of Newton that overlooked the meeting hall, he said, "Forgive us, Sir Isaac Newton. Your universe has been overturned....The deflection by 1.75 seconds of the Sun's length... is verified by the eclipse observations." Newton, a former (1703-1727) president of the Royal Society and probably its most renowned member, surely would have approved.

Many were still not convinced, including the Nobel committee, which passed over Einstein in 1919, 1920, and 1921. Having heard conflicting opinions from Campbell and Eddington, the world of science anxiously awaited the next solar eclipse, which would allow astronomers to gather more data and settle the question one way or the other.

The next total solar eclipse was going to take place in Australia in 1921. Seven expeditions set out to confirm either Einstein and Eddington, or Campbell. Several teams had no luck and lost out to clouds or faulty equipment, but Campbell's team, which had clear days and excellent equipment, gathered data that resoundingly corroborated Eddington's earlier findings. Campbell personally sent a telegram to Einstein congratulating him on the success of his theory.

The Nobel committee was still cautious. In 1922, they finally awarded Einstein the Nobel Prize, but not for his general theory of relativity. They chose to recognize him for his first paper in 1905 on the photoelectric effect. The Nobel committee was correct in its judgment of the importance of this work. Einstein's treatment of the photoelectric effect was a precursor of the quantum mechanical approach that came to dominate physics in the twentieth century and is still on the cutting edge today.

Following the First World War, a wave of anti-Semitism swept Germany as Jews became the scapegoats for Germany's defeat. The anti-Semitism helped propel Adolf Hitler and his Nazi party into power. When Hitler became chancellor of Germany in 1933, Einstein was in America visiting Cal Tech. Einstein decided then and there that he would not return to Germany. Having received offers from London, Paris, Madrid, and Jerusalem, Einstein chose to remain in the United States and relocate to Princeton, New Jersey, where the world-renowned physicist became the very first scholar at the newly formed Princeton Institute for Advanced Studies. He lived and worked there for the rest of his life, striving to formulate a grand "theory of everything" that would combine relativity and quantum mechanics. As of this writing, such a theory has still not been elucidated.

The President of the United States relied on Einstein's advice. As World War II unfolded, and Americans worried that the ascent of Nazism threatened the way of life and the very existence of the American nation. Few knew about the possibility of nuclear weapons. Einstein, upon urging by other scientists, such as Leo Szilard, Edward Teller, and Eugene Wigner, signed a letter to President Franklin Delano Roosevelt urging him to develop a nuclear bomb immediately. Roosevelt took this advice and instituted the Manhattan Project, an all-out effort to create the bomb before the Germans did. The project was successful and culminated with the use of nuclear weapons against the nation of Japan in 1945. However, Einstein had never stopped being an ardent pacifist. He later admitted that if he had known that the Germans were not close to making an atomic bomb, he would never have made his influential personal appeal to Roosevelt to move ahead with the project.

Einstein became exceptionally famous. He was a true celebrity, so famous that in later years, the nation of Israel invited him to become its president. In rejecting the offer of the presidency of Israel, he said that he lacked "the natural aptitude and the experience to deal properly with people."

Einstein's iconic image is still recognizable everywhere. He was the first person to describe two interwoven processes: how a gravitational field acts on matter and how matter generates gravitational fields in spacetime. Today, thanks to Einstein, we know that gravity is not a force of nature, but the result of matter deforming the fabric of spacetime, and that Newton's laws fail when we are very close to an extreme source of gravity.

In addition, Einstein was a kindly man who unswervingly stood by his friends when their need arose, and he always remembered those who had helped him. Doctor Emmy Noether left Germany for the United States in 1933 after Hitler took power. Once in the United States, she taught at Bryn

Mawr until she died in 1935. When Albert Einstein saw a terse notice of her death in the New York Times, he was outraged and wrote the following in a letter to the editor on May 1, 1935:

> In the judgment of the most competent living mathematicians, Fräulein Noether was the most significant creative mathematical genius thus far produced since the higher education of women began. In the realm of algebra, in which the most gifted mathematicians have been busy for centuries, she discovered methods, which have proved of enormous importance in the development of the present-day younger generation of mathematicians…
>
> Her unselfish, significant work over a period of many years was rewarded by the new rules of Germany with dismissal, which cost her the means of maintaining her simple life and the opportunity to carry on her mathematical studies. Farsighted friends of science in this country were fortunately able to make such arrangements at Bryn Mawr College and at Princeton[1] that she found in America up to the day of her death not only colleagues who esteemed her friendship but grateful pupils whose enthusiasm made her last years the happiest and perhaps the most fruitful of her entire career.

Einstein never lost sight of the fact that Fräulein Noether had done the mathematical work that provided an essential foundational element for his general theory of relativity—work that even today bears the name "Noether's Theorem."

The general theory of relativity required a paradigm shift in the thinking of scientists around the world. Thanks to Einstein, we now know that we live in a four-dimensional universe with three spatial dimensions and a fourth dimension of time. Einstein was able to solve this riddle with pure thought, and his theory has led to modern explanations for the origin of the universe. Paul Dirac, Nobel laureate, said that it was "probably the greatest scientific discovery ever made." Max Born, another giant of twentieth-century physics, called it "the greatest feat of human thinking about nature, the most amazing combination of philosophical penetration, physical intuition, and mathematical skill." Neil deGrasse Tyson, director of the Hayden Planetarium in New York City and host of the TV series "Star Talk," described Einstein's general theory of relativity as "…one of the greatest achievements of the human mind. It is beautiful, and simple, and profound." Michio Kaku, City College of New York and CUNY Graduate Center physics professor and frequent commentator on astronomy and science, says about Einstein, "Here was a man, because of the power of his intellect, the sheer power of his mind, was able to rise above poverty, rise above war and strife, to become one of the great figures in all of human history….You can achieve by the sheer power of intellect and for me that's always been a shining beacon."

Einstein once explained his fame to his curious son by invoking a picturesque analogy. He likened the curving of the fabric of spacetime to the plight of a wandering sightless insect: "When a blind

---

[1] Einstein is referring to Noether's work at the Institute for Advanced Study, located in Princeton, New Jersey. The Institute is separate from but maintains close ties with Princeton University.

beetle crawls over the surface of a curved branch, it doesn't notice that the track it has covered is indeed curved," he said. "I was lucky enough to notice what the beetle didn't notice."

Einstein's theories predict results that have only recently become measurable. As an outgrowth of the general theory of relativity, Einstein proposed the existence of gravitational waves. Direct evidence for gravitational waves was discovered on September 14, 2015, and reported shortly before February 12, 2016. The waves, streaming from a distant pair of violently colliding black holes, behaved just as Einstein predicted. This is another spectacular case of Einstein's ideas being proven correct by astronomers long after he conceived them in his fecund imagination. After the initial confirmation, scientists around the globe have continued to detect gravitational waves.

Because electromagnetic waves (the visible light that reaches our eyes, for instance, but also X-rays, gamma rays, radio waves, etc.) encounter interference as they travel across the universe, they can only tell part of the story. Gravitational waves experience no such barriers, meaning they offer a wealth of additional information. Black holes, for example, do not emit visible light but can be studied via gravitational waves. A new era of gravity-wave astronomy is under way, and each new finding will expand our understanding of the universe.

Einstein had profound mathematical ability; he held his own with David Hilbert, one of the best mathematicians in history. Einstein loved physics and used astronomy to advance his theories, and these two subjects have enthralled acolytes since their youth. It is possible for even young students to catch at least a glimpse of what Einstein experienced and feel the same joy (hopefully without the heart palpitations) as the wonders of science unfold before their questing minds, and challenging physics problems capitulate, as they so often do, to persistent mathematical attacks. Students' eyes still light up as Einstein's must have when, after protracted effort, a problem reveals its deepest secrets.

Einstein was a man who could comprehend the universe and explain it to the layperson. But most importantly, Einstein is admired as a humble man with the common touch. He would talk with anyone. He did not care about a person's age, status, or abilities. There exist many anecdotes of his amiability and willingness to help others. The name Einstein perennially conjures an image of a tousle-haired, compassionate, and congenial scholar of an earlier era. However, behind the great achievements of Einstein, we must recognize that he devoted his life to curiosity, concentration, and perhaps most of all, hard work.

World-changing thinker, grand theorist, the towering intellect of the twentieth century, he was also a humanist, a champion of peace, and an exemplar of kindness. By any criteria, this was a truly great man: knowledgeable, creative, and wise—a sage for all time.

6
Stonehenge

...a lifetime burning in every moment
And not the lifetime of one man only
But of old stones that cannot be deciphered.
T. S. Eliot

The most beautiful thing we can experience is the mysterious. It is the source of all true art and science.
Albert Einstein

Out of the misty, verdant expanse of the windswept Salisbury Plain in rural south-central England rises a mysterious and imposing assemblage of rough-hewn, weathered, gargantuan stones. This edifice has been celebrated and marveled at for over 5,000 years. Millennia ago, this awe-inspiring shrine was already attracting visitors from as far away as Spain and the Mediterranean Sea. Even today, thousands of visitors arrive annually to bask in the morning sunlight and gaze in fascination at one of the Earth's most unusual sights—an iconic, prehistoric, megalithic monument that dominates its flat, treeless, pastoral setting. This magnificent curiosity of the ancient world is known as Stonehenge.

Before the era of scientific archeology, Stonehenge was sometimes imagined to have been a temple for Druids or Romans. The Druids, a caste of Celtic priests, were contemporaneous with the Roman Empire around 2,000 years ago. One of the first known descriptions of them was given by the Roman commander Julius Caesar in his account of his military campaigns.

It is now known that Stonehenge was in existence for thousands of years before any Roman or Druid laid eyes on it. For historians, archaeologists, and tourists alike, Stonehenge is an enigmatic reminder of a lost Stone Age civilization. Although many of its stones have fallen from their original heights, the monument's grandeur still generates wonder, captivates the imagination, and mutely poses the questions: "Why did those Neolithic people build this?" and "Why, out of England's 50 million acres, did they choose this particular five-acre plot?"

Archeologists and other scientists face daunting obstacles when attempting to answer these questions. The people who built and used Stonehenge left no written records to reveal their methods, intentions, or beliefs. Therefore, students of Stonehenge have delved into the Earth, examined the stones, and tried to puzzle out the meanings behind the configuration of the monument's different parts. Unraveling the mysteries of Stonehenge has proven exceedingly difficult, often relying on guesswork and supposition. However, the situation is not hopeless.

Stonehenge endures in the middle of the Salisbury plain awaiting examination by a modern army of archeologists and anthropologists wielding all their knowledge, experience, and technology. Scientists survey the site from above with helicopters, peer under the topsoil with metal-detecting magnetometers pulled behind tractors, inspect deeply buried mysteries using ground-penetrating radar, and even get down on their hands and knees to dig gently in the chalky soil with trowels and scrape away odd bits of loam from their priceless finds with slender brushes. Researchers employing computer graphics software have mapped the environs of Stonehenge and constructed a virtual model of the monument that can be studied and examined from all angles. A scientist can enter a date and time, and the computer program will provide a panoramic view of the heavens along with the rising and setting of the Sun as seen from the exact center of Stonehenge on that day. Scientists routinely utilize radiocarbon dating to determine the age of organic substances, mass spectrometers to determine the elemental ingredients of materials, and microscopes to scrutinize all artifacts for pertinent characteristics and anomalies. These tools of the archeologist's trade have proved invaluable in the evaluation of artifacts and have enabled researchers to make many thought-provoking discoveries which otherwise would have remained beyond their reach indefinitely.

Archaeologists now surmise that Stonehenge was just one of many prehistoric structures, collectively called henges, built of earth, river gravel, timber, or stone. Like Stonehenge, most henges are circular in arrangement. More than 1,000 stone circles remain in the British Isles, but Stonehenge is unique. No other circle has such massive stones, nowhere else were the stones smoothed into regular shapes like giant building blocks, and no other circle has stones topped with nine-ton stone lintels.

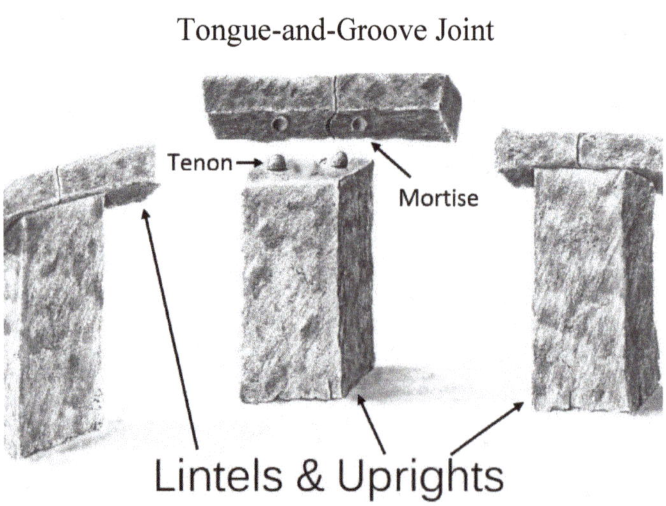

Stonehenge was clearly erected by people who were skilled in constructing large structures out of timber. The ancient engineers' woodworking genius was set in stone as evidenced by the joints employed. For instance, the stones of the trilithons, despite being harder than granite, were fitted together with mortise-and-tenon joints like those used by expert woodworkers to join the parts of a cabinet or dresser, and a tongue-and-groove joint was used to link each lintel in the outer circle to its neighbor. A bare tenon can be seen protruding from the upper surface of one of the upright stones in the picture on the previous page.

These ancient stonemasons of the English Isles were a devoted, hardy, and capable folk. They were a hardworking agrarian culture consisting of people who were cerebrally, anatomically, and spiritually just like us. These manipulators of 20-ton blocks were modern humans in every sense; all they lacked was modern technology and education.

Construction of Stonehenge began more than 5,000 years ago, around 3100 BCE—about 500 years before the building of the great pyramids of Egypt—and continued intermittently through three stages over 15 centuries. The builders of Stonehenge were in the midst of the transition between the Old Stone Age and the New Stone Age. The word "stone" refers to the material from which tools were made during those periods, while the dividing line between the two stone ages was the invention of agriculture. In the Old Stone Age, also called the Paleolithic period, humans procured their foodstuffs by hunting and gathering, while during the New Stone Age, or Neolithic period, they abandoned their nomadic practices to adopt a more settled lifestyle involving the growing of crops and the tending of domestic animals. A huge undertaking like Stonehenge would have required a reliable and plentiful food supply of the kind associated with the advent of agriculture. By the final stage of construction, around 2000 BCE, the Bronze Age would have commenced and the people of that time and place would have learned how to fabricate metal tools. Although nowhere near the quality of today's tungsten-carbide steel tools, these tools of alloyed copper and tin, which combine to form bronze, would have been considerably more effective at digging in the chalky soil or dressing stones than the tools fashioned from stone or deer-antler that preceded them.

In the centuries since Stonehenge's original construction, many of its stone blocks have toppled from their original positions. Furthermore, the stones that remain have endured millennia of harsh weathering. Also, many of Stonehenge's stones have been carted off over the years. Medieval and early modern builders eagerly raided the site for material to construct houses, barns, and churches because there is no natural building stone within 13 miles of Stonehenge.

A 17th-century depiction of Stonehenge illustrates most of the remaining stones connected and in upright positions. A photograph taken in early 1877 shows very few of the original blocks in the correct positions with most of the slabs disconnected and strewn on the ground. However, a picture taken in 2008 from a similar angle reveals that part of the stony formation had by then been reconstructed to provide visitors a better indication of how the original structure appeared. Several of the larger stones have been set in concrete to hold them in place and ensure that they will remain upright for generations to come.

The architecture of Stonehenge is characterized by concentric rings of megaliths of different sizes, materials, and configurations. The centermost ring consists of 19 smaller stones, known as bluestones, in a horseshoe arrangement. The bluestones are spotted dolerite, an igneous rock that shows a bluish hue when wet or freshly chipped. Geologists have proven that the bluestones originated in the Preseli Mountains in southwestern Wales. The years-long search for the quarry where the bluestones originated, ultimately resulted in a location that matched the geology of the bluestones superbly. A partially excavated bluestone was even discovered at the quarry. Their transportation from Wales, a straight-line distance of over 120 miles, to the site of Stonehenge was no trifling feat considering their weight of two to four tons. It is likely that the bluestones were transported to Stonehenge by sea, river, and overland, a distance of some 240 miles. An ancient boat aged within a few hundred years of the proposed transport of the bluestones has been found preserved in mud; it would have been suitable for carrying the massive bluestones. The bluestones were presumably the first stones to be set up at the site.

Trilithon

Immediately outside this inner bluestone ring stands a horseshoe-shaped ring of five trilithons. A trilithon is a group of three stones in which one of the stones rests crosswise atop the other two. The middle trilithon is the largest and is known as the great central arch. It is oriented so that its opening is positioned in an east-west direction. The stones in this inner ring of trilithons are up to 30 feet long and weigh about 40 tons apiece. They are identified as sarsen stones, a kind of sandstone. As part of the construction process, they were painstakingly pounded until their outer surfaces were almost perfectly smooth.

Around the trilithon ring, more bluestones were erected in another ring. Initially, there may have been as many as 61 bluestones in addition to the 19 in the inner ring. Today, however, only 43 remain in both rings combined. It is likely that the bluestones were conveyed to the site in two batches widely separated in time and have been rearranged several times.

Outside the outer ring of bluestones is situated a ring of 25-ton slabs of sarsen stone arranged in a post-and-lintel construction 108 feet in diameter. The stones of this circle were so carefully aligned by the builders of Stonehenge that, on average, the stones are less than three inches out of place.

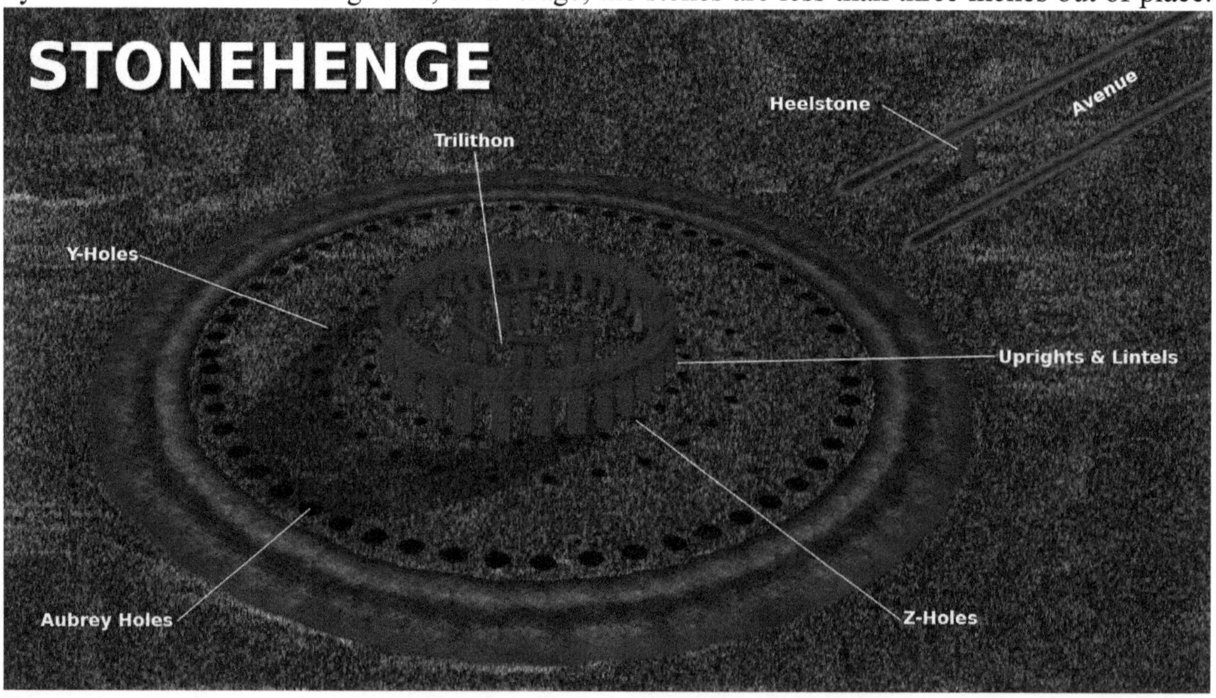

Formerly, the ring would have been unbroken, consisting of 30 standing stones capped with 30

lintels, but today many of the stones are missing. The illustration on the previous page shows how Stonehenge may have looked with the outer ring completed. Unfortunately, only 17 of the uprights are still standing. The sarsen stones were originally transported from the Marlborough Downs 20 miles to the north. The massive crosspieces were probably lifted into place by alternately raising one end and then the other and placing blocks of wood under each end as it inched upward. Modern researchers reenacted this process with volunteers and found that a small group of people could raise a twenty-ton stone slab eight feet in a single day.

Around the post-and-lintel formation, and very near it, lie two concentric rings of holes. The inner ring consists of 29 holes and the outer ring contains 30 holes, called the Z and Y holes respectively. Large wooden posts originally jutted upward from these holes.

Farther from the center, Stonehenge's builders dug, and apparently almost immediately refilled, a circle of 56 shallow holes. They were named Aubrey Holes after their discoverer, the 17th-century antiquarian John Aubrey. It is surmised that workers used deer antlers as picks to loosen the chalky bedrock because archeologists have found the worn pieces of antler and determined their age by radiocarbon dating. Recent evidence indicates that the Aubrey holes were formerly burial sites with a bluestone capping each one as an outsized tombstone.

Almost touching the ring of Aubrey holes, but just outside of it, are four stones called station stones. The station stones, if connected by lines, would form an almost perfect rectangle. On the east and west side, these lines would be tangent to the post-and-lintel circular formation. Imaginary lines drawn diagonally between the station stones would intersect in the exact center of Stonehenge.

Outside the last ring lies a circular ditch and dike. At a distance of 256 feet from the center of Stonehenge stands the heelstone, which is 16 feet tall with an additional four feet buried in the ground. It is eight feet wide and seven feet thick, and has a naturally pointed top. It is unworked, which suggests that it was set in place earlier than the other stones.

Leading directly to the heelstone is part of an ancient path called the Avenue. The Avenue runs towards Stonehenge from the Avon River about a mile and a half away. Nearing Stonehenge, this broad lane curves and then abruptly becomes a perfectly straight path which, if continued through the midst of the standing stones, would precisely bisect the monument.

It was recently discovered that the Avenue runs exactly between two deep channels that were carved into the terrain by ice-age freezing and thawing. These channels, which predate Stonehenge and are no longer visible from ground level, are aligned with the rising of the Sun on the summer solstice and the setting of the Sun on the winter solstice. This revelation is significant and very unlikely to be a coincidence. The presence of these naturally occurring grooves running parallel to the sunrise or sunset on these special days was probably the original reason why Paleolithic hunter-gatherers held the place sacred. Humans of the Stone Age could have interpreted these fortuitously aligned channels as a message from the gods.

Within two miles of Stonehenge are the remains of the ancient wooden village of Durrington Walls. This settlement, the largest known in northern Europe at that time, flourished for about a half-century around 2500 BCE. Signs of ancient wooden houses have been detected there, along with vast amounts of bone and pottery. The village was home to as many as 4,000 people and most likely served as a headquarters and staging area for Stonehenge when the sarsen stones were being transported and erected. Over 2,000 tons of sarsen stones were moved at this time. The bluestones were also rearranged, and it is estimated that all of this work took less than 50 years. For the people of that epoch, the arduous labor of quarrying, erecting, and smoothing the huge stones would likely have been a labor of religious devotion, a welcome opportunity to serve a power greater than themselves. Durrington Walls was inhabited by people of the newly arrived Beaker Culture, which represented a major shift in local pottery styles and burial practices (away from cremation). The Beaker People seem to have quickly displaced the peaceful Neolithic farmers who preceded them. However, it is evident that the newcomers must have shared some of their predecessors' beliefs because the Beaker People adopted Stonehenge as a sacred site, presumably suitable for the interment of their political and religious leaders.

Explaining Stonehenge

The elaborate constructions at Stonehenge, composed of extraordinarily heavy and gigantic rocks, inspire curiosity and admiration but refuse to explain themselves definitively. Although innumerable interpretations have been proffered, most likely the original purpose or purposes for which this monumental formation was created will never be divined, nor whether those purposes changed during the ensuing 15 centuries. We do know, however, that Stonehenge was still being used as late as 1100 BCE. Furthermore, we can be sure, setting aside all speculation, that Stonehenge was the great communal project of that time and place. It represents a very serious undertaking, one that required intelligence, purpose, strategy, and persistence.

Although we remain unsure of the actual purposes of Stonehenge, there are observable relationships between the stones of Stonehenge and astronomical phenomena. On the morning of the summer solstice, an observer standing in the center of the inner circle can witness the Sun rising to the left of the distant heelstone until the glowing orb seems to alight perfectly atop it like a flawless golden jewel adorning the fingertip of a sun god, or perhaps a golden sphere resting atop a 16-foot golf tee. On the summer solstice the rays of light passing over the heelstone also pass through the great central trilithon. Likewise, on the evening of the winter solstice in December, the sunset appears precisely framed between those same uprights.

The station stones seem to mark the travels of the moon on its monthly cycle of moonrise in the east and moonset in the west. The gap between moonrise and moonset varies over an 18.66-year cycle, which corresponds to 235 lunar months. This cycle is called the Metonic cycle, after an ancient Athenian astronomer who studied it, although the builders of Stonehenge evidently knew about it 2,000 years before Meton did. From one cycle to the next, the distance between moonrise and moonset increases and decreases, but never passes beyond the limits marked by the station stones. This cycle also predicts lunar eclipses. It has been suggested that the ring of 56 holes has a similar relationship with the 18.66-year lunar cycle. After all, the phases of the moon repeat every 18.66 years, and three of those cycles equal 56 years.

The correlation of so many architectural features with astronomical phenomena surely reflects planning on the part of the builders. Stonehenge is aligned to the solstices within 1 degree, which is astounding for that time. This spectacular edifice must be a manifestation in massive stones of ancient humanity's expertise in astronomy. But was it more than a solar and lunar observatory? Without being able to read the minds of those long-dead ancient people, we can never know Stonehenge's purpose with certainty. However, knowledge of the yearly cycles of the seasons would have been vitally important to a primitive society, especially one experiencing a newly acquired reliance on agriculture. Crops must be planted and harvested at the correct times of year to optimize yield. Stonehenge could have been a source of astronomical and therefore agricultural knowledge that bolstered and aided food production throughout the British Isles.

There is evidence that large groups of people journeyed to Stonehenge at certain times of the year. At Durrington Walls, over 10,000 shards of pottery have been unearthed, which is in itself a marvel as most finds of Neolithic pottery are but a few pieces. Animal remains uncovered nearby include the bones of domestic livestock. The bones show that the animals were almost all the same age when slaughtered—about nine months—and this age places their death around the time of the winter solstice. Strontium levels in teeth produce a sort of isotopic fingerprint that reveals where an animal spent most of its life. Thus, we know that some of the animals who died at Stonehenge were brought from as far away as the Orkney Islands, a distance of 700 miles. The animals, 90% of which were pigs, were probably either consumed as part of the festive celebrations or offered as sacrifices, in either case serving a socio-religious purpose. Researchers have found animal remains in Durrington Walls that were discarded with meat still on the bones, strongly suggesting that revelers were fully satiated. That such waste occurred indicates that these were feasts of epic proportions on a scale not seen in ordinary daily life.

The fact that Stonehenge was constructed of stone is itself further indication of the societal importance of the site. Scholars have asked why people would choose to construct their most important monument out of stone when they were so adroit at building with timber. The answer may lie in the stones themselves. Stone will last for ages, while wood rapidly rots in the English weather. Thus, stone was the natural choice for a people who wished to erect an eternal cenotaph to their dead, express lasting devotion to their gods, or imbue the locale with enduring spiritual significance.

A modern-day researcher, Mike Parker Pearson, has been given special permission by the British government to investigate the complete site, including the Aubrey holes. Pearson was guided to Aubrey hole 7 by a cryptic scrawl left in a notebook by an archeologist's assistant near the dawn of the twentieth century. There, two feet beneath the grassy surface, the team found a slightly corroded lead plaque under which were some 500,000 charred bone fragments. These bones represented an anthropologist's worst possible scenario—they were the hopelessly jumbled remains of individuals who had died thousands of years before. After a sorting process requiring more than a decade of almost unbelievable patience and meticulous attention by an expert osteoarchaelologist, it was found that the bones belonged to at least 63 individuals, including women and children. The most common bones found were ear bones, and there were also numerous occipital bones, which was fortunate because anatomists can distinguish the occipital bone of a male from that of a female. By carbon dating, the remains were dated to between 3000 BCE and 2500 BCE. The crystal structure of the exhumed bone material provided clear evidence that the bones had been subjected to cremation at surprisingly high temperatures—over 1000° F. A strontium analysis revealed that a third of the bones came from western Wales where the bluestones were quarried. Although it is possible that ancient residents of Wales journeyed to Stonehenge to die, it is more likely that Stonehenge was a burial place for those who had died elsewhere. Tribes would have cremated the remains of important individuals wherever they died, constructing a tall funeral pyre for the purpose, and afterward transported the recovered bones, including even the tiniest, to Stonehenge for burial under one of the sacred bluestones. With the bones in their pit were two mysterious artifacts: a mace-head of polished volcanic rock likely used as a scepter-like symbol of authority and a clay disk that was possibly a shaman's incense burner.

Chalk residues from the other Aubrey holes were detectable on the bone fragments, and Pearson was able to determine that the bones had once occupied different Aubrey holes. The bones had been removed by researchers in 1919 and placed together for some reason in Aubrey hole 7. It was also found that the bluestones had once resided in the Aubrey holes; the chalk particles were crushed in such a way that only objects the size and weight of the bluestones could have been responsible.

The area surrounding Stonehenge, outside the 56 Aubrey holes, was also a burial site of the Stone Age elite, but for a different, later people. The culture responsible for these burials is referred to as the Beaker People after their distinctive pottery consisting of wide-mouthed vessels with incised lines around the rim. Unlike those whose bones were buried in the Aubrey holes, the bodies buried outside Stonehenge were not cremated; they were interred individually, and the bodies were buried in round barrows, or mounds, often with artifacts. Many such burial mounds have been found close to Stonehenge. Some of the interred individuals appear to have been rulers because they were buried with golden objects of exquisite workmanship. For instance, the body of an apparent chieftain was accompanied by a geometrically incised, lozenge-shaped, solid gold ornament the size of a human hand. Another burial was accompanied by a bronze dagger with a wooden hilt that had been adorned with 140,000 amazing golden micro-studs, each more slender than a human hair. It is hardly surprising that the period when these burials took place is called the "golden age" of Stonehenge.

In addition to the rich and powerful, there are some surprises among the burial mounds that ring Stonehenge. An otherwise healthy male in his late 20s was shot in the back and sides by three arrows at close range before being buried in a ditch near the monument. Was he executed, or sacrificed? A fifteen-year-old boy buried at Stonehenge around 1500 B.C. grew up near the Mediterranean. One interred man was discovered to have come from what is now Germany; this individual, known as the "Amesbury Archer," was buried about three miles from Stonehenge with almost a hundred artifacts, including golden hair ornaments and a ceremonial copper dagger.

The presence of travelers from distant lands suggests that Stonehenge could have been a place of pilgrimage. Perhaps Stonehenge was considered a place of healing. Some of the burials were of people who had suffered traumatic injury; they may have traveled to Stonehenge in hopes of being miraculously healed. On the other hand, some of them might have sustained their injuries at Stonehenge.

The size and majesty of Stonehenge, the presence of the deceased, evidence of ritualistic worship, and the correspondence of astronomical phenomena with the layout of the stones—all these facts point to a ceremonial purpose for the site. Whether an observatory, a computer, or a cathedral, it has been suggested that such a gargantuan and impressive construction would also have served as a source of cultural pride and a force for cultural unification for the ancient peoples of Great Britain.

Around 1500 BCE, however, the site ceased to be a significant place of worship, burial, religion, or astronomical observation. The great stones had served as the focal point of a little-remembered culture's greatest technological, astronomical, and spiritual ambitions, but after its heyday of a thousand years, it was abandoned and began the slow and dreary process of falling into ruin.

Today the monument, despite its missing and fallen stones, retains an aura of mystical power that still captivates visitors. The true purpose of Stonehenge remains an enduring mystery. We may never divine the original intentions of the industrious people who built Stonehenge. Nonetheless, we can be confident that this imposing structure inspires today the same awe and wonder that resonated in the souls and bosoms of these ancient people dwelling in the land that would one day become England.

* * *

On the longest day of the year, as the Sun Spirit begins to ascend the sacred sky, she lovingly surveys her serene surroundings. First, she spies a standing spectator, the heelstone, a splendid solitary sentinel. Second, she spots her trilithons made of sarsen stone surrounded by a set of huge stone slabs that shade their diminutive brethren the bluestones. Behind this noble vanguard, she sees her humble human servants advancing down a broad avenue to join the privileged acolytes, leaders, and shamans of the tribe who are already standing in rapture, watching her intently from the inner circle. Soon, all her devotees will be worshiping, praying, and asking for her blessing. She is pleased, and this particular assemblage she adores. She silently says to herself: No heavenly tears this day. For, after all, she is a blissful, fruitful, and benevolent goddess. She will bless all marriages with children, will ensure that rainfall is plentiful so that there will be sufficient food to last her people through the coming winter, and will honor the dear departed so that they may join her in the heavens. She will continue to communicate her decrees in signs that only the shaman and leaders may interpret as they study the astronomical signals in which her commands come veiled. She beams with pleasure and smiles radiantly.

Six months later, on the shortest day of the year, the Sun Spirit soundlessly descends beneath the horizon, secreting herself with the recently deceased. A long line of torch-wielding men, women, and children solemnly walk in twilight to the center of Stonehenge where they chant with deep emotion, pay their respects to the departed, and fervently wish for a swift return of the love, care, and warmth of their spiritual mistress. To her admiring multitude, she is everything. She is the All-Mother, the creator of life, and tonight, the herald of spring. This time of year is cold and dreary, yet filled with hope, promise, and portent, for her followers know that from this moment forward, if she so wills it, the days will once more grow longer and engender a rebirth of plant, animal, and human life. All the flora and fauna will flourish, and humans too may experience resurgent fertility. Later that night the Sun goddess hears the sounds of a grand celebration resounding among the stones and across the plain. She rejoices with the knowledge that her admirers are jubilantly celebrating, marking the beginning of a new year with raucous escapades of revelry, feasting, and dance, and she smiles inwardly.

Horsehead Nebula

7
A Gentle History of Astronomy Part I

An Ancient Astronomer of Alexandria

...the ways by which men arrive at knowledge of the celestial things are hardly less wonderful than the nature of these things themselves...
Johannes Kepler

The history of astronomy is a history of receding horizons.
Edwin Hubble

Since time immemorial, humanity's horizons have defined its existence. Our primitive ancestors' world revolved around the herds of ungulates thundering over the grassy veldt, the sprightly antlered quadrupeds rushing through evergreen forests, and the schools of fish swimming beneath the gleaming waters of an aquamarine sea. Only with the onset of night was humankind able to exceed the limits of daily existence and catch a glimpse of life beyond the mundane routines of survival, as the reddish-orange sunsets steadily morphed into lovely rotating starry panoramas. Studying the stars opened up new avenues of thought. Over time our ancestors discovered that the recurring star formations were reliable predictors of the future. It was as if the gods who resided in the starry heavens were communicating to humanity: "This is when the migrating herds will travel. Soon the floods will arrive to quench the thirsty roots of the sprouting plants. This is when the birds take flight to warmer climes." Thus, the cloudless skies of night that were humanity's first picture book became, with time, its first textbook—a textbook not only enriching our developing minds with abstract ideas but energizing the ongoing search for better ways of life. From humble origins around a campfire, humankind marshaled all the ingenuity, tenacity, and perspicacity that it possessed and commenced the ascent towards a better society and a better life. The upward journey of humanity, which began ages past and continues today, is like climbing a mountain path; with each step upward, wider and grander vistas open all around.

Succeeding generations honed their mental abilities, and for thousands of years, the sages of our species sought to positively shape their world and humankind's future. These individuals focused all their resources and intellect upon improving their vantage point. They discovered hidden passes through mountains of dogma to develop sounder theories about nature, crossed seemingly impassable crevasses of ignorance with profound leaps of intuition that allowed them to solve apparently insurmountable problems, and surpassed the physical limitations of their senses by developing increasingly more powerful tools to extend their range of vision. In a process that could be considered the epitome of the human experience, humankind reached ever higher vantage points physically, intellectually, and imaginatively.

Astronomy was born the day humankind realized that the way to a better future was through a deeper understanding of the stars. In astronomy, each advancing step is unique. One might be the invention of a new piece of equipment with which to examine the heavens, such as Galileo's telescope or the Large Interferometric Gravity Wave Observatory (LIGO). Occasionally it is the sudden arrival of a new insight like Newton's apple or Einstein's equating the force of gravity to a ride in an accelerating elevator. Sometimes a key step is the development of a new theory, such as Copernicus' world-shifting cosmology or Newton's elegant gravity equation. Regardless of their form, each step leads to the next, and together these steps constitute the uplifting of humanity to ever higher realms of understanding from which the view is more magnificent than before.

In recent days, we have learned to take flight. The millennia-old question about the flatness of the Earth has been resoundingly answered. The Earth has been viewed, photographed, and filmed from space, which has utterly removed all doubt about our planet's spherical shape. Thanks to courage, diligence, and genius, we gained the vantage point of the Moon; tomorrow, our vantage point may reach to Mars and beyond. Humanity will always surpass its previous boundaries as it unremittingly pushes against the confining envelope of our limited vision and knowledge.

Humanity is not alone in utilizing star-knowledge. Migrating birds steer by stars, dung beetles maneuver in harmony with them, and mammals set their waking and sleeping cycles according to what is happening overhead. Grunions synchronize their mating schedule by the phases of the Moon, as does South Pacific coral. Animals such as insects, grunions, and corals, however, never seek wider horizons. Only humankind, with its superior mental faculties, has been able to lift itself to never before imagined heights from whence it can survey ever expanding fields of knowledge and pursue ever-greater pinnacles of achievement. This is more evident in astronomy than in any other discipline. Truly, astronomy, as the physicist Henri Poincaré said, "raises us above ourselves." This uplifting process began with the ancients.

Ancient Astronomers

Tens of thousands of years ago, near the time when the sparkling diamonds of the firmament first captured human attention, our ancestors began to rigorously study the sky. Ancient astronomers did not possess telescopes, so they were limited to observations made with the naked eye and rudimentary sighting devices. With only these tools, ancient astronomers made a number of profound discoveries, and, after the invention of writing, left detailed records, diagrams, and accounts of their sightings. It is certain that they not only looked but also sought explanations for what they observed. For those early humans, it was not enough to know that the Moon appears to change shape; they were curious to know why. When they did not know the answer, they fabricated one: The days get shorter because the goddess of springtime is being held captive in the underworld. The Great Bear never dips beneath the horizon because Hera, Queen of Mount Olympus, has decreed that it will never be permitted to bathe. The river floods because it is reacting to the heat cast upon the Earth by the rising of the Dog Star after its long sleep. These are tales of the type that might have been told to explain astronomical phenomena people could see but not understand.

The first astronomers left no written records because human beings had not yet learned how to write. However, they left something much more impressive than books or inscriptions. They bequeathed to their descendants imposing yet cryptic monuments—the ones that survive are made of stone—that reveal to experienced and perceptive experts a surprising amount of what the preliterate ancients knew about astronomy. An outstanding example is Stonehenge, a circle of huge standing stones in southern England. On the morning of the summer solstice, the rays of the rising Sun bisect the henge perfectly, as does the last glint of sunset on the winter solstice. Features of Stonehenge's design strongly suggest that its unlettered builders were able to predict lunar eclipses.

For a preliterate society, interest in astronomy would not have been a "pure" or abstract pursuit. Knowledge of astronomy conferred powerful survival advantages. Being able to time the solstices and equinoxes went hand in hand with expertise in the recently developed agriculture of the New Stone Age, as well as helping to predict the migrations of the great herds. The rotating canopy of the stars taught us the very concept of time. The luminous gems that punctuated the night sky were punctual in their movements. The rising and setting of the Sun defined the day, the phases of the Moon delineated the month, and the movements of the stars communicated the passing seasons. Stars revealed the passage of time, and the more prominent stars taught the first tillers of the soil when to sow and when to reap. From out of the mists of time, between myth and history, circumpolar stars, constellations, and asterisms guided early navigators as they plied their trade. They shepherded shrewd spice merchants and their caravans of dromedaries laden with redolent resins through blistering desert siroccos. The stars steered discerning dealers in precious metals and ivory ferrying their wares in nimble dhows across treacherous wind-tossed Eastern seas. They directed savvy silk traders and their Bactrian beasts of burden over narrow, Khyber-like mountain passes.

With each successful passage, newly learned concepts were spread across the continents, and from this wealth of hard-earned knowledge our ancestors' traditions developed, explanations proliferated, and cultures expanded. It was as if merchants, traders, and the bankers who financed them entrusted their safe passage, and thus their wealth, to the impassive vault of the heavens.

Astronomy was being carried out in a variety of regions in the ancient world. Petroglyphs on the American continent record astronomical observations made there. Astronomy was a part of ancient Chinese culture. Finally the Mesopotamians, who most scholars believe were the first to invent writing, left the oldest astronomical records that have survived until the present.

The Mesopotamians

The Fertile Crescent is a region of the world that stretches from the present-day countries of Iraq and Kuwait in the east to northern Egypt in the west. The eastern portion of the Fertile Crescent contains two mighty rivers, the Tigris and the Euphrates, which flow through Iraq and eventually join before draining into the Persian Gulf. The land between these two rivers is called Mesopotamia and is often referred to as the cradle of civilization. Mesopotamia has a rich history that began approximately 8,000 years ago. This wonderfully fertile land was the most densely populated region of the world for thousands of years and boasted a succession of powerful city-states which gave rise to influential civilizations including the Sumerian, Akkadian, Babylonian, Assyrian, and Chaldean. Evidence indicates that these groups interacted often and shared much. All were agrarian societies that started as small villages. All employed the same underlying technologies including the wheel and the bridle. All were polytheistic, and their building projects were comparable to each other in purpose, scale, and style. Among the magnificent structures they erected were shrines, temples, ziggurats, and massive walls. They all shared similar political structures (autocratic governments), schools, writing, legal codes, literature, mathematics, astronomy, and astrology. Since they had so much in common, these various early civilizations are often collectively referred to as Mesopotamian.

The ancient Mesopotamians wrote on clay tablets, which when baked become hard as stone and can endure for thousands of years. Thus, today, we can pluck these tablets from where they have lain in sand-covered oblivion for millennia and read records of observations of the planet Jupiter that were pressed into wet clay with a pointed reed stylus over 4,000 years ago. These clay tablets are the oldest written records still in existence and possess both archeological and artistic value. They reveal, among other things, that before 2000 BCE the astronomers of Akkad had a zodiac almost exactly like ours today.

An ancient library has been preserved at Nineveh, from which we learn that star-charts were even then divided into constellations, many of which were known by the names which they bear to this day. The signs of the zodiac, already in that distant epoch, were used for determining the paths of the seven "dwellers of the sky": the Sun, the Moon, and the five visible planets Mercury, Venus, Mars, Jupiter, and Saturn. Every early culture that we know of studied the signs in the heavens, and the knowledge they accrued served to guide sailors, farmers, and rulers alike.

Later Mesopotamians erected astronomical observatories and tracked the movements of the stars and planets with awe-tinged attention. They believed that the stars that did not twinkle were the homes of the gods, or were the gods themselves—divinities with the power to dispense life or death depending upon their benevolent or malevolent nature and how well they had been propitiated. The ancients felt they were catching a glimpse of a deity every time they saw bloodshot Mars or gleaming Venus.

The Babylonian civilization endured in Mesopotamia for thousands of years, and by 900 BCE, this superstitious culture had compiled lists of over 7,000 different omens. One of their preferred methods of divination was the interpretation of sheep entrails, but they also looked to the heavens for guidance. They felt that the stability of their political system depended on the interpretation of omens, because a ghastly one could portend the death of a king or even the toppling of a kingdom. Concerned in particular at how comets, which were considered unusually grave omens, appeared and disappeared unpredictably in the sky, the Babylonians systematically sought ways to improve their ability to predict the future. Therefore, from 900 BCE until about 150 BCE they kept meticulous astronomical records (e.g. the study of a Babylonian tablet revealed that Halley's Comet was observed during 163 BCE), which were used exclusively for astrological purposes. Most serious thinkers today dismiss astrology as a mixture of delusion, superstition, and wishful thinking, but to ancient humans, it constituted a plausible and seemingly useful set of beliefs. Their longing to understand astrology proved to be a boon to future generations as it led directly to increased knowledge of astronomy.

Pythagoras

From their meticulous records, the Babylonians were able to extract details of some of the very subtle and hard-to-detect movements of the Sun, Moon, and planets. The Babylonians kept their reputation as the world's foremost experts in astronomy and mathematics for centuries, so that many of the greatest Greek mathematicians, such as Pythagoras (left), for example, were said to have studied at the feet of the Babylonian mathematicians. In Roman times, more than a millennium after the heyday of the Babylonian savants, mathematical calculations relating to astronomy and astrology were still being referred to as "Babylonian numbers." Even today, when we divide the hour into sixty minutes or the circle into 360 degrees, we are following the base-60 system used by the Babylonians. The Babylonians made impressive discoveries such as the Saros cycle, which is an 18-year cycle of solar and lunar eclipses.

It was precisely because the motions of the planets could be analyzed mathematically that astronomy made such rapid advances in Babylonia and later in Greece. Physics, on the other hand, languished until the 1600s, when scientists finally started using an advanced mathematical approach. Astronomy also benefited from its close connection with religion and philosophy, which provided a social value that other sciences simply could not match.

The Egyptians

Egyptian civilization arose shortly after that of Mesopotamia and grew to greatness at roughly the same time. A green ribbon of life hugged the ancient Nile River, stretching from the almost impassable cataracts of the Nubian Desert to the Mediterranean Sea. Upon this luxuriant strip of plenty arose the fabulous civilization of ancient Egypt. The mighty Nile River served this civilization as a highway upon which slow-moving barges sailed carrying textiles, food, exotic animals, and building stones. The Nile was also a plentiful source of water for irrigation, and its annual floods brought fresh layers of rich silt each year to reinvigorate the depleted soil. The Nile was the very lifeblood of Egypt and still is today.

The star Sirius was connected to the annual flooding of the Nile River, for the heliacal rising of Sirius portended the coming inundation. In addition, Sirius anchored the Egyptian calendar. Their twelfth month always had Sirius rising, and if Sirius did not rise in that month (as sometimes happened due to the gradual accumulation of errors over the course of decades or centuries), then the calendar would be changed. In other words, the erroneously constructed human calendar was adjusted to match the obvious "perfection" of the stars.

The fundamental religious belief of the Egyptians was that their Pharaohs were living gods. At death, these godlike beings were transported to the afterlife to reside with their kin. They needed, of course, all the accouterments of a living god: ornate furniture; treasures of gold, silver, and iron; weapons; chariots; games; and even figurines of servants to attend them throughout the afterlife. No expense was spared in supplying the deceased with sure access to eternity, and this included the construction of tombs so awe-inspiring that to this day they remind us of the ancient glory of the civilization that made them rise above their windswept plateau and point toward the hallowed heavens. As tombs for their pharaohs, the Egyptians of the Old Kingdom built massive stone structures, one of which, the Great Pyramid of Khufu on the Giza Plateau, dwarfs the others and was the tallest building on Earth for thousands of years.

The great pyramid of Khufu is almost perfectly aligned with the north geographic pole. It is more accurately positioned than the Greenwich observatory, built thousands of years later. The precise alignment, off by less than one-twentieth of a degree, shows the skillfulness of the Egyptians at astronomical observation, because only by careful observation of the sky and the movements of the circumpolar stars can one find true north. Today, the star Polaris is located at almost exactly true north, but when the pyramids were built, about 4,500 years ago, it was not there. Instead, the star Thuban was near the celestial north pole (nearer, in fact, than Polaris is today) and most likely served the people of that era as their North Star. Thousands of years from now, Vega will be the pole star; the cycle, brought about by precession, repeats itself every 25,868 years.

The physical arrangement of the three pyramids on the Giza Plateau matches the configuration of the three stars of Orion's belt. Furthermore, the southern shaft of the King's chamber in the Great Pyramid aligns with Alnitak, the brightest of those stars. The stars of Orion's belt symbolized for the Egyptians their god Osiris, who was associated with resurrection and rebirth, concepts of vital importance to Pharaohs obsessed with immortality. The northern shafts penetrating the Great Pyramid pointed towards Kochab, the second-brightest star in Ursa Minor, which was associated by the Egyptians with the immortality of the soul. The southern shafts were said to point towards Sirius, which was equated with the cosmic mother of the kings of Egypt.

The alignments that researchers have noted may be no more than coincidences. The tantalizing connections between the interior passageways of the Great Pyramid with Sirius and the stars of Orion's belt, while relating the myths of Egyptian afterlife with the Egyptian gods Osiris and Isis, may be an example of finding connections that are not really there. All we can be sure of is that the pyramids, their shafts, and the stars are aligned in several intriguing ways. However, a people who could build the greatest and most enduring of the seven wonders of the ancient world (the world's tallest building for almost 4,000 years); a people who could erect 2,300,000 stone slabs averaging 2.5 tons in 20 years and then cap it off with 144,000 casing stones of highly polished limestone, and place it with such precision; a people who could do all this could indeed have aligned the inner passageways of their Great Pyramid with the stars if they so chose.

The Chinese

Thousands of miles from Mesopotamia and Egypt, a civilization flourished between forested mountains on a broad fertile plain through which the Yellow River (Huáng Hé) flowed in what is now China. By about 2000 BCE, the inhabitants of that region had established a prosperous society with agriculture, metalworking, ceramics, and writing.

The Chinese felt there was a close connection between their imperial rulers and the heavens. One of the titles of the Chinese emperor was Son of Heaven. This may explain one possibly apocryphal account of two Chinese astronomers who were executed in 2159 BCE for failing to predict an eclipse. Besides giving an insight into the harshness and brutality of that ancient society, this incident also reveals that astronomers in China at that early date were already in possession of a method of eclipse-prediction that was expected to be perfectly reliable. Failure to predict an eclipse would have reflected poorly on an emperor, calling into question his attunement with the heavens.

Although the Chinese invented their system of writing around 4,000 years ago, records from that era are spotty because of societal instability and the perishability of their writing materials. Recordkeeping improved by 776 BCE, and it is clear from the records that the Chinese were by that time quite adept at astronomical observation. In particular, they did a better job of documenting comets than their European counterparts of the same period. Visits of the comet now known as Halley's Comet to the inner Solar System were recorded by Chinese observers in 240 BCE and 12 BCE. These visitations either escaped detection by Europeans or were noticed but not recorded. The Chinese were also the first to discover sunspots.

The Mayans

The Mayan civilization of Central America wove religious, astronomical, and calendrical aspects of their culture together. Celestial events were considered messages from their gods, and cities were laid out on the consideration of celestial phenomena. The wise men who studied the heavens could seemingly predict the future and held great power. The most important object in their multi-layered universe was the Sun, which they acknowledged as the giver of all life. Its symbol was a red eagle with a large, all-seeing eye.

The Mayans demonstrated the sophistication of their astronomical knowledge by building impressive monuments. These Mesoamerican constructions show that their builders, just like the Egyptians, were interested and

Head of Kukulcán

El Castillo (Mayan Temple)

knowledgeable about the timing of the Sun's movements. In the Mayan city of Chichén Itzá, one of the sacred pyramids, called El Castillo or the Temple of Kukulcán, was constructed in such a way that on the vernal and autumnal equinoxes, shadows cast by the Sun onto the pyramid create the illusion of a 120-foot long snake slithering down the steps (visible in picture on the lower right of the preceding page), terminating in a dramatically carved stone head (preceding page, lower left). The descending snake was their god Kukulcán. El Castillo is a step pyramid with a square base and a flight of 91 steps on each of its four sides. An extra step on top makes a total of 365 steps, representing the days in a year.

El Caracol (Mayan Observatory)

The "observatory," called El Caracol (the snail) by the Spaniards because of its spiral staircase, is a partially collapsed tower in Chichén Itzá (left). It still boasts a square base with its corners aligned to the equinoxes and windows that track the movements of the planet Venus.

Some of the Mayans' astronomical calculations were impressively accurate. They calculated the length of a lunation (the period between two full moons) at 29.53086 days, not differing from todays accepted value of 29.53059 until the fourth decimal place. We know of only a tiny portion of the Mayans' astronomical knowledge because Spanish conquistadors burned almost all of the books and records the Mayans left.

The Greeks

Greece is a country of rocks, mountains, small islands, jagged valleys, and narrow inlets of the sea. Early city-states were separated by walls of mountains. Short on arable land, the Greeks developed a seafaring way of life, utilizing trade while also seeking sustenance from the sea. Because of their land's harsh topography, the early Greeks became a hardy folk capable of withstanding invasions by vastly superior forces, and they maintained their freedom for centuries. This heritage of political independence also fostered a spirit of intellectual independence.

The Greeks inherited their astronomical information from the Mesopotamians, the Egyptians, and the Phoenicians (another advanced seafaring society of that era). Having taken possession of the invaluable knowledge of their predecessors, the Greeks built upon and expanded it. Like earlier civilizations, they harnessed knowledge of agriculture to the rhythms of the stars. Hesiod, a Greek poet of the eighth century BCE, recorded in his poem *Works and Days* the astronomical events—specifically the heliacal risings of the Pleiades, Orion, Arcturus, and Sirius—that were used by Greek farmers to time annual activities such as plowing, sowing, pruning, and harvesting.

Around the seventh century BCE, the Greeks were the originators of a dramatic shift in the way humans sought knowledge of the world around them. Until the Greeks, the default explanation for astronomical phenomena, and all others for that matter, was that gods or other supernatural entities were responsible for them. Beginning with Thales of Miletus (624-547 BCE, at right), the Greeks sought more naturalistic explanations. This set the stage for a blossoming of scientific knowledge, including knowledge of astronomy. According to one account, Thales, who earlier had studied with the Egyptians, was able to correctly predict a solar eclipse in 585 BCE, which caused such fear among the Median and Lydian armies that they laid down their weapons in mid-battle and fled the field in terror at that clear sign of divine anger. Thales also reportedly aided Greek navigators by disseminating the knowledge that the Little Bear was a better pointer to true north than the Great Bear.

Thales

Anaximander

Anaximander (611-547 BCE, at left) was a disciple of Thales. His accomplishments include the introduction of the sundial and measuring the amount of tilt in the Earth's rotation. He began the Greek tradition of cosmology by writing a detailed book in which he proposed a perfectly reasonable, yet erroneous, disposition of the Sun, Moon, and stars within the universe.

Early cosmologists believed that the Earth's position in the cosmos was maintained by some supporting object or beast, such as, for instance, an immense turtle. Anaximander was the first person to propose that the Earth required nothing to support it. This conceptual leap, enshrining one of the axioms upon which modern astronomy is built, was lauded by the philosopher Karl Popper as "one of the boldest, most revolutionary, and most portentous ideas in the whole history of human thought."

During the fourth century BCE, Eudoxus of Cnidus (c.390-c.337 BCE) recorded the first known theory of planetary motion. The book he wrote is no longer extant, but we have some quotes from it in works by other Greek authors. Eudoxus regarded each celestial body as being embedded in its own transparent sphere, with the concentric spheres all nested one inside another like Russian dolls, a view which remained popular for many centuries.

Eudoxus

Eudoxus (left) wrote at least two books: *The Mirror* and *The Phenomena.* Both are lost, but a poet known as Aratus of Soli versified *The Phenomena* around 270 BCE, and that version is still extant. The poetic form made it easier to memorize. Because of its popularity and accessibility, this work helped spread useful astronomical knowledge and promoted the study of astronomy. Aratus' poem records some of the earliest known names of stars and constellations and is the oldest known description of the heavens.

Eudoxus was an exponent of the idea that all heavenly bodies move in perfect circles. This principle possibly originated with Plato (428-348 BCE) and was later enshrined in Aristotle's work *On the Heavens.* Plato taught that the heavens are perfect and operate according to geometric principles. Since the circle is the perfect geometric shape, Plato reasoned, heavenly movements must be circular. Plato did not have a time-lapse camera, but he surely would have been delighted with the image on the next page (top) which shows the motion of the circumpolar stars over several hours. Plato strongly believed that geometry could be the means to a complete understanding of

the universe. Tradition has it that engraved at the doorway of his Academy was the warning: "Let no one ignorant of geometry enter." Plato issued a challenge to all astronomers to explain the movements of the stars and planets in terms of circular motion. Eudoxus was, as far as we know, the first to accept Plato's challenge and make exclusive use of circles in his cosmology.

Aristarchus of Sámos (310-230 BCE) was remarkable for deducing that the Earth revolves around the Sun, although this view never became popular in his time. Authorities such as Aristotle (384-322 BCE) argued, quite cogently, that if the Earth orbited the Sun, then the stars would appear to shift their positions over time, and no such shift was visible. Aristarchus' view made little headway against the voice of such an esteemed authority as Aristotle, who seemed to have conclusive evidence on his side. The shift that Aristotle was referring to is called "parallax," and it is an observable phenomenon, although the Greeks could not detect it. Our accurate instrumentation, much more sophisticated than anything the Greeks had, can measure the minute movements resulting from parallax. The first stellar parallax to be accurately measured was that of the star 61 Cygni, in 1838, by the German astronomer Friedrich Bessel. Aristarchus's heliocentric idea would not be widely accepted for almost 1600 years. He was truly ahead of his time.

Aristarchus (right) also devised a way to measure the distances between the Earth, the Sun, and the Moon. The method he developed involved observation of the angle between the Sun and the Moon with the Earth at the vertex when the Moon was half-full. His estimate was inaccurate due to lack of precision in the measurement of the angle, unavoidable due to the primitive nature of the instruments available to him. The inaccuracy of his measurement, however, detracts in no way from the brilliance of his method.

Aristarchus

Eratosthenes

Another prime example of the ingenuity and astronomical acumen of the ancient Greeks was the measurement of the circumference of the Earth by Eratosthenes of Cyrene (c. 276-c.194 BCE, at left) in the third century BCE.

He was aware that a vertical post in the city of Aswan, in Upper Egypt, cast no shadow at noon on the day of the summer solstice, while such a post in Alexandria did cast a shadow at the same moment. Knowing how far apart the two cities were and the length of the shadow cast in Alexandria, he was able to use simple geometry to calculate the angle between the two cities with the center of the Earth as the vertex of the angle, and then deduce the circumference of the world. His estimate was probably off by only about five percent, an amazing accomplishment, although there remains some uncertainty about the exact length of the unit he used. Note that there was no question in Eratosthenes' mind about whether the Earth was a sphere or not. In fact, Aristotle had previously presented compelling evidence of the Earth's spherical shape, pointing out that during a lunar eclipse the Earth's shadow upon the Moon was always circular. Also, it had long been observed that the last visible part of a departing ship was the tip of its mast, as would be the case if the vessel were traveling away on the curved surface of a sphere.

Apollonius

Apollonius of Perga (262-190 BCE, at left) made the contribution of the off-center circle which would account for the observed speeding up and slowing down of a planet. The "great geometer" is best known, however, for his books on conic sections—what he did for conics has been compared to what Euclid did for geometry.

Conic sections are curves that can be created by slicing a cone with a plane. (right). Apollonius investigated the curves in the spirit of pure inquiry and the joy of discovery. Astonishingly, these very curves proved many centuries later to be the key to understanding how celestial bodies actually travel through space.

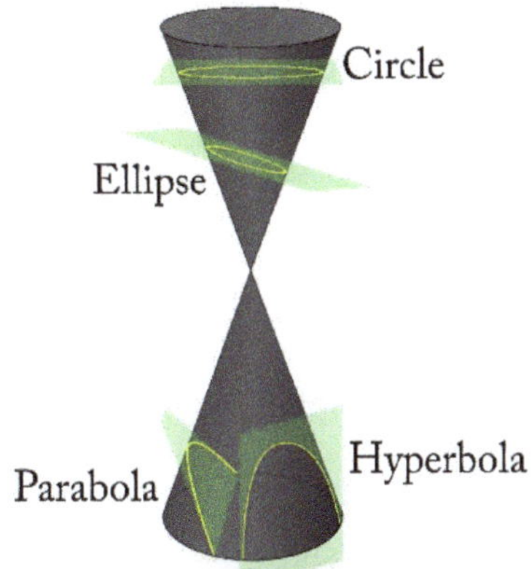

Conic Sections

Hipparchus

The astronomer Hipparchus of Nicaea (c. 190-c. 120 BCE, at left), revered by many as the greatest of the Greek astronomers, compiled the first known catalog of stars in the second century BCE. He included reports of eclipses from Babylonian records dating back to 800 BCE. He developed a system for classifying the brightness of stars, and the system we use today is based on his method. Hipparchus improved on Aristarchus' measurement of the distance from the Earth to the Moon (59 Earth Radii), and he was off by only one-sixtieth of the true distance. Amazingly, his estimate of the length of the year was accurate to within about 200 seconds of the value accepted today. Hipparchus also discovered the wobble in the Earth's axis of rotation, a truly awe-inspiring achievement as this wobble, called precession, proceeds at the rate of less than one minute of arc per year. In the meantime, Hipparchus invented plane and spherical trigonometry, an excellent example of how research in one field (in this case, astronomy) can go hand in hand with discoveries in a related field (mathematics).

Arguably the most important of the Greek astronomers was Claudius Ptolemy (85-165), known as the "prince of astronomers." Although his nationality is unknown, he worked in Alexandria, then in the Roman Empire, and wrote in Greek. His lasting influence sprang from his masterwork, *η μαθηματικη συνταξις* (*Hē mathēmatikē syntaxis*). This was translated into Arabic as *Almagest* (*The Greatest*), although the original Greek title meant "Mathematical Collection." This compendium of astronomical knowledge was so useful, thorough, and authoritative that historians rarely find astronomy books from earlier periods. The practice of the time was to discard older, outdated books (which were a nuisance to store and maintain) when better ones became available, and the *Almagest* was recognized for over a thousand years as the epitome of excellence.

Ptolemy

Ptolemy (left) based his research on the idea that the movements of heavenly objects could be expressed mathematically. Using careful observation and reams of data, he concluded that the planets, the Sun, and the Moon move in circular orbits around the Earth. To account for phenomena such as retrograde motion (the appearance of a planet changing direction in its movement from night to night), Ptolemy added additional circular movements to the orbits. Unfortunately, these additional circles proliferated uncomfortably as more and more were added to account for the data. The resulting complicated system did manage to closely match the observed movements of the heavenly bodies, and over a thousand years passed before it was finally rejected. This system, in which Earth is at the center of the Solar System (or the universe), is called the geocentric theory and is still referred to as the Ptolemaic system. Ptolemy should not be harshly criticized for being wrong. He used the data at hand to create a reasonable hypothesis, and he did not have accurate enough instruments to disprove it. His mathematical ability was exceptional, he was skilled at trigonometry, and, perhaps even more importantly for his influence on subsequent generations of astronomers, was a prolific writer. In addition to his *Almagest*, he penned works on optics, trigonometry, and astrology.

Ptolemy's geocentric theory was mistaken, but the tables in his *Almagest* were accurate enough for practical use. Christopher Columbus relied on data from Ptolemy to successfully navigate to the New World. By finding the relevant numbers in the correct table in the *Almagest*, a navigator or astrologer could determine the location of any of the known planets or major stars, which, along with measurements taken on site, would reveal the traveler's position on the surface of the Earth (or at least the latitude; the measurement of longitude proved to be a much thornier problem). The *Almagest* contained a catalog of about 1,022 stars, arranged into 48 constellations.

The Renaissance sage Bacon was undoubtedly referring to Ptolemy when, giving credit to early humans for the first star-catalogs, he went on to say that our present listings of the stars were "as they come to us blown softly through the flutes of the Grecians." When Bacon said this, he was likely thinking of the lyrical language of the Greek poets and their love of music as contrasted with the prosaic, everyday speech used in cosmological and other treatises of their predecessors. Bacon was assuredly right to acknowledge the contributions of the Greeks, originators of so much of what we think of as Western culture, including science, mathematics, and democracy. He clearly understood the role the Greeks played as recipients, guardians, and developers of astronomical knowledge.

The Arabs

After the great civilizations of Greece and Rome passed from dominance, the cultures of Europe in the following age were little interested in the study of astronomy. This shift in emphasis began in Roman times, since the Romans, notwithstanding Ptolemy and the great accomplishments of the scholars of Egyptian Alexandria, were more interested in administration and conquest than activities like astronomy, which they considered effete. When the Roman period ended around 500 CE, all of Europe descended into the "Dark Ages," so called because life in the time of the Black Death and continual feudal warfare led to a focus on survival with little time to devote to intellectual pursuits.

With Europe in this state, then, it fell to the Arabs to become the guardians of ancient knowledge. Many of the ancient Greek books were translated into Arabic at this time, to be retranslated into Latin centuries later during the era of the European Renaissance. This is the only reason we possess these ancient works; without the Arab conservators, much of the scholarship of the Greeks and their forebears would have been lost to us forever.

During the Abbasid dynasty (750-1258), people came from near and far to study and do research in the capital, Baghdad, at an institution known as the House of Wisdom. Astronomers at the House of Wisdom studied the works of the ancient Greeks and made many discoveries of their own. Their observations were so painstaking that they were able to detect subtle changes that were taking place very slowly in the heavens. For example, in the ninth century, the Baghdad astronomers observed that the obliquity of the ecliptic (the angle by which the ecliptic deviates from the celestial equator) had decreased from the value given in Ptolemy's *Almagest*. From Ptolemy's time to theirs, the obliquity of the ecliptic, which is related to the Earth's tilt, has changed by only a tiny fraction of a degree, but the Arab savants detected this minute change.

Alhazen

Ibn al-Haytham, latinized as Alhazen (c. 965-c. 1040, at left), was born in Basra in what is now Iraq, where he made significant contributions in mathematics, astronomy, and physics. Recognized by his peers and later commentators for his aptitude, abilities, and acumen, this Arabic scholar was called Ptolemaeus Secundus (the second Ptolemy) due to his phenomenal range of interests and fascinating discoveries. It was a rare researcher in those days who used experimentation to test theories, but Alhazen worked according to the principle, "Prove what you believe!" He was not afraid to question accepted wisdom if it was not supported by the evidence. His philosophy about old knowledge can be summed up by the following quotes attributed to him: "If learning the truth is the scientist's goal…then he must make himself the enemy of all that he reads." "Blindly following ancient customs and traditions doesn't mean that the dead are alive, but that the living are dead."

Alhazen authored over 100 books, half of which are still extant. A quarter of his books were astronomical works that discussed a variety of technical issues like how to make an accurate observation or determine the location of the Milky Way. However, he also wrote books on physics, such as a seven-volume study of optics in which he gave experimental proofs of the nature of light, the structure of the eye, and principles of reflections in mirrors. Alhazen also invented the camera obscura, a room that functions as a pinhole camera, which produced an inverted image.

Alhazen was suspicious of Ptolemy's model of the universe and wrote an entire book titled *Doubts Concerning Ptolemy*. He hoped to discover the "true configuration" of the planets and create a model based on a complete geometric description that would explain the motions of the planets without the complications and difficulties that seemed to proliferate in Ptolemy's model. Alhazen accepted the geocentricity of Ptolemy and his epicycles, but he never accepted the equant, an imaginary point invented by Ptolemy that he used to explain a planet's movement as its speed changed in its orbit. Alhazen abhorred what he considered arbitrary mathematical inventions to describe reality.

Omar Khayyam

Though not an Arab, the Persian Omar Khayyam (1048-1131, at right) was a great astronomer of the widespread Islamic culture that originated in Arabia. More famous today as a poet, author of the famed *Rubaiyat of Omar Khayyam*, he was also a mathematician, astronomer, and government official. His knowledge of astronomy and mathematics was so advanced that the calendar he devised, in which eight leap years occur every 33 years, is more accurate than the calendar we use today.

At its peak, the Arabic civilization spread westward to the Iberian Peninsula, where Muslim and Christian nations existed side by side until the fifteenth century. In the city of Toledo, Spain, in the twelfth century, a school of translators flourished, where Moors, Greeks, and Europeans worked for centuries translating Greek and Arabic books, including books on astronomy, into Latin. One of the most important translators was Gerard of Cremona. He had heard of the *Almagest* and the wonders it contained, but he could not read it because no Latin version was available. Therefore he traveled to Spain and learned Arabic in order to be able to read and translate it. Residing in Toledo the rest of his life, Gerard translated Arabic versions not only of the *Almagest* but also Aristotle's *On the Heavens*, Euclid's *Elements*, and about 20 other works of astronomy and geometry. Thus, in one man's lifetime, most of the key works of ancient astronomy once again became available to Western civilization.

Despite the influx of knowledge from the Greeks by way of the Arab civilization, European astronomy did not regain the level of the ancient Greeks until 1496. Cardinal Bessarion requested Georg von Peuerbach, a mathematician, astronomer, and author, to write a synopsis of Ptolemy's *Almagest* in 1460. Unfortunately, Peuerbach died after completing six of the thirteen chapters of the abridgment. Fortunately, his student and colleague Johann Müller von Königsberg, known by his Latin name of Regiomontanus, completed the *Epitome of Ptolemy's Almagest* in about 1462. Subsequently, a highly acclaimed movable-type edition of the *Epitome* was published in 1496. This highly influential tome was a translation, redaction, and summary of Ptolemy's work that was widely disseminated among European astronomers and navigators.

Despite the influence of Ptolemy, whose monumental work was not only wrong but authoritatively so, a huge change was imminent. With the dawn of a new age of seeking and learning, the Renaissance, astronomers would begin to question the long-held dogmas of Ptolemy and embark on a new path to discovery and a fuller understanding of the cosmos.

8
A Gentle History of Astronomy Part II

Galileo Addressing Ecclesiastics

The publication of my writings will dispel the fog of absurdity by most luminous proofs.
Nicolaus Copernicus

The knowledge whose content makes up astronomy is the gain from more than 2,000 years' work on one of the most abundant objects of human knowledge, in which the foremost minds of all times have summoned up all the resources of genius and diligence.
Carl Friedrich Gauss

The Renaissance

The fifteenth century in Europe marked the end of the Middle Ages and the beginning of a more enlightened and open-minded age. A great awakening swept across Europe, in which intellectual curiosity and a desire to rediscover the lost knowledge of the ancients surged through academia and irrevocably altered the intellectual and cultural landscape. This effervescent period, known as the Renaissance, or rebirth, marked a rapid and sustained increase in technological, artistic, and scientific attainment, and advancements in astronomy began to bubble forth quickly. Science took its first tentative steps toward supplanting superstition, and human reason began its long march toward an ultimate triumph over dogma—a victory which, unfortunately, has not yet been fully realized today.

During the Renaissance, there was renewed interest in the ancient knowledge of the Greeks and Romans. It became fashionable for every educated person to learn Greek and Latin in order to sample the literature of the ancient sages, which included works on mathematics, science, medicine, and astronomy in addition to poetry and history. It was as if the Greeks, who thousands of years earlier had given birth to a new age of intellectual discovery, were inspiring the world for a second time. Progress accelerated as all fields of human endeavor were invigorated. A plethora of new ideas, many of them revolutionary, were advanced in biology, mathematics, engineering, and astronomy. The seeking of naturalistic explanations was once again in ascendancy and became the foundation for scientific progress. Like Archimedes' pulleys that allowed him to pull a fully laden three-masted vessel singlehandedly out of the ocean, the reawakened Greek influence pulled Europe and the world out of the Dark Ages and set us on the path which we have been following ever since—the path that could one day lead us to the stars.

Copernicus (1473-1543)

Of all the astronomers of the Renaissance, the Polish clergyman Nicolaus Copernicus (right) was possibly the most instrumental in bringing astronomy into the modern age. He achieved fame, after years of studying and calculating, by reviving and promoting Aristarchus's old and unpopular cosmological system in which the Sun is at the center of the Solar System. He reported that he had found in the writings of the Roman orator Cicero a reference to a fourth-century Greek philosopher Hicetas, who claimed that the Earth moved. Copernicus presented as his primary argument the observation that his theory placed the planets in a reasonable and logical order by distance from the Sun, distances that he was then able to calculate. This logical ordering had been lacking in geocentric theories. Copernicus based his theory on data published by Ptolemy; the essential clues had been there in the *Almagest* all along, and anyone could have found them. Despite this availability, Copernicus's system was still not widely accepted until a century or more after his death.

Copernicus was reticent about sharing his theory, and he delayed its publication, perhaps for fear of reprisals by religious authorities. He must have been concerned about attacking centuries of rigid consensus, and he knew that his scheme was not the only possible explanation. Gravity, for example, which according to Aristotle was a force directed toward the center of the Earth from every point in the universe, would need to be rethought entirely. Also, if Earth were just another planet orbiting through space, how could it have weather, earthquakes, and irregularities in its shape when the other bodies in space supposedly did not possess such imperfections? Copernicus claimed that he delayed publication for 36 years, and the book *De Revolutionibus Orbium*, containing the full explanation of his theory, did not come out until the year he died.

Brahe (1546-1601)

Tycho Brahe (right) became fascinated with astronomy as a youth when he witnesses a solar eclipse. He made his first recorded observation, a conjunction of Jupiter and Saturn, while still a teenager. When he compared what he saw with the existing reference books, he found that the accepted authorities were inaccurate by as much as a month. Then and there, smitten with zeal for astronomy he decided to become a meticulous and dedicated researcher, devoting himself for the next 58 years to accurate observation of the heavens.

The king of Denmark gave him an island and an observatory, and Brahe filled the observatory, which he named *Uraniborg* ("fortress of the Muse of the heavens") with new, larger, and more precise instruments of his own invention to aid him in his observations. Although it had nothing to do with his activities as an astronomical researcher, an interesting detail about Brahe is that he possessed a prosthetic nose, which he wore to conceal a disfiguring injury suffered in a rapier duel during his college days.

Brahe never owned a telescope because it had not yet been invented, but that did not stop him from gathering huge amounts of accurate data. He considered the updating and correcting of existing astronomical manuals an almost sacred duty. His discovery of a "new star" in 1572 was particularly important. This finding weakened the traditional contention that the heavens were perfect and unchanging, preparing the way for the emergence of new theories and a deeper understanding of the cosmos. The new star, located in the constellation Cassiopeia, was brighter than Venus. The remnant of this *nova* (Latin for "new") can still be seen with a telescope today.

Brahe's model of the heavens differed from that of Ptolemy and from the new theory promulgated by Copernicus. From observing the path of a comet, he deduced that the planets were not held in crystal spheres as the ancient Greeks had proposed. In Brahe's view, the Sun and Moon revolved around the Earth while all the other known planets orbited the Sun.

Copernicus and Brahe were the last significant exponents of the idea that heavenly motions were circular. Together, these two formed the fulcrum upon which the world of astronomy could begin to pivot away from the rigid philosophy that demanded perfect circles to the plasticity of a mathematically oriented science that acknowledged the possibility of ellipses. With the acceptance of ellipses, the Sun could now assume its rightful place at one of the foci of a conic section whose equations not only enabled the prodigious work and wonderful insight of Kepler, but also made the future sublime achievements of Isaac Newton possible. The combination of exact mathematics and a better understanding of the physical realm of matter and space enabled science to step out of the shadows of dogmatic tradition, cast a light upon the world of reality, and lift humankind onto a loftier path of discovery.

Kepler (1571-1630)

Johannes Kepler (right) worked for a time as an assistant to Brahe. In 1600, Brahe assigned Kepler the task of determining the path of Mars's orbit while withholding all his other data for reasons that remain unclear. Perhaps Brahe felt that Kepler had not yet earned access to the data. Very soon after this, in 1601, Brahe died, and a month later Kepler was appointed Imperial Mathematician. Unlike Brahe, Kepler accepted the Copernican (heliocentric) theory. Kepler continued to work on the Mars orbit problem and was able to use Brahe's own data to disprove the hybrid theory to which Brahe had clung. Kepler demonstrated, making use of Brahe's precise observations, that the orbit of Mars was not a circle as the Ptolemaic system ordained. Instead, Kepler found that an elliptical orbit with the Sun placed at one focus of the ellipse fit Brahe's data best.

Kepler ultimately formulated the three laws of planetary motion that bear his name. The first law states that planets orbit in ellipses, as he had shown for Mars. The second law asserts that a line segment drawn from the Sun to a planet sweeps out equal areas in equal lengths of time. Kepler's first two laws were published in 1609. Nearly a decade later Kepler published his third law, which holds that there is a fixed mathematical relationship between a planet's orbital period and its average distance from the Sun, and this relationship is the same for all planets. Kepler's dogged determination over twenty years of arduous calculation and testing resulted in his three laws, which represent an awe-inspiring achievement and remain the basis of our understanding of planetary motion today. Kepler's laws proved instrumental to Isaac Newton for verification of his law of gravitation since bodies moving under the influence of a force like gravity can be shown mathematically to orbit in ellipses and obey the other two laws of Kepler as well. Surely Isaac Newton was referring to Kepler, among others, when he said that in having seen farther than others, Newton had "stood upon the shoulders of giants."

The year 1609 proved to be a bellwether year for astronomy. That year saw not only the publication of Kepler's proof that planets orbit in ellipses, but also the invention of a device that would augment the power of human vision and allow searchers of the skies to see more than they had ever seen before. That device was the telescope.

Galileo 1564-1642

Throughout most of history, astronomers investigating the heavens relied upon the power of their own eyesight. Rudimentary instruments for measuring the locations of stars and planets existed, such as the quadrant and sextant, but the advantage went to the sharp-eyed.

Early in the seventeenth century, surely the most noteworthy event in the history of astronomy occurred. Some resourceful tinkerers in Flanders inserted a pair of spectacle lenses into a tube (eyeglasses had been in use for over 300 years), and a new era in astronomy dawned. The Flemish craftsmen had just invented the telescope, a device that would revolutionize the world of astronomy.

When news of the invention reached the Italian mathematician Galileo Galilei (above), he immediately realized that the new "spyglass" would be of immense value to astronomers. He set to work building his own telescopes and in one day surpassed the expertise of his Flemish predecessors. The discoveries Galileo made when he turned his telescope toward the sky almost immediately revolutionized astronomy, upending millennia of accepted dogma and forcing astronomers down new paths of discovery that would call humankind's place in the universe into question and forever alter the way we think of our planet, our universe, and ourselves.

Since the era of Claudius Ptolemy and his Plato-inspired belief that heavenly objects must move in perfect circles, the ancillary belief that everything about the heavens was perfect and immutable had held sway for a millennium and a half. Galileo's telescope quickly and irrefutably debunked that belief. His discovery of mountains on the Moon showed once and for all that a heavenly body could have imperfections and did not have to be a perfect sphere.

His greatest discovery was that the planet Jupiter possesses its own moons, which orbit Jupiter just as our Moon orbits the Earth. Adherents of the Ptolemaic model believed that everything in the heavens orbited the Earth, so Galileo's observation of the moons of Jupiter was an epochal event.

No longer could any reasonable person hold to the antiquated view that everything orbits the Earth, because anyone with a telescope could see four moons orbiting Jupiter. Galileo's revolutionary discoveries provided clear evidence in favor of the Copernican view, the heliocentric model.

Galileo's discoveries famously fell afoul of the Catholic Church. A scriptural passage recounting one of the military exploits of the biblical hero Joshua refers to the Sun miraculously standing still in the sky. This implies that under ordinary circumstances the Sun moves around the Earth. According to the Church leadership of Galileo's time, anyone claiming that the Earth moves around the Sun is thus contradicting holy writ and must be corrected, by force of judicial action if necessary. After many vicissitudes, Galileo was punished by being placed under house arrest and had to recant his claims. He doubtless counted himself lucky to have avoided being burned at the stake, a punishment which befell no small number of those who disagreed with the Church's teachings, among which was Giordano Bruno, who espoused several heretical beliefs, including that the stars were enormous balls of fire and had planets orbiting them. In 1979, the leader of the Catholic Church admitted that Galileo had suffered injustice at the hands of the Church, but no formal pardon has ever been issued.

The man who invented the astronomical telescope also made impressive discoveries in physics, such as the acceleration of objects due to gravity and the timing of the movements of a pendulum bob. He even made a serious attempt to measure the speed of light. Galileo was a true giant of science, one of those upon whose shoulders Isaac Newton would stand.

Newton 1643-1727

On Christmas day of the year after Galileo died, a child was born who would become the greatest scientist the world had seen up to that time, and possibly the greatest theorist and mathematician of all time. That child was Isaac Newton (below).

Newton's work changed astronomy forever by stating a law in the form of a mathematical equation that governs how bodies on Earth or in space attract each other. In other words, he quantified the force of gravity. His simple formula for this force was one of the crowning achievements in the history of human thought. Researchers were now able to explain the movements of heavenly bodies according to this gravitational force.

Using his simple mathematical law, Newton was able to calculate the masses of Jupiter and Saturn. The huge numbers his calculations yielded were surprising yet absolutely correct. Newton was also able to explain ocean tides, as well as those minute irregularities in the Moon's orbit known as nutation and libration. He ascertained that the Earth is an irregularly shaped spheroid, bulging at the equator and flattened at the poles. A key to his discovery was that the masses involved could be treated as if all the mass were located at the exact center of the body, which dramatically simplifies the use of his formula.

It was Newton who began the rigorous study of the spectrum. He let light into a room through a small hole and split the resulting beam into its constituent colors by use of a prism. Newton's study of the spectrum laid the groundwork for later studies of the spectra of starlight, which permitted astronomers to determine the chemical constituents of stars, including the Sun. Light spectra also are used to measure the Doppler shift, which reveals how fast distant galaxies are moving away from us.

The type of telescope used by Galileo, in which light passes through lenses enclosed in a tube, is called a refractor. Newton invented a telescope in which the light does not pass through lenses but is reflected by a curved mirror. The vast majority of large telescopes today are of this type and are referred to as reflectors. This invention alone would have made Newton an immortal hero of astronomy. Upon seeing Newton's reflecting telescope for the first time, the members of the Royal Society immediately elected him a member of that august body of English science. Newton also described how moons and planets move according to his law of gravitation. Furthermore, he invented calculus, a system of mathematics which is indispensable not only to astronomy but to all branches of science.

Newton was the "Lion of Cambridge," whose mathematical claw all others recognized. The intellectual achievements of this British icon surged through the world of science and mathematics, forever removing us from the dark ages and hurling us down the path to the Moon, the planets, and beyond. The breadth and power of his contributions are unparalleled.

Newton and his contemporaries established the fertile bed on which modern science could begin to flower. New discoveries blossomed forth in rapid succession like a profusion of wildflowers emerging in springtime on a mountain meadow.

Halley (1656-1742)

Halley's Comet

Edmond Halley (right) was a close associate of Newton. He gained fame by predicting the 1758 appearance of the comet now known as Halley's Comet (left). He thus proved that the orbits of some comets are periodic. Another very important contribution of Halley's was that he facilitated the publication of Newton's discoveries, a pivotal moment in the advancement of knowledge. Halley was an extraordinary man; not only was he important for astronomy in his own right, but he was an intellectual giant who taught himself enough Arabic to translate Apollonius of Perga's treatise on conics into Latin.

Doppler (1803-1853)

Christian Doppler of Austria described how the frequencies of light and sound waves are affected by motion. He deduced that since the pitch of sound from a moving source varies for a stationary listener, the color of the light from a star should change depending on how fast and in what direction the star is moving. This phenomenon became known as the Doppler shift and conveys essential information about the nature of the cosmos. For example, Edwin Hubble based his discovery of the expanding universe on this effect, and the search for dark matter was undertaken because anomalous galactic rotation speeds were detected from Doppler shifts.

New Planets

Herschel

Since time immemorial, humans have taken note of the presence in the night sky of five planets: Mercury, Venus, Mars, Jupiter, and Saturn. They are distinguishable from stars because they varied in their configuration from time to time. Whereas stars remained fixed in the same constellations, the planets would wander from one constellation to the next over a few weeks, sometimes even appearing to change direction and reverse course.

It was not suspected that additional planets might be lurking out in space somewhere, too dim and unobtrusive to be seen and identified. However, a part-time musician named William Herschel (1738-1822, right top), who indulged in an expensive and painstaking hobby of making large telescopes (right), discovered a light which was not where it belonged as he was surveying the sky one night. He soon realized that it was a planet, the one we now call Uranus. This discovery elevated Herschel to prominence and knighthood. (Though German, he lived in England and was knighted by the English queen.) He also discovered moons of Saturn, although he was not the first to do so, and detected the existence of ice caps on Mars. Yet another astonishing achievement of Hershel was the discovery of infrared radiation. He detected it by placing a thermometer just beyond the red band of light on the outer edge of the colored spectrum dispersed by a glass prism.

Herschel's Giant Telescope

Once astronomers knew about Uranus, there was a rush to see if there might be still more undiscovered planets. Making use of Newton's law of gravitation, researchers found irregularities in Uranus's orbit that could be accounted for by a massive additional planet perturbing Uranus from its expected path. Calculations led to a prediction of where to find the dim, mysterious body, and it was duly observed, by Johann Galle, in 1846. This was the planet that became known as Neptune.

Clyde Tombaugh

Clyde Tombaugh (1906-1997, at left) was the discoverer of Pluto, which was considered a planet for many years before being demoted to dwarf-planet status. He found it in 1930 by looking at photographs taken by the telescope at the Lowell Observatory in Arizona.

Astronomical Distances

Until the twentieth century, astronomers had little idea of the size of the universe or even the galaxy. Newton had surmised that the distances between stars must be immense, calculating a distance to Sirius that proved to be reasonably accurate, considering the state of astronomical equipment in the 1600s, and was off by only about 50%. However, even centuries after Newton, most believed that the galaxy was the universe. The strange and beautiful spiral nebulae (Latin for "clouds") that were visible among the pinpoint stars were part of our own neighborhood, as far as anyone knew.

The problem of astronomical distances seemed to be intractable because all one could do was to measure, as Newton did, the apparent brightness of an object, which supplied no information about its distance. Although it might be tempting to suppose that a bright object is closer to us than a dim object, this is by no means certain. The bright object could be much farther away from us but possessing greater intrinsic brightness, so that even from its greater distance it shines more brightly than the closer, dimmer object.

As the Earth orbits the Sun, we view the distant stars from different angles at different times of the year, and we ought to be able to make use of this effect, called parallax, to gauge how far away they are from us. Even the ancients were aware of this. Unfortunately, even for the nearest stars, the parallax is exceptionally tiny. The ancient astronomers were not able to detect it at all, and even for modern astronomers, the method of parallax is useful only for finding the distances to the nearest stars.

Henrietta Swan Leavitt

The key to finding the distances to the stars and, eventually, the size of the universe was the discovery of a "standard candle," a star whose intrinsic brightness was known and could thus have its distance from us calculated from its apparent brightness. The first standard candle was a type of star called a Cepheid variable, named after a variable star in the constellation Cepheus. Cepheid variables were studied by Henrietta Swan Leavitt (1868-1921, at right top), a researcher at Harvard University. A variable star pulses predictably in brightness and size over a period of a few days; Leavitt discovered that the intrinsic brightness of a Cepheid variable is predictable depending on how long the star takes to move through its entire cycle of dimness and brightness. This watershed discovery meant that, for the first time, astronomers could tell the distances to individual stars, and thus they could also deduce the distances of nebulae or galaxies of which those stars were members. This information proved immensely useful to Edwin Hubble in his calculation of the size and age of the universe.

Fraunhofer

New Waves

Astronomers studying the heavens with the unaided eye or through a telescope are making use of light, that is, electromagnetic waves in the visible portion of the spectrum. By the nineteenth century, however, scientists knew that electromagnetic radiation occurred in a broad range of wavelengths beyond the visible spectrum. These wavelengths could be investigated just as the emanations of visible light from stars were being examined.

Annie Jump Cannon

One result of this expanded understanding was the beginning of the study of spectra. Stars and nebulae emit a nearly complete spectra of light, but their spectra display gaps where a few of the wavelengths are missing. The missing wavelengths proved to be unambiguous markers that indicate which elements are present in a star or nebula. Joseph von Fraunhofer (1787-1826, previous page, bottom) pioneered the research into the spectrum of our own Sun, and this work was later extended to other stars. Using an extensive collection of stellar spectra at Harvard, Annie Jump Cannon (1863-1941, at right) developed the system of stellar classification, based on surface temperature, which is still in use today. Cecilia Payne (1900-1979, at left), also working at Harvard, arrived at the surprising conclusion that the Sun consists mostly of hydrogen and helium. (Until then, astronomers had assumed that stars were made of the same materials as the Earth.) She made her discovery by studying the strengths of the absorption lines in the Sun's spectrum.

Cecilia Payne

Einstein (1879-1955)

No one doubts that Albert Einstein (right) was the greatest scientist of the twentieth century. His name became synonymous with brilliance, and his contributions to astronomy and physics are legendary. Not only did he revolutionize our understanding of astronomy and the universe by expanding and enriching our interpretation of gravity; his ideas still drive physics today. His predictions continue to be borne out as scientists conduct more and more investigations into the minutest phenomena.

Einstein showed that gravity is not a force. Instead, he found, the presence of a massive object curves spacetime so that objects move as if they were under the influence of a long-distance force. The curving of spacetime was proven in 1919 by observation of starlight that passed through curved space near our Sun before reaching us. Thus, Einstein's theory owes its acceptance to the careful work of astronomers, without whose painstaking observations Einstein's theory would have been nothing but equations on a piece of paper.

Hubble (1889-1953)

Scientists today know how large the universe is and how old it is because of the work of Edwin Hubble (right). Making use of the discoveries of Henrietta Leavitt, he was able to measure the distances to far-away galaxies. This alone was a huge discovery, but he went far beyond this simple measurement. Carefully examining the spectra of stars in distant galaxies, he made another startling breakthrough. Stars containing hydrogen, for example, have a characteristic gap in their spectrum, as explained above. Hubble saw, however, that in more-distant galaxies, this gap was shifted. He deduced that the shift was due to the Doppler Effect, from which he drew the astonishing conclusion that the galaxies are moving away from us.

Further study revealed that the more distant a galaxy is, the faster it is receding from us. If the distance to a galaxy, its speed, and the rate at which its speed is changing are known, a truly amazing result is possible. The age of the universe can be calculated, that is, the date when the distance between all the galaxies were zero. The calculation is, in fact, remarkably simple once the theoretical background is understood. Thus, we can confidently state today that our universe is a little less than 14 billion years old.

The Big Bang

The idea that the universe is expanding, as shown by Hubble, and that its age can be determined, implies that there must have been a time in the distant past when it was very young and small. The theory that the universe began from a singularity (a dimensionless point containing all the mass of the cosmos) originated in the 1920s with the Belgian astronomer Georges Lemaitre (1894-1966). He referred poetically to the time of the origin of the universe as "a day without yesterday." The Russian-American physicist George Gamow (1904-1968) became an ardent proponent of Lemaitre's theory and formulated an explanation of how the lightest chemical elements arose in the first seconds after what came to be widely known as "The Big Bang" (the beginning of the universe). A striking piece of evidence in support of the Big Bang was unexpectedly discovered by Arno Penzias and Robert Wilson while working at a small radio telescope in New Jersey in 1964 (left). They detected a strange, feeble radio signal that was present no matter where in the sky they pointed the radio telescope. Unknown to the two men, George Gamow had already proposed that background radiation should remain from the Big Bang, and that this radiation would permeate the universe. Penzias and Wilson had discovered this radiation, and it was almost exactly as energetic as Gamow had predicted in his 1948 paper.

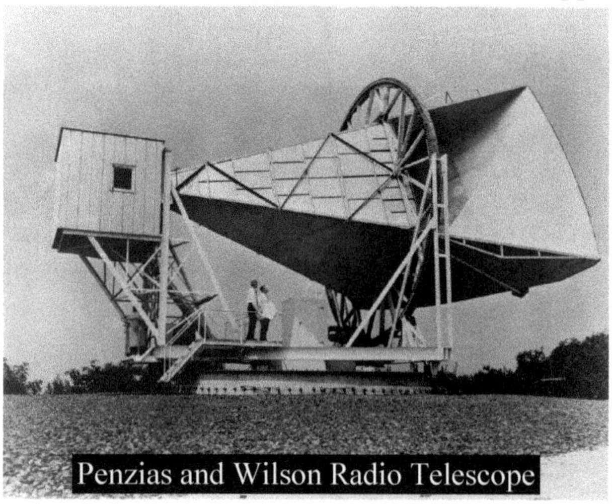

Penzias and Wilson Radio Telescope

A New Vision of the Solar System

Gerard Kuiper

The Dutch-American astronomer Gerard Kuiper (1905-1973, at right) revolutionized our idea of the Solar System by proposing the existence of a band of asteroids and planetoids beyond the orbit of Neptune. At the time of his proposal, in 1949, the only body known to occupy this region was Pluto, then considered a planet, but other bodies were discovered in the Kuiper Belt, as it came to be called, starting in the 1990s.

Jan Oort

Jan Oort (1900-1992, at left), proposed in 1950 that the inner Solar System, containing the orbits of the eight planets, is surrounded by a huge, diffuse, roughly spherical cloud of icy comets left over from the formation of the Solar System. The collection of comets he posited is known as the Oort Cloud.

Shedding Light on Dark Matter and Dark Energy

In the early 1930s, the Bulgarian-Swiss astronomer Fritz Zwicky (1898-1974), working at the California Institute of Technology, made observations of a cluster of galaxies that were orbiting each other. By measuring the galaxies' speeds utilizing the Doppler Effect, he was able to calculate their masses. Surprisingly, the mass he calculated did not match the observed mass of the visible matter in those galaxies. He suggested that a large amount of matter in those galaxies must be invisible, or perhaps hidden. The finding did not initially generate much interest in the astronomical community, probably because scientists were just coming to understand that the cosmos was larger than the Milky Way, and many exciting discoveries were being made in other areas, such as the Big Bang, relativity, and supernovae.

Vera Rubin (1928-2016) was responsible for the explosion of interest in "dark matter" in the second half of the twentieth century. Starting in 1963, she examined how galaxies rotate, and she found that the outer edges of galaxies are not rotating at the speed they should if the matter we can detect there is obeying Newton's Law of Universal Gravitation. One explanation for this anomaly is that the rotation is affected by additional matter. This matter, invisible and mysterious, has come to be known as "dark matter." Most scientists accept the existence of dark matter as the explanation for the otherwise inexplicable galactic rotations curves. However, extremely rigorous and costly searches for the identity of dark matter have consistently failed to turn up any direct evidence beyond a sort of gravitational silhouette which hints at the presence of this enigmatic phenomenon.

Dark energy, on the other hand, is related to the energy responsible for accelerating the expansion of the universe. Saul Perlmutter (1959-) found evidence for this acceleration by studying Type IA supernovae, a type of very dependable and consistent nova which could be used as another standard candle. The problem with the accelerating expansion of the universe is that there is no apparent source of energy to fuel it. "Dark energy" is nothing more than a shorthand phrase to refer to the hypothesis that there is some undetected cause which is making the universe expand ever faster and faster. Scientists have been unable to make much progress toward determining what that cause might be.

Gravity Waves

Ordinary telescopes are limited to visible light, but later inventions enabled astronomers to explore other wavelengths of electromagnetic radiation, including radio waves and x-rays. In the twenty-first century, scientists have expanded their exploration of the heavens utilizing an entirely different type of wave—gravity waves.

In the second decade of the twenty-first century, a colossal interferometer was built, spanning the North American continent, which is capable of detecting minute flexing in the fabric of spacetime. Known as LIGO, the Laser Interferometry Gravity-wave Observatory, this device is phenomenally accurate and is capable of detecting changes smaller than the diameter of a proton. Work at this observatory will surely usher in a new era of gravity-wave astronomy.

Exploration by Spacecraft

Starting in 1957, humans began to launch satellites into space. Today, many thousands of these devices orbit the Earth. They gather data on all kinds of physical phenomena, ranging from weather patterns in the Earth's atmosphere, to the presence of nuclear research facilities in rogue nations, to the existence of planets around nearby stars, to quasars in the most distant galaxies of the visible universe.

Exploratory spacecraft also have been sent far beyond Earth's orbit. Trying to learn about the universe by gazing through the glass of telescope lenses and the interfering impurities and irregularities of the Earth's atmosphere is nowhere near as clear, revealing, or effective as sending robotic messengers to gather data at close range. Therefore, starting in the 1960s, probes have been dispatched to every planet in the Solar System as well as many moons and asteroids, and their achievements have been breathtaking: Rovers traverse the rust-red sands of Mars. A probe has darted nimbly among the rings of Saturn. Probes have braved the scorching, thousand-degree temperatures and crushing pressures on the surface of Venus. Others have visited the farthest reaches of the Solar System beyond the orbit of the most distant planet. Two intrepid machines, the Voyagers, are even now streaking through interstellar space more than two billion miles from Earth.

The Future of Astronomy

The irresistible quest to expand our horizons will continue to motivate the human species as it searches the heavens with enhanced eyes and ears. Eventually, we will launch ourselves physically into the cosmos in ever larger and more advanced spacecraft as we search for new realms to explore. We cannot know what discoveries will materialize in the far future, but we can be certain that discoveries will undoubtedly occur.

Dark energy and dark matter will be pursued relentlessly in the coming years and decades until they are either fully elucidated or utterly rejected in favor of hypotheses that better fit the data. Humans will continue seeking new horizons, other Earths, and extraterrestrial life. Space travel, even if accomplished only or mainly by robots, will accelerate and become ever more commonplace.

By the end of the 1800s, some of the world's top scientists felt that scientific discovery was nearing its end. With Newton's theory of gravity and Maxwell's eloquent electromagnetic equations, it seemed that the few remaining enigmas of the universe were on the verge of being fully solved. New scientists were warned not to enter the field of physics because it was almost entirely known and just a few more loose ends needed to be tied up. Then, lo and behold, things we did not know, phenomena we had never measured, began to intrude upon our reasonings. The twentieth century saw some of the greatest discoveries in the history of the human race: galaxies, quasars, supernovae, relativity, and the expansion of the universe. The twenty-first century saw no slowdown.

This trend of discovery can and must continue. As we look further and further, we will inevitably see more. Our telescopes see back in time, and our intellects can envision the Big Bang and reconstruct its details. Will we ever be able to see before the Big Bang? Most scientists say no—there is no way to cross the singularity—but many scientists also said we would never fly and that we would never travel in space, and the presumptions of those forlorn and misguided predictors of the future are now little more than smears of ink on the discarded leaves of forgotten history books.

We are inventing new ways of probing the cosmos, and we will continue to invent. Starting with telescopes that scanned the visual spectrum of light and progressing to radio telescopes that pored through the longer wavelengths, now we have advanced to the point where we are catching the subtlest of waves, the elusive gravity waves, minute visitors from a billion light years away. Instruments of future astronomy may make the capturing of neutrinos routine or uncover the so-far-invisible dark matter particles.

While our physical bodies may be perpetually bound to a planetary gravity source, our minds will remain forever unbound, endlessly seeking new vantage points from which to see farther and ceaselessly refining our understanding of the cosmos. We will not stop until we have described and explained absolutely everything, and until that day comes, we will continue advancing astronomy.

Will humans or their machines one day span the length and breadth of the cosmos? Will we travel from solar system to solar system or galaxy to galaxy in immense space arks? Will creatures of the future be electrical entities streaming through the universe without loss of energy, living forever? No matter who humanity's descendants are, how they think or how they live, they will still be studying astronomy.

Woodcut engraving of Halley's Comet in 1690.

9

Curiosity: More Powerful than Rocket Fuel

Nature, in the form of man, begins to recognize itself.
Victor Frederick Weisskopf

Man…is not a figure in the landscape—he is a shaper of the landscape. In body and in mind he is the explorer of nature, the ubiquitous animal, who did not find but had made his home in every continent.
Jacob Bronowski

The important thing is not to stop questioning.
Albert Einstein

Albert Einstein once wrote, "I am neither clever nor especially gifted. I am only very, very curious," The great physicist was expressing a truth about all human beings, for all are endowed with curiosity: curiosity about our immediate surroundings, about the far horizon, and about the universe beyond that horizon. We are natural explorers who, after eons of maturation on our home planet and millennia in the history of astronomical investigation, are taking our first tentative steps into the mysterious void of space. We continually broaden our explorations, yearning to add to our knowledge and build more wings onto our House of Wisdom. Whether on our own world or beyond it in the cosmos, we instinctively long to climb any mountain, whether physical or metaphorical. Then, upon reaching the summit, we espy a distant, higher peak, and we say to ourselves, "This too must be climbed." Inquisitiveness is a hallmark of our species, and it, along with our inherent drive to overcome all barriers, has formed the foundation for our species' ongoing success. This inquisitiveness has conferred upon us a survival advantage throughout our existence as a species. As a growing cub leaves the den and ranges farther and farther as it matures, so humans have explored caves and continents, oceans and deserts, and have now ventured into space. Our inquisitiveness has been distilled into its purest form in us by evolution. It is as if we labor under an irresistible impulse to cast light into darkness and uncover what is hidden.

We are always striving for an advantage, trying to discover a new resource or create a new tool that will help us thrive and shape our future. Space is one such resource. It is a vast, unexplored territory that holds great risk but also great potential. Space is inevitably the next step in our quest for a successful and thriving future, and we are already beginning to beam our light into its immense darkness. Our curiosity shines in the darkness like the rays from a towering lighthouse that illuminate unknown paths through the ocean of space. Like sailors of old, we are creating vessels of exploration capable of embarking on missions of discovery and penetrating the great unknown. We will eagerly follow those revealing rays to unimagined destinations and new harbors in the cosmos.

* * *

About 75 million years ago, a dome of magma pushed up the plains of Ethiopia. Then, over millions of years, complex movements of the plates of the Earth's crust caused a system of immense valleys to form, chief among them a 4,000-mile long trench called the Great Rift Valley. In this valley are several calderas that filled with fresh water from the surrounding highlands. By about 26 million years ago, these lakes and their environs were home for the predecessors of humans.

Our ancestors lived a primarily arboreal existence, seldom descending from the trees. They found relative safety high above fierce predators that roamed the jungle floors. While in the trees, they expanded their territories in search of fresh supplies of food and mates. Over time, they learned how to use hammer-stones. These first tools were useful for any number of purposes, from cracking open oil-palm nuts to defending oneself against prowling predators.

Around this time, the climate began to change rapidly in this region of Africa. Tectonic plates shifted, splitting apart huge landmasses and forcing sections of the crustal plates upward to form lofty mountain ranges that have lasted up until today and are still known as the "Roof of Africa." The climate cooled and became drier, and the rainforest gradually changed into open savanna. These geologic and climatic processes triggered a turning point in humankind's history. The descendants of those ancient primates were forced from the safety of trees onto the ground-level killing zone where vicious quadrupeds armed with dangerous power, speed, claws, and teeth held sway. This new peril compelled our progenitors to change and adapt. The survivors were those who underwent multiple modifications in body type.

Standing more upright allowed us to peer over high grasses for both dangerous predators and edible game. Since our hands were no longer required for locomotion, they were freed to perform other vital tasks. With their opposable thumbs, our hands, which were useful for grasping stones, could now wield more effective weapons. Even more importantly, those hands allowed parents to carry their helpless young, transporting them rapidly from place to place as food supplies and threats from predators waxed and waned. We became a bipedal species that could sprint across savannas in search of safety among the diminishing groves of trees. We also learned to travel in groups for mutual protection or in pursuit of food. We learned that the social unit is greater than the individual. However, the full benefits of acting as a unified group are unavailable unless the members of the group can communicate with each other effectively. Thus was born the need for an ever-increasing range of vocalizations, which eventually developed into full-blown language. The birth of language was contemporaneous with the dawn of technology.

Our journey toward dominance of the environment and the globe relied on technology. We improved our technology a little at a time, tool by tool and method by method. One of the very first technologies we seized upon was fire. Taking ownership of fire, we transformed it from a mystery and a danger into a powerful asset.

The importance of fire is made clear in the mythologies of cultures around the world, which commonly include tales explaining how humans came into possession of fire. In these stories, fire was usually bestowed upon humankind by a god. Sometimes the gift was bestowed as an act of rebellion in disobedience to still higher powers because the power of fire was recognized as too fearsome to be entrusted to mere mortals. Such fanciful accounts provide us with insight into the thought processes of early humans. However, no matter how control over fire was learned, it quickly became an essential part of human life and a key to human survival. A dependable source of warmth, even through protracted ice ages; a protection against marauding beasts; a tool for cooking fibrous roots and fresh meat to make them more digestible and palatable; fire was all these things and more.

With fire eventually came the power to work metals. Humans laid aside their stone tools once they had enough heat to smelt copper and then bronze. With further advances came the ability to forge tools and weapons out of the even stronger metals, iron and steel.

Eventually, maturing knowledge of science, mathematics, and engineering ignited the industrial revolution. Fire enabled a lifestyle that allowed us to become smarter and safer, bringing us together around numerous common hearths to form great communities. Today, fire enables us to dream bigger than ever. This seminal technology has carried us forward to breathtaking new frontiers. The burning of gasoline in automobile engines propels us at breakneck speeds across continents. Fire shoots jet airplanes through the sky so fast they outrun the noise of their own motors. And fire will permit humankind to achieve someday its ultimate exploratory destiny, the expansion of human civilization to occupy new worlds, new planets, new star systems, and new galaxies. When people of the Earth ride into space on that great journey, the vehicle they ride in will be balanced on a bright pillar, a brilliant, blinding cone of that foremost ancient power that brought us down out of the African trees: fire.

The chemical process we call fire, in which oxygen reacts rapidly with fuel and gives off heat, light, and other products, is basically the same reaction that our bodies perform in breaking down our food into usable energy. The difference is that within our bodies the reaction is catalyzed, modulated, and held under control. The reaction transpires in the interior of each living cell, and a major proportion of the energy so released is committed to satisfying the massive caloric needs of our outsized brains.

Our brains were required to make ever-more-important decisions for survival, and they responded by rapidly increasing in size and complexity. Technology, language, mutual protection, decision-making, and better senses required a larger, more capable brain. Education became vital as transmission of knowledge to the next generation increased in importance. Our children, instead of transitioning to adulthood in one year as some primates do, came to spend long years in a state of near-helplessness as they slowly matured and mastered increasingly demanding tasks. Rearing, nurturing, and protecting the young now required a greater investment of time and attention. A larger brain and increased time as a pre-adult consumed additional calories as well. The tradeoff was superior probability of survival.

The human brain became, and remains today, our most crucial survival attribute. Human beings are not the largest, fastest, or most agile of animals. Neither do we possess the great strength, sharp fangs, protective armor plating, or other physical attributes that are so useful to other species. What we do have is a brain. Ours is not even the largest of brains, but it is the most complex. A brain, the seat of curiosity, is a vast assemblage of neurons that grow and intertwine with each other, forming patterns of almost unimaginable complexity. The 100 billion neurons of a human brain form approximately one quadrillion connections with each other (as many as 10,000 connections per neuron). From this astonishing number of connections arises the miracle of our minds.

A 130,000 year-old specimen of *Homo sapiens* was recently unearthed in the southwestern part of Ethiopia and had a braincase equivalent to that of a modern man. The humans of that time were us: A child from that era, if transported to our time and offered the opportunity of a modern education, could become a scientist, a diplomat, or an astronaut. Our brains have figured out how to build spacecraft that enable us to travel to, and even take up residence in, that harshest, darkest, and deadliest of environments—the cold and seemingly barren vacuum of space. And yet those are the same brains which once guided our hands in the fashioning of spear points on the plains of Africa over 250,000 years ago.

Today, wild primates still thrive in the highlands of Ethiopia. Although these primates have had the same amount of time to evolve, learn, and explore, they are still living much as they did millions of years ago, while humans have launched themselves into space. The Ethiopian primates have a complex multilevel society and possess a diverse repertoire of vocalizations, but their "speech" does not even remotely approach the complexity and variety of human language. Humans, in contrast to the primate denizens of the Ethiopian highlands, have traveled. We have spread across the seven continents, descended to the bottom of the deepest ocean trenches, and climbed the highest mountain peaks. We are still finding new places to explore while our simian cousins continue to haunt their seemingly timeless habitat. Exploration of our environment is hard-wired into our DNA. It is an instinctual survival mechanism bestowing superior chances of success.

To more effectively explore, one builds tools. Tools are not unique to humans. A species of corvid, the Caledonian crow, can use a stick to pry food out of a crevice. Chimpanzees prepare sticks for use as termite-mound probes by stripping off leaves and twigs, and chimps have also been observed to use rocks as hammers. But constructing complicated tools such as boats, nets, and spaceships is a skill unique to humans. This ability is thought to have emerged in the hills around the many lakes in the Ethiopian Rift Valley and surrounding highlands thousands of generations ago. Around the ridges and calderas of Ethiopia, outcroppings of volcanic glass, or obsidian, are found. These outcroppings are of no small importance for the history of humanity. The early humans who lived here chipped pieces of the obsidian to form spear points as well as knives, scrapers, and other sharp implements. Today, numerous specimens of these ancient hunting tools have survived the erosion and neglect of time and can be found strewn around the region.

The spear points were often fastened to throwing sticks with strips of animal hide or directly inserted into a cleft at one end of a shaft. Creating spear points was an arduous task that required patience, practice, and, most of all, knowledge. The method for crafting a lethal spear point could not have been perfected even by an adroit individual in a single lifetime but required countless trial and error experiments over multiple lifetimes. Then the process of efficiently transferring knowledge from one generation to the next requires that elders show, demonstrate, and, most of all, explain the process to the young. However, this process requires outstanding communication skills. In other words, to transmit this knowledge and technique from one generation to the next, our ancestors needed language. Thus, a simple obsidian spear point found in an African valley constitutes concrete evidence that in that bygone epoch, there existed minds similar to ours. We can say that at that time we left our ape-like relatives behind and started along the road leading to the stars. With the power of verbal communication, so eloquently revealed by those carefully crafted artifacts found scattered across the Ethiopian terrain, it is but a small step from spear points to spaceships.

Modern humans began to trickle out of Africa in small groups around 125,000 years ago. Their entry into the rest of the world was like the first rays of light from a sunrise presaging the glorious light of day to come. By 60,000 years ago, the migrants were streaming out of Africa in waves. Genetic markers track the global spread, indicating that human populations flowed north through the Middle East and into Europe and Asia, eventually crossing the Bering Strait and reaching the southern tip of South America approximately 16,000 years ago. Living as nomadic hunter-gatherers for millennia, these primitive peoples began to experiment with agriculture about 12,000 years ago. Around 6,000 years ago, those hardy ancestors of ours began to produce the world's first surplus of food, which freed a few of them to devote their time to thinking and the pursuit of knowledge. When sufficient fundamental knowledge accrued, early cultures began to construct cities with temples, palaces, citadels, walls, aqueducts, and roads. Social complexity and division of labor contributed to a continual increase in expertise in all the activities that contribute to survival. Now they had time to create a vital innovation that proved crucial on humanity's pathway to space: writing.

Initially, writing arose for purposes of law, taxation, and record-keeping, but with it came those nobler fruits of the human intellect: science, mathematics, and engineering, as well as that written record of a culture's shared tales known as literature. Writing almost instantly became one of the most important aspects, indeed, a defining characteristic, of human civilization.

Writing conferred a huge advantage upon any society that possessed it. It represented a true paradigm shift in the evolution of data storage. It motivated and enabled humans to become more exact in their thinking, and it fostered a new growth of mental acuity. It opened new floodgates of creativity and gave rise to unprecedented opportunities for the creation and appreciation of a new type of beauty. The majesty of the written word was acknowledged for its inherent worth, and over time, more and more information was written down. Time and again, cities and nations vied with each other to amass huge libraries because wise leaders recognized the value of written knowledge. Writing exponentially expanded the storage capacity of humanity's collective memory. With the invention of writing, huge amounts of information could be transmitted from one generation to the next and also from city to city, and from culture to culture, across oceans, across worlds, across time. Knowledge became widely available, easily accessed, and persistent. It was rarely lost and could always be added to. Writing ensured that the advance of knowledge would be inexorable, and irreversible. As a repository of knowledge and a generator of learning, writing became the vehicle for continued proliferation of the known, which ultimately propelled us to the stars.

Less than a single lifetime ago, sending a vessel with human passengers into space would have been impossible; however, now, in the 21st century, it is so routine that it often goes virtually unreported. As of 2019, astronauts make the journey from Earth to the International Space Station four times a year. This very limited and short-distance space travel is the preliminary step on a much grander journey—the exploration of the universe. After almost 13.8 billion years of change in the universe, including 4 billion years of life on Earth, a species arose that was characterized almost purely by curiosity. Over the span of just 250,000 years, that species has transformed itself, built impressive civilizations, accumulated necessary knowledge, and developed awe-inspiring technology. From swinging through the treetops of prehistoric African forests, this descendant of apes has become a builder and pilot of spacecraft.

Our journey has occupied but a moment in cosmic time, from the rising of a volcanic ridge in the Great Rift Valley of Africa to today, when capsules regularly dock at the International Space Station. In the Rift Valley, the environment shaped us. Now, we have learned enough to actively shape our environment. In past days, we formed mud and clay from the ground into bricks for our dwellings. Today we take iron, aluminum, and titanium from that same ground and fashion wondrous structures that make space itself inhabitable. Just as we shape our tools, shape our houses, and shape our dreams, now, indeed, we have started to shape our destiny. Creatures of Earth have begun their inevitable journey into the vastness of space. In doing so, we are answering the call of our most basic nature; we are following the impulse that has guided us since the inception of our species. It is the impulse to survive, yes, but much more than mere survival, it is the impulse to learn and discover, to illuminate the darkness of the unknown and the undiscovered with the light of knowledge. This impulse has always contributed to humankind's survival and will continue to do so. Yet great cosmological and astrophysical forces that we do not fully comprehend are inexorably moving in our universe, and they are far greater than the forces that created the Great Rift Valley millions of years ago. We will adapt or perish. Only by continuing to learn and explore will we be able to achieve that future transformation. But regardless of what transformations and discoveries the future may hold for us, there will be one constant, one unchanging characteristic that defines what it means to be human. If human beings are still exploring the cosmos a million or a billion years from now, our endless, expanding existence will still be guided by one thing: curiosity.

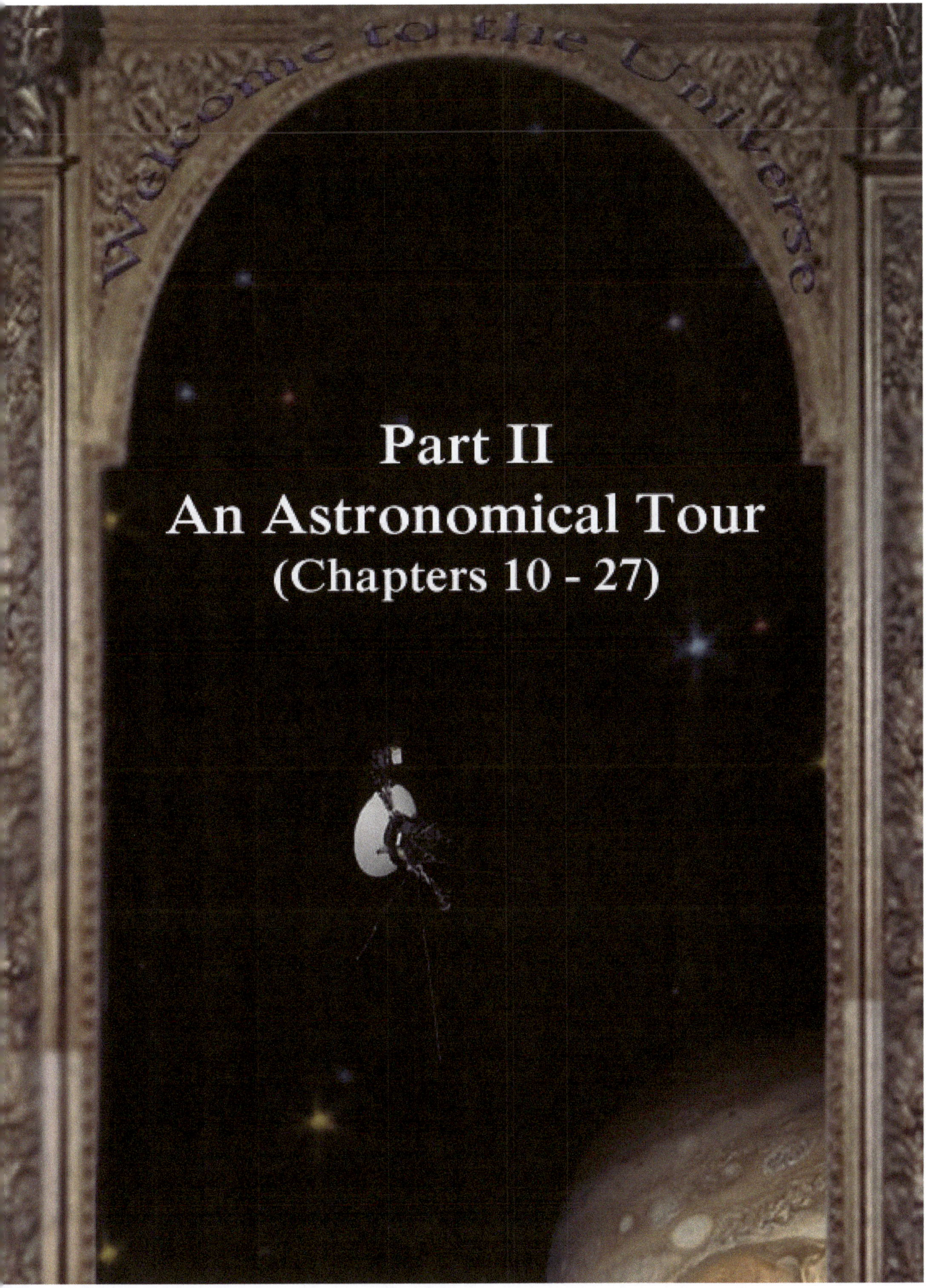

Part II
An Astronomical Tour
(Chapters 10 - 27)

10
Earth: Our Special Home

Home is where the heart is.
Pliny the Elder

We love our planet Earth. We should - it is our home, and there's no place like home. There can't ever be a better place than Earth.
Dimitar Sasselov

That I am part of the earth my feet know perfectly, and my blood is part of the sea.
David Herbert Lawrence

Living here on Earth, we breathe the rhythms of a universe that extends infinitely above us. When resonant harmonies arise between this vast outer cosmos and the inner human cosmos, poetry is born.
Daisaku Ikeda

Among the magnificent array of planets in our Solar System there exists a unique world of water reflecting light in a dazzling assortment of blues. This spinning sanctuary of sapphire is an oasis of life awash with water, the most precious fluid known to humanity. Surrounding this aquamarine haven is a barren, frigid, obsidian space populated with…virtually nothing. The nearest neighbors, except for a large, well-lit moon, are many tens of millions of miles away and consist of a paltry scattering of dark boulders, a few multi-hued gaseous balls, and a relatively large, whitish star. As viewed from space, this liquid jewel reflects a marvelous sheen of cobalt blue intermittently wrapped in cream-colored clouds. A few umber and viridian continents thrust upward between the waves of the planet's broad cerulean oceans, but the landmasses are secondary to the water. There is so much water that from a distance the planet resembles an azure droplet floating in space. This gyrating gem is the birthplace and home of human life: Earth.

Let us turn back time to examine Earth's turbulent infancy and its metamorphosis into the mild, mature, aquatic planet we know today. Proto-Earth was molten, forged in the ferocious, fiery maelstrom of the early Solar System. It took tens of millions of years for it to cool and become the early Earth, a strange, severe place with six-hour days and a Moon so close that it dominated the sky, turning night into day and creating violent tides. During this time, Earth's pioneer organisms gained a footing and commenced to manipulate their environment. These nascent harbingers of the explosion of life to come helped transfigure the planet into the relatively placid and peaceful domicile we know today. Here broad streams of air gently propel far-ranging currents of water, and minuscule dust particles softly jitterbug upon rays of sunshine while a modest, upright creature speculates about the origin of the universe and Earth itself. With the passing of time and the unceasing inquisitiveness of the human mind, knowledge of our world has advanced to the point that scientists feel confident in positing the following series of events that led to the origin of the Earth and, ultimately, to the rise of humanity.

Our Solar System came into existence about five billion years ago in an undistinguished region 30,000 light years from the center of the Milky Way galaxy. A tenuous cloud of dust and gas was drifting through space when a nearby stellar explosion shook the surrounding region and fired out a shock wave at 70 million miles per hour. This shock wave rippled through the sparse dust, churning the particles and creating regions of slightly increased density, where the cloud began to coalesce. Tiny, delicate clumps of matter took shape, at first drawn together by electrostatic attraction and then, as they grew, by gravity. The burgeoning cloud collapsed to form a twirling pancake-like structure. The resulting solar nebula flattened into a revolving disk with a swirling sphere of condensed gas and dust at its center. The core of that sphere grew so hot and compressed that nuclear fusion began. As the Sun started converting hydrogen into helium through fusion, huge quantities of additional heat and light were produced, as well as energetic streams of charged and uncharged particles.

Around the newborn Sun, a multitude of planetesimals hurtled haphazardly through space. These planetesimals experienced countless collisions over a period of several hundred million years in a process called accretion that resulted in the formation of the planets and moons of our Solar System. About 4.6 billion years ago, one of these planetary bodies had grown by gravitational attraction into a waterless, moonless, molten mass. This world would become the rotating rock we call Earth.

In this red-hot, liquid world, the lighter elements floated outward while more massive substances, such as nickel, iron, and heavy radioactive elements, sank toward the center. There crushing pressure compressed them into a solid core. Farther away from the center, there was less pressure, resulting in a layer of liquid iron around the solid core. Earth's rotation caused this iron to flow, which continuously produced Earth's protective magnetic field. Above the liquid metal, a zone of magma-like mantle oozed like putty. The outer crust, in contact with the thin film of water we call oceans, took the form of massive, independent plates floating on an underground sea of magma. On that unstable base, these plates, some bearing entire continents, traveled independently across the surface of the planet carried by ponderous convection currents. These meandering plates were subjected to bouts of violent shaking and upheaval along their edges as they continually smashed into one another. Their shapes and elevations changed as the enormous forces involved gave rise to mountain ranges, rift valleys, plateaus, and chains of islands in a process that even today is ongoing. Meanwhile, above the planet's oceans, continents, and islands, an atmosphere formed, a blanket of gases which would prove vital for the development of life.

The Earth came into being in a part of the Milky Way galaxy that just happened to have the right conditions for life. The new Earth was not too near the center of the galaxy, where it would be roasted in radiation and bombarded with so much light and heat that life as we know it could not evolve. Neither was it too far from the center of the galaxy in a region where the necessary ingredients for life would be too sparse. It was, to use a phrase that in almost every respect seems astonishingly apt in describing the Earth, "just right."

After the Earth's initial formation, a sequence of events occurred which was both surpassingly crucial for the later history of our planet and, as far as we know, phenomenally improbable. First, a planet the size of Mars collided with the early Earth. Astronomers have dubbed this interloper Theia after the mythological mother of the Moon. The colossal violence of the collision of Earth and Theia hurled a molten ring of debris around the Earth. Some of the material from this ring possibly fell back as a glowing rain of molten lava upon the Earth while the remainder gradually condensed to form the Moon. As if that was not enough cataclysmic disaster for one planet's lifetime, the Earth was also pelted by a million or more meteors in an event known as the Late Heavy Bombardment. This bombardment by asteroids and comets lasted for several hundred million years and delivered water, surface metals, and atmospheric gases to the Earth's surface. The Earth cooled, a crust formed, waters gathered, and an anaerobic mixture of gases covered the surface. Within about a billion years of the Earth's formation, the building blocks of life began to assemble themselves into the self-replicating entities that transformed the planet over the ensuing billions of years. The Earth eventually settled into a semi-stable state with broad expanses of water, continents of dry land, and an atmosphere of indispensable gases.

Upon and within the early Earth, molecules of increasing complexity emerged. These included amino acids, lipids, simple sugars, and nucleic acids. Energy for the chemical reactions that haltingly constructed these molecules came from the Sun, which bathed the Earth in heat and light. The most important of the molecules was deoxyribonucleic acid (DNA). This delicate and intricate molecule, with its famous double helix, could store massive amounts of information. It possessed the astonishing abilities of replication and adaptation as well as the capacity to increase its storage capacity with the addition of simple polymer subunits. The genetic language, the DNA-based code in which the diverse traits and propensities of every living thing on Earth are written, became the blueprint for all life. From the interior of every living cell, molecules of DNA steadily adapted and modified our ancient progenitors, allowing them to better survive in their ever-changing environmental niches. This subtle yet powerful process continues unabated today. The DNA that shapes humans is ultimately responsible for the creation of our minds, memories, and consciousness as well as our physical makeup. Our own bodies, from within, have transfigured us into superbly creative beings capable of altering our environment and thriving upon planet Earth. Billions of years of evolution have culminated today in the species we call *Homo sapiens*.

Water

Scientists posit that the single most critical factor for the evolution of life on Earth was water. Our warm, wet, whirling world possesses copious yet modulated amounts of liquid water: not too much, which would leave the Earth covered with a global ocean, nor too little, in which case the Earth would resemble a vast Saharan desert of shifting dunes. Sometimes the sea level has fallen due to increasing amounts of polar ice while at other times it has risen due to warming trends. Over millions of years, the oceans have repeatedly risen and fallen thousands of feet. During the last two millennia, the rates of rise and fall have been minimal, less than three inches per hundred years. However, in the twentieth century, sea levels rose by almost half a foot. The melting of polar ice is due to increasing average temperatures around the globe. According to NASA, the last time the Earth was as warm as it is today occurred 125,000 years ago. Since 1880, when scientists began keeping track of global temperatures, the record for the highest annual average temperature has been broken numerous times, including, ominously, in 2014, 2015, and 2016. In 2017 and 2018, average temperatures decreased slightly but were still higher than in any year before 2014.

The illustration at right shows a hot spring emitting steam amid pools of liquid water in front of mountains adorned with snow. This is not entirely atypical; on Earth, water exists in all three of its phases: gas, liquid, and solid. This is because water changes phase across a relatively narrow range of temperatures. Some organisms survive in boiling water, and some thrive on the frozen undersurfaces of Antarctic ice sheets, but all of these organisms require water. From polar ice caps to steaming geysers, from ice-covered lakes to deep-sea hydrothermal chimneys, the humble water molecule possesses surprising attributes that make it indispensable for life on Earth.

One of water's extremely useful and vital characteristics is its superior dissolving power. With the ability to dissolve countless inorganic substances, water constitutes an excellent transportation system for moving necessary ions and nutrients into living cells and removing wastes. Another very important and highly unusual characteristic of water, which is due to the shape of the water molecule and its slight polarization, is that water is less dense in its solid form than in its liquid form. One side of the molecule bears a slight negative charge and the other side carries a slight positive charge. Since opposite charges attract, in any sample of water, the molecules will tend to line up and form regular geometric shapes. If the water is liquid, there is so much movement and jostling that the links between the molecules are fleeting. As ice freezes, however, the three-dimensional structure becomes static, forming a lightweight hexagonal crystal less dense and taking up more space than the liquid from which it formed. Thus, ice floats on water and does not sink to the bottom of an ocean or lake. Over the course of evolutionary time, this has been of extreme importance. Layers of ice forming on top of bodies of water serve to protect organisms and, paradoxically, keep the liquid water underneath a bit warmer than it would be if it were not so protected. If frozen water were to sink to the bottom of water basins, then in a short time, all bodies of water would freeze completely, killing the living things therein.

For our water to be so useful to us in all its versatile roles, the most important factor is temperature. For the temperature to be favorable, we need to be the proper distance from the Sun. As when standing close to a hot campfire on a cold night, we do not want to be too close or too far away. We need it to be just right: the Goldilocks zone. A difference of a few degrees could convert the Earth into a broiling hellhole or a frozen snowball. Our nearest planetary neighbors, Venus and Mars, barely missed the Goldilocks zone. The latter is a rusted, barren, presumably lifeless rock, while the former has become a scorching, caustic oven. Elevated temperatures would also cause chemical reactions to occur too energetically, which could have destroyed the delicate building blocks of incipient life, while temperatures too cold would have slowed down those initial chemical processes so much that life might never have evolved.

The water cycle began four billion years ago. As water showered the Earth's surface, it dissolved the land's salts and carried them via rivers to be deposited into the oceans. The resulting brackish water may have aided in the development of life. Earth's oceans today teem with marine life that thrives in seawater containing about 3.5% salt by weight. However, salt water is not suitable for human consumption. The human body cannot get rid of the salt fast enough, so drinking salt water leads to organ failure and death. In fact, very few terrestrial creatures can consume ocean water and live. Salt-water crocodiles, for example, have specialized salivary glands that eliminate excess salt. Humans must consume fresh water, which is effectively devoid of salt. Approximately 2.5% of the water on Earth is fresh water, and only 1% of that is readily available; most is inaccessible in immense ice sheets and glaciers. (Interestingly, ice is one of the purest forms of water on this planet because, as water freezes, it forms a crystal lattice. During the formation of ice, the solidifying water forms a matrix which forces out the salt.) Humans must seek out streams, rivers, lakes, and wells as sources of fresh water. Even today, seafarers must carry fresh water with them or possess devices capable of distilling or deionizing water, or else lament, like the ancient mariner who sailed through lonely seas surrounded by water as far as the eye could see and uttered the bitter truth: "Water, water, everywhere, nor any drop to drink." For humanity, fresh, pure water is the most precious of all of Earth's liquids.

The Atmosphere

Earth has been able to hold its oceans of water for billions of years, partly due to its atmosphere, which exerts pressure and prevents the oceans from boiling away. The Earth also has sufficient mass that its gravity is able to maintain our oceans and atmosphere on the surface, preventing those precious molecules from escaping to space.

Earth's early atmosphere was rich in hydrogen, carbon dioxide, and water vapor, which means that the greenhouse effect, the warming due to certain gases in the atmosphere, was much stronger then than it is now. However, the Sun was also dimmer, producing only a fraction of the energy it produces today. The Sun and the atmosphere have been involved in a four-billion-year balancing act in which changing gas concentrations regulate the rising heat of the Sun. The greenhouse effect is extremely important for Earth as it adds approximately 60°F to the planet's average temperature. Without our greenhouse gases, Earth's average temperature would hover around a frigid 0°F, and our cherished globe would be a frozen snowball covered in sheets of solid ice thousands of feet thick. There would be little, if any, life, as in Antarctica, where temperatures never surpass 10°F.

On the other hand, the highest recorded air temperature on the planet was observed in the Libyan Desert at 134°F. Add another 80°F and water would boil. Only a very few single-celled microbes can survive at boiling temperatures. Increasing the temperature another 51°, to 265°F, would cause even these microbes to perish within two hours.

The greenhouse effect was not the only mechanism by which the early Earth kept warm. Internal heat created by radioactive elements decaying inside the Earth contributed, and there was also some residual heat from the bombardment of Earth by the meteors that originally formed the planet. In this suitably warm environment, the earliest microbes—cyanobacteria—learned to use the Sun's light for photosynthesis. In photosynthesis, carbon dioxide is taken in and oxygen is excreted. The unpretentious cyanobacteria were thus undertaking the herculean and vitally important task of converting the Earth's anaerobic atmosphere into one rich in oxygen capable of supporting later plant and animal life.

Furthermore, the increased oxygen cooled the planet. Some of the oxygen formed ozone, which ascended high into the atmosphere and began carrying out the function for which we still value it today—the absorption of potentially harmful ultraviolet radiation. Unfortunately for those ancient cyanobacteria, the oxygen they gave off was poisonous to them, and they found themselves in a world grown toxic. They were so successful that they destroyed their own environment.

The atmosphere those doomed cyanobacteria helped create possesses as one of its major components a gas that is remarkable for its reactivity and usefulness. Free oxygen, which is plentiful in this planet's atmosphere, reacts with almost everything to form compounds. Fortunately, Earth's supply of free oxygen is constantly replenished by the descendants of those ancient bacteria: plants and photosynthetic bacteria. Also present in our atmosphere is a varying quantity of water vapor, and there is also much nitrogen. The tiniest pieces of the mixture of gases surrounding this planet are of great consequence. For example, one might be tempted to discount the significance of a rare, minuscule molecule that occurs in the atmosphere at a proportion of .04%, but this is carbon dioxide, which, in its way, is at least as necessary to this planet as oxygen. Carbon dioxide contributes to regulating the temperature of the planet and, amazingly, supplies the carbon atoms that constitute the building material for the physical structures—“bodies”—of all the plants and animals that thrive on the Earth’s surface.

The storage of carbon, in whatever form, is called “sequestration.” Most people are unaware of the staggering amount of carbon that lies sequestered in mineral deposits such as the calcium carbonate of the white cliffs of Dover as well as peat bogs, ocean coral, hydrated crystals on the ocean floor, and petroleum. If even a small portion of this enormous reserve of carbon were to be suddenly released into the atmosphere, the result would be catastrophic for all the planet’s animals, including humans.

The atmosphere from which all that carbon was extracted is much more than a chemical warehouse. It is a vast protective blanket that regulates the planet’s temperature. By spreading heat around the globe, it prevents extreme temperatures from building up. Excessive heat from the equator continuously flows to the colder poles, warming them and cooling the equator. Constant heat distribution by wind and water produces Earth’s large temperate zones. These moderate regions stimulate the existence of an abundant diversity of life.

Earth’s atmosphere also acts as an asteroid shield. Every day hundreds of tons of space dust and rocks rain down upon the planet. Most of these particles, pebbles, and stones are incinerated high in the Earth’s atmosphere due to friction with the air. If large enough, these appear in the night sky as shooting stars. However, if Earth’s atmosphere were 100% oxygen, then even a tiny meteor would ignite a global fire. Fortunately, Earth’s atmosphere is only 21% oxygen and 78% nitrogen. Nitrogen does not readily combine with anything and therefore acts as a fire-suppression system. Thus, the animals of Earth have a plentiful supply of oxygen to meet their needs, but not a dangerous excess.

In the distant past, there were times when the atmosphere held a higher percentage of oxygen, and this has led to some bizarre results. For instance, some species of prehistoric insects grew to be giants. During the carboniferous period three hundred million years ago, when the supercontinent Pangaea first took shape, the denizens of the Earth’s jungles and forests were mainly insects. *Meganeura*, the stealth fighter jet of its day, was a carnivorous dragonfly that had a wingspan of over two feet. The largest known land-dwelling insect of all time, *Arthropleura,* was an herbivorous millipede that grew to a length of over six feet and was about a foot and a half wide.

Earth, then and now, has always been a dynamic planet. Weathering, erosion, volcanoes, rain, glaciers, wind, and plate tectonics all act collectively as Earth’s intricate recycling system for the creation of new rocks. Plate tectonics is the name given to the process in which great plates of the Earth’s crust slip and slide over the planet’s surface. Scientists believe that water makes plate tectonics possible, as it lubricates the plates. Without plate tectonics, our planet would be geologically dead and uniformly dull. We are a much more geologically interesting world than, for example, Venus or Mercury. Plate tectonics creates rising mountain ranges and plunging ocean trenches. This allows a carbon-silicate cycle that is yet another contributor to temperature regulation and has helped maintain a life-zone of temperatures on Earth for billions of years.

One of the most important of the life zones is the intertidal zone, where land, water, and atmospheric gases interact simultaneously. Tidal zones allowed species that evolved in the oceans to transform themselves by stages into land-dwellers. If the Earth were just a water world, the transition to flying through the air would have been much harder if not impossible, and nothing would be able to walk on land. Organisms destined eternally to live, grow and develop underwater would not have discovered fire, electricity, or metallurgy. It is one thing to have great intelligence—dolphins and octopi are extraordinarily intelligent—but how would they write? What would they build? What tools would they use? Would they ever dream of space or of other worlds? Would there be a word in their language for galaxy, star, or moon?

The Moon

Intertidal pools depend for their existence on our nearest celestial neighbor—the Moon. Its gravity draws the tides systematically inward and outward, twice daily creating the miniature marine basins which may have been instrumental in the evolution of life. Part of the reason that these basins are so large is that the Moon is close to Earth and exerts a strong gravitational pull. Compared to all the moons in our Solar System, our Moon is the closest in size to the planet it orbits. Therefore, scientists refer to the Earth-Moon system as a binary system.

Another effect of the size, mass, and nearness of the Moon is that it gyroscopically stabilizes our planet's axis of rotation, preventing wild, drastic swings of the poles such as hapless Mars experiences. Another planet with an unstable tilt is Uranus, where night and day last for decades instead of days or hours. When a planet swings chaotically upon its axis, continents can rapidly change from frozen wilderness to desert wastelands within periods of thousands of years instead of millions of years as on Earth. Rapid, intense, worldwide alterations in temperatures would hinder the ability of life to evolve into complex creatures.

The Earth is tilted on its axis at an angle of 23.5° at present, and a line extended perpendicularly from the North Pole points almost directly at the star Polaris. These alignments are in flux, however. Although the Moon does an excellent job of stabilizing the Earth's tilt, the amount of tilt still varies over thousands of years, and the line from the North Pole sweeps out a circle through the sky. Thus, five thousand years ago, in the time of the ancient Egyptians, the pole star was Thuban in the constellation Draco; twelve thousand years from now, the pole star will be Vega.

Since the Earth's tilt is responsible for its seasons, especially in temperate climates, these subtle movements may have substantial effects on the Earth's weather. Some scientists, like Brian Cox, theorize that such climate changes during the evolution of early humans were of great importance in creating selective pressures for the development of increased brain capacity. At any rate, the presence of the Moon has kept these fluctuations from being so ferocious that they would have prevented or delayed the development of life.

In addition to stabilizing the Earth and preventing wild shifts in the planet's tilt, the Moon has also shielded us from a shellacking by meteors. Each of its 80,000 impact craters represents a meteor strike that might have hit the Earth with potentially extinction-causing force.

Electricity and Magnetism

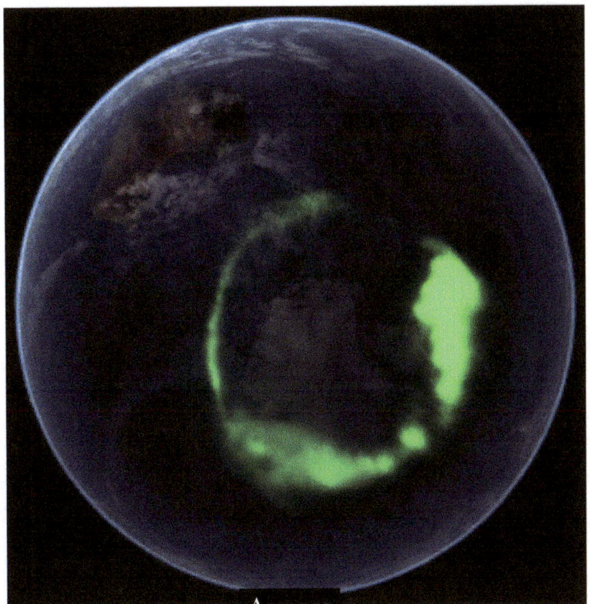
Aurora

The Earth has a powerful magnetic field, known as the magnetosphere, which is relatively stable. However, during the last 20 million years, according to evidence found in ancient rocks on the ocean floor, the Earth's magnetic poles have reversed themselves about once every 200,000 years. The protective power of the Earth's magnetic field is vital, and its importance can hardly be overstated, because it deflects deadly solar and cosmic radiation. The radiation is in the form of charged particles, and charged particles naturally follow circular paths when passing through magnetic fields. As a result, many of the particles streaming toward us from the Sun are deflected, but those which are caught become trapped and follow a spiral path, as dictated by the laws of magnetism, towards the Earth's magnetic poles. There they ionize atmospheric gas particles, which emit light we can observe as scintillating, colored auroras (above, viewed from the IMAGE satellite). Without this magnetic safety net, our atmosphere would be effectively ripped away from the Earth and blasted into space by the relentless impacts of the solar-wind particles. The presence of a magnetosphere is one of the critical factors necessary for life on Earth.

In the nineteenth century, scientists began to realize that electricity is a manifestation of the same fundamental, universal force that also accounts for magnetism. Thus, it is no surprise that in addition to acting as a huge magnet, the Earth is a world that flashes with electricity. Lightning strikes kill dozens of people each year, yet this "bolt out of the blue" might have provided the initial spark for life. By jolting molecules of methane, water, carbon dioxide, and other ingredients of the primordial soup, lightning (as shown by the famous experiments of Miller and Urey starting in 1953 and Sagan in the early 1960s) could have started the Earth on the path to life. The scientists' experiments showed that amino acids, sugars, and even nucleic acids similar to DNA could be created by an electrical spark interacting with the simple chemicals known to have existed on the primordial Earth. Similarly, pungent ozone can be created by an electrical discharge, from a lightning bolt to a sparking electrical cord; the electrical discharge splits oxygen molecules to produce solitary atoms of oxygen that then recombine with free oxygen molecules, which contain two oxygen atoms apiece, to form the three-atom ozone molecule. In the Earth's upper atmosphere, a layer of ozone created by the ultraviolet portion of sunlight forms a vital protective barrier against harmful radiation.

Conclusion

The varied and improbable characteristics of the Earth have arrayed themselves in harmonious unison to give rise to and sustain life and human beings. Many observers experience astonishment and joy at the wondrous complexity of the interacting factors and forces that collaborated in such a consummate way that the result was our existence as living organisms on this planet.

As our planet has changed over time, so have we. Our nervous system has evolved and can detect specific shapes, colors, and textures. Since beauty is in the eye of the beholder, it is only natural for our species to perceive the Earth as the most attractive of all possible planets. The exquisite value of the tranquil Earth's bountiful lands, generous oceans, and plentiful flyways populated with abundant life is self-evident. Earth is home for millions of species and provides sustenance, warmth, and shelter for all. For humanity, our world is like a self-sustaining kitchen garden, factory, and storehouse all rolled into one.

We renew our bodies with every morsel we consume, we revive our spirit with every breath we take, and we refresh our vigor with every sip of water we drink. We draw inspiration, derive pleasure, and gain perspective from the visual magnificence of our majestic world. From the perpetually snowcapped Himalayas to the Challenger Deep in the Mariana Trench; from the humid Amazon to the arid stretches of the Gobi; from the barren Antarctic ice sheets to the lush coniferous forests of Siberia, from the wide-open grasslands of North America to the Congo River basin of Africa where light barely penetrates the canopy of trees—this is our planet, our home.

Our home is awash with wonders and bejeweled with ornaments. An open seam, almost 50,000 miles long, forms the longest mountain range system in the Solar System. This is the mid-ocean ridge system, of which 90% lies underwater and which forms a backbone in every ocean on the planet. Running down the spine of those mountains is a great rift valley, where new molten crustal rock continually emerges from the Earth's interior. The Great Barrier Reef, half the size of Texas and located off the northeastern coast of Australia, is a wonderland of life, home to myriads of species of fish, mollusks, and corals in bright and iridescent hues of blue, red, and orange. At thundering Victoria Falls in south-central Africa, clouds of water vapor rise a thousand feet on the great Zambezi plain as water plummets into a 350-foot rift over a mile wide. This mist is *Mosi oa Tunya*, "the smoke that thunders." Hypnotic auroras in blues, greens, and pinks adorn the poles. They shimmer like undulating sheets that billow in magnetic winds. These are the indicators of an active world, a life-supporting planet brimming with energy. This dynamic, vigorous, self-sustaining, royal-blue planet powerfully signals to the universe that here lives a vibrant, healthy, robust biosphere.

Over the past 4.6 billion years, our planet has rearranged its continents and oceans several times, tenanted its seas with a multitude of fishes, colonized its land with legions of animals, and filled its skies with flocks of flying creatures. It has withstood the assault of moons, planetesimals, comets, and meteors. It has experienced enormous outpourings of lava, titanic volcanic eruptions, deadly pyroclastic flows, and massive earthquakes. Meanwhile the scars of these events are eventually smoothed by weathering, erosion, and the recycling of the Earth's crust by unrelenting tectonic forces.

We live on planet Earth, and our descendants will most likely continue to thrive here as long as conditions allow. Furthermore, Earth will be our only home until we become able to voyage safely elsewhere. Knowledge of how our planet functions is indispensable if we wish to live here indefinitely. Furthermore, understanding how our planet and all the other celestial objects interact within the vastness of space is imperative if we wish to sojourn among the distant planets, stars, and galaxies.

Even though we cannot alter the movements of tectonic plates or the orbit of our planet, we can do much to influence the evolution of our planet. We can promote rational discourse, work cooperatively for our mutual benefit, practice sustainable forms of agriculture, plan for the beneficial future usage of our planet's resources, harness renewable energy supplies, and, of paramount importance, take care of our planet, for we only have this one.

Among the varied species that populate our world, we occupy a unique position. Humanity has the ability to anticipate and influence its own future. We sprang from nature, and until now we were practically powerless in its grip, buffeted by the turbid ebbs and flows of immense forces beyond our ken or control. Today we can dramatically alter nature, to destroy it irremediably or remold it for the benefit of all life. Therefore, we must make a choice. Are we disinterested passengers aboard a mighty ocean-going vessel, blithely ignoring the Ocean of Truth all around us as we distract ourselves with short-term pleasures? Or will we learn to focus on the long-term health of our all-important vessel, recognizing that we are the custodians of a delicately balanced ecological paradise?

Protection of the Earth means the preservation of natural systems and the maintenance of a world suitable for human survival. The Earth has been in a state of flux since its first years, and we humans are merely a sparkle in the eye of geologic time. We are part of the Earth's dynamic processes, collectively a tick on the timeline of Earth's existence—but a tick with a conscience. It behooves us to maintain at the least, enhance if possible, and thoroughly cherish this one-of-a-kind, revered rock we call Earth. We are the captains of our ship, this spacefaring sphere, the planet Earth—our special home.

11
The Moon: Faithful Companion and Springboard to the Stars

Three things cannot be long hidden: the sun, the moon, and the truth.
Buddha

The Moon! Artemis! The great goddess of the splendid past of men! Are you going to tell me she is a dead lump?
D. H. Lawrence

Everything has a natural explanation. The moon is not a god, but a great rock, and the sun a hot rock.
Anaxagoras

A mysterious, mesmerizing companion orbits the Earth, tugging at human hearts as it constantly pulls at our planet's substance, powerfully massaging its surface and core. Partners forever, the Earth and Moon perform a magnificent round dance as they spin through space together. By far the most luminous light in the night sky, the enthralling lunar sphere at times appears as a solitary, spectral schooner sailing over a dark sea. At other times, this floodlit orb resembles a shining pendant of shimmering pearl that silvers the Earth and acts as a radiant beacon that summons the gaze of dreamers, lovers, astronauts, and scientists. With amazing consistency, this mottled, iron-gray globe waxes and wanes, producing an elegant monthly display of horn-shaped crescents, curves, and circles as it passes through its spellbinding sequence of phases. Throughout history, this eternal beauty has captivated humanity, inspired occultists and admirers to invoke its putative mystical powers, and motivated experts to direct the force of their intellect toward the fathoming of the cosmos. The evocative, glowing presence of this ghostly galleon has enriched cultures, birthed traditions, and stimulated affections.

Humans have rarely realized that as the Moon maintains its continuous, lonesome vigil over the Earth, it unobtrusively provides priceless protection. Like Sancho Panza who dutifully aided his knight-errant, Don Quixote, the Moon unfailingly watches over the Earth. Our invaluable squire constantly stabilizes Earth's rotation, shields the Earth from wayward comets and asteroids, waits as a steppingstone to the exploration of space, and beckons to humanity to participate in the greatest exploration of all—the discovery of self and our place in the universe.

The Moon's majestic domination of the night sky makes it easy to understand how any tribe, culture, or people seeking something to worship might have chosen the Moon. In the past, the Moon was often revered as a goddess who controlled the daily tides and governed the cycles of female fertility. Witches seeking to enlist the forces of magic have long been known to stir their boiling cauldrons, chant their bizarre rituals, and cast their malevolent spells under the light of the Moon. For millennia, humans observing the Moon's ashen beauty have longed for control over its darker side. The Moon is the acknowledged ruler of the night, like the triple-bodied goddess Hecate of ancient Greek myth, who went everywhere accompanied by baying canines. The Moon has long been imbrued with macabre power over all beasts that howl, and even today, cringing youngsters thrill to spine-tingling stories of shape-shifting monsters who arise to do mayhem under the influence of the baleful cyclopean stare of the mist-shrouded Moon. Fearsome creatures of the night emerge from peaceful repose to scour their neighborhood seeking prey under the Moon's pale light. Real wolves howl to frighten their quarry or to summon the pack in pursuit of an intended victim. In our imaginations and in reality, terrifying events transpire under the Moon's watchful gaze.

Scientists, instead of trading weird tales, gathering shadowy folklore, and elaborating on ancient mythologies, choose to follow a different approach as they scrutinize, investigate, and analyze the Moon. They bring to bear all the tools, techniques, and insights of science to ascertain all they can about our captivating neighbor. As a result, today we know many things about the Moon. For instance, the Moon orbits approximately 239,000 miles from Earth, a mere 60 hours' journey via Apollo spacecraft. Earth's only permanent natural satellite, the Moon is the fifth largest moon in the Solar System and is larger than the dwarf planet Pluto. The Earth and its abnormally large Moon form a remarkable double planetary system, for the Moon is the largest planetary satellite relative to its primary anywhere in our Solar System.

The Earth and the Moon are so large and so close to each other that they exert powerful gravitational forces on each other. One effect of gravity is that the two bodies orbit each other. It is not technically correct to say that the Moon orbits the Earth. Instead, both the Earth and the Moon oscillate around the center of mass of the combined system. Since the Earth is so much more massive than the Moon, the center of mass of the system (the barycenter) is located inside the Earth, though fairly near its surface on the side facing the Moon. The slight weaving of the Earth as it orbits this point inside itself is quite measurable, although not as clearly evident as the Moon's wobbling orbit around the Earth.

Phases of the Moon

No one who studies the Moon for more than a few days can fail to notice the striking transformation it undergoes as it progresses through its phases. The Moon's light comes originally from the Sun and is reflected off the Moon's surface. The phases occur because different portions of the Moon's surface are illuminated as the Moon orbits the Earth. Surprisingly, the Moon does not reflect light very well. Its albedo is low—only 0.07, which means that only 7% of the light that hits its surface is reflected. If the Moon were just slightly more reflective, the Earth would in effect have two suns in its sky.

The Moon is dusty, desiccated, and devoid of life. An airless body about a quarter the diameter of Earth, the Moon has no atmosphere to moderate its heat. While there are occasional outgassings of various volcanic fumes, these do not last and are quickly dissipated by the solar wind. Therefore, the temperature on the outermost surface of the Moon can fluctuate between 260° F during the heat of the day and −280° F during the cold of the night. Astronauts' space suits must be well designed because the temperature can rise or fall hundreds of degrees very quickly at sunrise or sunset. The Moon has almost no tilt, so points near the poles in the bottoms of craters do not receive any direct sunlight and lie in regions of permanent shadow. There, temperatures can remain as low as -365° F for millions of years.

The Moon has long concealed the mystery of its origin underneath layers of dust and rock. Like the Mona Lisa, who hides her secrets behind a subtle, knowing, yet unfathomable smile, the Moon has guarded its secret through the eons. It is a mystery that ancient cultures and astronomers have contemplated for millennia. Recently, however, having visited the Moon six times, humans have been able to bring back to Earth and extensively analyze some 1828 pounds of its rocks. We have also carried out thorough photographic surveys of our celestial neighbor's near and far sides. Our extraterrestrial expeditions have produced reams of data, which scientists must consider when attempting to deduce the details of the formation of the Moon. One of the key clues to the Moon's origin is that Earth rocks and Moon rocks have the same isotope ratios, implying a common origin. Furthermore, the Earth's spin, the Moon's spin, the Moon's orbital rotation around the Earth, and the Earth's rotation around the Sun are all counterclockwise as viewed from the direction of the star Polaris. Thus the theory that the Moon was a visitor from afar which happened to be caught in the Earth's gravity is very implausible.

The far side of the Moon has a thicker crust than the near side, and there are virtually no plains, or maria, on the far side of the Moon. (The word "maria" is Latin for "seas" and was adopted by early observers of the Moon who thought the large, smooth expanses resembled bodies of water. These sharp-eyed observers of the pre-telescopic era even bestowed upon the maria imaginative names like "Sea of Storms.") Rock samples indicate that the Moon once had a molten surface. The Moon contains relatively little iron, a lack which contributes to the Moon's lower density. Evidence exists of collisions in other star systems that result in debris disks. The highlands of the Moon are older than the plains. Finally, the Moon has over 80,000 craters. How can all these disparate observable physical characteristics be merged into a persuasive, logical, and coherent theory of the Moon's origin?

Like a Renaissance painter crafting a masterpiece, who begins with a "vignette" or preliminary drawing, astrophysicists have begun to sketch in broad strokes the story of the Moon's origin. Approximately 4.5 billion years ago, the proto-Earth was one of many planetesimals that occasionally collided with each other as they spun around the Sun. Sometime during this period, a Mars-sized trespasser called Theia struck the Earth a glancing blow. This cataclysmic event nearly destroyed the Earth and turned its entire surface into a global ocean of molten rock, while shattering Theia. Chunks of Theia and parts of Earth's crust intermingled and were hurled thousands of miles into orbit around the early Earth in a planet-circling loop of superheated fluid. The heavier metals of Theia remained on the Earth and descended to become part of Earth's solid core. Lighter silicate rocks reformed the Earth's surface. During the next few hundred million years, Earth's ring of molten, metal-poor rock coalesced into the Moon. This explains the density difference between the two bodies, the small iron core of the Moon, and the unquestionable chemical similarities in the surface rocks of the Moon and the Earth. It also explains why the surface of the Moon shows signs of having once been molten.

Because the newly formed Moon was much closer to Earth than it is today, it quickly became tidally locked and settled into a one-to-one resonance. One-to-one resonance means that the Moon's period of rotation on its axis is equal to the period of its orbit around the Earth, which is about 29 ½ days. Most of the moons in our Solar System are tidally locked with their parent bodies. Thus, the surface of the Moon we see today, called the near side, is the same side of the Moon that the mastodons, dinosaurs, and trilobites saw when they walked and swam upon the Earth. It amounts to slightly more than half of the Moon's surface, due to an effect called libration, which results from the elliptical shape of the Moon's orbit. Although the Moon's orbit is almost perfectly circular, it is elliptical enough to afford periodic glimpses of areas amounting to nearly 60% of its surface.

If the Moon formed as the result of a collision and quickly became tidally locked with the Earth, as suggested by sophisticated computer simulations, then it would make sense that the Moon's far side has a thicker crust than its near side. The Earth, being more massive than the Moon, would have retained its great heat far longer and radiated this heat into space. The near side of the Moon, since it was facing the Earth, was warmed, while the far side cooled much more quickly, resulting in a thicker crust as compared to the near side.

After the bodies cooled and formed a crust, hundreds of thousands of meteors collided with the two bodies during the subsequent 100 million years, a period known as the Late Heavy Bombardment. This protracted barrage created the heavily cratered four-billion-year-old highlands we see today.

The largest impact crater on the Moon is the Aitken Basin, located on the far side near the Moon's South Pole. This immense feature stretches across 1,600 miles of the lunar surface and is 3.7 miles deep in some places. It has been theorized that this basin was created during the Late Heavy Bombardment approximately 4 billion years ago. It is believed that the impactor that created the basin came in at a shallow angle, digging this massive hole yet not penetrating to the still-molten interior of the Moon.

The thicker crust also shielded the inner molten magma of the Moon from the countless meteors that bombarded the far side. Those meteors gouged great holes in the face of the Moon but did not penetrate to the molten interior. The near side was struck by meteorites too, but since its crust was thinner, some of these cosmic projectiles penetrated to the molten center. The resulting volcanic vents spewed massive outpourings of magmatic basalt for millions of years. Magma filled in the vast holes and lowlands of the near side of the Moon, and these lowlands became the great seas, or maria, that we see on the Moon today. The younger rocks of the maria have been dated between 3.2 and 3.9 billion years of age. With its far side of older, thicker crust and countless craters, and its near side of younger, thinner crust along with craters and maria, the Moon is a facially asymmetric satellite.

Since the Moon is so small, it quickly lost its atmosphere to space and cooled too quickly to allow plate tectonics. Today, it is likely that the Moon no longer has a molten interior, especially since it possesses no magnetic field. These factors together mean that the Moon possesses no mechanisms for erosion, weathering, or the creation of new rocks. The Moon has therefore maintained its distinctive appearance with little change since the Late Heavy Bombardment four billion years ago, although every meteorite that hits it alters the surface somewhat. This relatively untouched state means that the Moon's rocks preserve clues to otherwise inscrutable events that happened in the early Solar System. From these clues, scientists theorize the order of events in the creation of the early Solar System and the Earth itself. This part of the theory explains why the Earth's spin and the Moon's orbit have similar orientations, as well as why the far side of the Moon has a thicker crust and no maria. It also explains the 80,000 craters and the fact that the highlands are older than the plains of the Moon. Thus, the destruction of a proto-planet the size of Mars 4.5 billion years ago, its subsequent resurrection as a molten annulus of debris, and its finally becoming the glowing satellite we see today is a remarkable story—and it is a model that accounts for the known characteristics of the Moon today.

The formation of the Moon was also an extremely fortuitous event for the development of life on Earth. The Earth and the Moon together form a gigantic gyroscope that provides billions of years of stability to the rotations of both bodies. Without the stabilizing effect of the Moon, Earth's axis would wobble dreadfully, as Mars' does. Earth's axial tilt stays at an exceptionally steady 23.5 degrees, while Mars swings wildly from upright to 60 degrees of tilt, which creates havoc with its weather and most assuredly is not conducive to the evolution of life.

The exchange of tidal forces between the Earth and Moon has profound effects on each sphere. The Earth's mass causes movement of the crust of the Moon, and the Moon causes tides to rise and fall upon the oceans of the Earth, as well as deforming the land. Since oceans are liquids that flow more easily than relatively solid land, ocean tides are far more noticeable. Sea tides, depending on local topography, can advance over miles of beach; the tidal tug that the Moon exerts on the solid crust, by contrast, only deforms the ground by about eight inches. The Earth's atmosphere, too, which is a type of fluid, flows under the influence of the tidal forces and forms a bulge in the direction of the Moon.

One might expect the Sun, thirty million times more massive than the Moon, to exert an even greater tidal effect upon the Earth. However, the Sun is about 93 million miles farther away from the Earth than the Moon is. Newton's formula for gravity conclusively reveals that, by far, the most salient factor in gravitational calculations is the distance that separates the bodies rather than their mass. When both mass and distance are taken into account, the gravity formula reveals that the Moon's gravitational force upon the Earth is twice as great as the Sun's. Thus, tides caused by the Sun are only half as big as the tides caused by the Moon. When the Sun and Moon act in concert, their separate gravitational tugs combine to produce the largest tides, which are called spring tides and occur twice per lunar cycle. When the Sun and Moon act antagonistically, their opposing gravitational forces produce the smallest tides, known as neap tides, also occurring twice per lunar cycle.

The rotation of the Earth also affects the tides. The Earth's rotation has a tendency to hurl the waters of the ocean away from the planet, outward towards space, just as a ball rotating overhead on the end of a string appears to be experiencing a force directed away from the center of rotation. The outward force, often called a centrifugal force, is caused by the ball's tendency to continue to move in a straight line, also called inertia. (If the string attached to the whirling ball were cut, then the ball would stop circling and begin to fly in a straight line. A similar phenomenon would occur if the Sun were to disappear; the Earth would stop circling the now-empty center of the Solar

System and fly in a straight line outward. In fact, all the planets, asteroids, and comets would fly in a straight line to who knows where, becoming "rogue" planets and rocky pieces of debris. Let us fervently hope that this never happens.)

The gravitational force exerted by the Moon on the Earth acts most strongly on the part of the Earth that is closest to the Moon, and is weakest on the part of the Earth that is farthest away from the Moon. Meanwhile, the centrifugal force on the waters in the oceans remains constant. These two forces when acting in unison create a high tide on the part of the Earth that is closest to the Moon, but there is also a high tide on the side of the Earth that is opposite the Moon. There, the inertial force overcomes the Moon's gravitational force, thus creating an antipodal high tide. Elsewhere on the Earth, the two forces have a tendency to balance each other out. Thus, tides occur almost twice at any given point on the Earth during a 24-hour period, once when the point is nearest the Moon, and once when it is on the opposite side of the Earth. The time of the daily high tides occurs approximately one hour later each day because each day the Moon completes one twenty-eighth of its rotation around the Earth.

Billions of years ago, the Moon was much closer to Earth than it is today. The drifting away of the Moon is corroborated by measurements made using reflectors left on the lunar surface by Apollo astronauts. Scientists direct highly coherent light from Earth at these reflectors and accurately time the duration of the light rays' round trip. Over the years, the duration of the round trip has been observed to increase slightly, conclusively proving that the Moon is indeed moving away from the Earth.

It is estimated that when the Moon formed 4.5 billion years ago, it was only 25,000 miles away from Earth. Thus it would have been nine times closer than it is today. Being nine times closer would have made it appear about 81 times larger in the Earth's sky. Our Moon would have appeared as a gargantuan, swollen disk in the sky, reaching almost from horizon to horizon, and would have transformed night into day. Such proximity would have increased the Moon's gravitational influence on the Earth, resulting in colossal tidal waves a thousand feet high. These dramatic high tides would have occurred every three hours, since in that epoch the Earth was also rotating more rapidly than it is today, completing a revolution every six hours.

The Earth's rotation has slowed down since then because the Earth drags the Moon through space. Earth's rotation is still decelerating, but luckily this happens at an unhurried pace—only about a thousandth of a second per century. At that rate, we will all have an extra hour per day to run our errands…140 million years from now. Additionally, the elaborate binary dance of tidal effects has steadily flung the Moon farther and farther away from Earth. The effect is small: the distance between Earth and Moon increases by a mere one and a half inches per year. This is about the rate at which human fingernails grow. Despite the steady increase, there is no need to worry that the Moon will eventually sail far enough away to escape entirely from the Earth's gravitational hold. Long before that happens, the aging Sun will expand and engulf the inner planets, and the Earth and Moon along with them.

Exploration

For countless millennia, the Moon's beauty and brightness have enticed human viewers, luring them outward to learn and explore. According to ancient cosmologies, the Moon was held in place yet permitted to rotate by virtue of being imbedded in its own independently rotating crystal sphere.

This theory accounted for the Moon's motions even though the ancients who conceived this theory, in which the other heavenly bodies each had their own sphere, had no idea of the Moon's true nature (but see the quote from Anaxagoras, above). As was common among primitive peoples, the activities of the Moon were ascribed to deities. Among the Greeks, the goddess of the Moon was Artemis, who was also the goddess of hunting and of bears. She was the twin sister of Apollo, the god of the Sun, so together the two siblings ruled the principal lights of both day and night.

Even as the scientific worldview began to transcend the mythological, the Moon kept up its attraction. Who could gaze upward in the nighttime without longing to learn the truth of the nature and meaning of that brightest of the lights? Who could resist dreaming of someday, somehow, being able to journey upward into that night and stand on the surface of that glowing, mottled orb. Long before humanity could entertain realistic hopes of visiting the Moon, dreamers, and in particular, authors, laid out the path. Edgar Allen Poe and Jules Verne were just two of the imaginative geniuses whose fictional works provided the impetus for the quest which would come to such amazing fruition in the following century.

In 1960, the United States of America was locked in a tense "Cold War" with the Soviet Union. Citizens in both countries were terrified that the tension could, without warning, explode into a global nuclear holocaust. Meanwhile, every scheme, action, and maneuver on each side was calculated to demonstrate the superiority of that side's political and economic philosophy. Against the background of this historical context, in 1957 the Soviet Union had seized the initiative by launching the world's first artificial satellite, Sputnik.

Americans were terrified. Many feared that the Soviet Union would presently be able to rain death down upon them from space. With the memory of World War II seared into the national consciousness, this existential threat aroused the nation and spurred a rebirth of interest in the teaching of science, engineering, and mathematics. There was a clear sense that the nation had faltered and had a lot of work to do to catch up technologically and surpass its fearsome ideological opponent.

To inspire the nation and soothe its angst, newly elected President John F. Kennedy announced in his inaugural address that the United States would land a man on the Moon and return him safely to Earth "before the decade is out." Galvanized into action, Americans marshaled all their energy and determination to win the newborn "space race." President Kennedy was assassinated on November 22, 1963, and the quest to go to the Moon became a mission of honor to glorify the memory of the beloved late president. Although the task seemed impossible and the hurdles insurmountable, the American people accepted the challenge and worked with singleness of purpose to achieve the goal.

The task was assigned to the National Aeronautics and Space Administration (NASA). With the clock ticking, the scientists and engineers of NASA performed heroically in designing, building, testing, and perfecting the necessary new technologies. They worked in partnership with private industry as manufacturers across the country were contracted to produce the myriad of custom-made parts for the necessary equipment, apparatus, and vehicles that would successfully carry men to the Moon and back.

The name of the program to carry men to the Moon and back was recommended by NASA director, Abe Silverstein, who believed that "Apollo" captured the scope, spirit, and optimism of the ambitious program. In 17 missions filled with drama, suspense, and occasional despair, NASA succeeded beyond the expectations of most people, and their work culminated in the greatest achievement in the history of human exploration.

On July 16, 1969, the Apollo 11 mission was launched from the Kennedy Space Center. Four days later, on July 20, the Lunar Excursion Module, named "Eagle," landed on the surface of the Moon. Neil Armstrong stepped out onto the surface and uttered the famous phrase, "That's one small step for [a] man, one giant leap for mankind."

Harrison Schmitt with Flag

Six more Apollo missions followed Apollo 11. The picture at right is from Apollo 17 and shows astronaut Harrison Schmitt next to a United States flag on the surface of the Moon with the Earth in the background.

"Science has been a very important part of making our beloved United States of America what it is. We are an economic superpower and a military superpower primarily because we've dedicated ourselves to being leaders in science," said reporter Dan Rather. The exploration of the Moon by astronauts from the U.S.A. is a prime example of this.

Conclusion

The Moon has been the companion of the Earth since its fiery birth. It has proved to be a consequential ally and a friend of life and of human beings. The Moon has stood as a guardian and protector from meteors, and it has maintained and stabilized the Earth's spin and its seasons for billions of years. At the dawn of terrestrial life, it stimulated marine organisms to swim out of the oceans and crawl upon the land by way of tidal pools created by its gravity.

In our own era, this fellow traveler has been the site of one of humankind's most ambitious and impressive achievements. In the future, will it be a steppingstone to greater achievements? The first step toward exploration of the cosmos will have to be a return to the Moon to establish a permanently staffed base. For humans to one day walk on the surface of Mars or visit the planets circling distant suns, the Moon could be a necessary proving ground, way station, and supply depot—a veritable springboard to the stars.

Past, present, and future, the Moon is easy to take for granted. On a clear night, when it lights up the sky, remember that over the billions of years of its partnership with Earth, it has given much, much more than light.

Artist's conception of a collision of the type which hypothetically occurred between Earth and Theia and gave rise to the Moon.

Footprint left by a human visitor to the Moon

12
Eclipses

I have always read that the world, both land and water, was spherical, as the authority and researches of Ptolemy and all the others who have written on this subject demonstrate and prove, as do the eclipses of the moon...
Christopher Columbus

The like Eclipse having not for many ages been seen in the Southern Parts of Great Britain, I thought it not improper to give the Publick an Account thereof, that the sudden darkness, wherein the Starrs will be visible about the Sun, may give no surprise to the People, who would, if unadvertised, be apt to look upon it as Ominous, and to interpret it as portending evill to our Sovereign Lord King George I and his government, which God preserve. Hereby they will see that there is nothing in it more than Natural, and no more than the necessary result of the Motions of the Sun and Moon; and how well those are understood will appear by this Eclipse.
Edmond Halley[1]
(Published slightly before the eclipse of May 3, 1715.)

While the universe abounds with an incredible array of immense, impressive, and gorgeous phenomena, all the other astronomical wonders pale before the fascinating and stunning spectacle of a full solar eclipse. As human beings, we respond immediately, viscerally, and primally to the sight of our beloved, life-giving Sun being slowly obscured into blackness by the encroaching disk of the Moon.

During a solar eclipse, continent-sized spectral shadows race across the surface of the Earth faster than the speed of sound. Glimpses of the waves and ripples near the surface of the Sun become perceptible, and just at the moment of totality the typically invisible corona magnificently reveals itself. This stirring sight remains in view for only a few fleeting minutes before the Sun's glare

[1] The eclipse to which Halley is referring happened on April 22, 1715, according to the old Julian calendar in use in England at that time. The English did not adopt the modern (Gregorian) calendar until 1752. Under the new calendar, the date of the eclipse would have been May 3, 1715, and that is the date we recognize today.

once again overwhelms it. A total solar eclipse is an oddity, a marvel, and a revelation.

Eclipses are not uncommon. There is an eclipse of some kind at least twice annually, and some years may have as many as five. From any particular point on Earth, however, a total solar eclipse can be seen but once every 375 years, so it is easy to understand why eclipses have consistently been considered significant events and have had a lasting impact on the course of human history. Today, eclipses represent research opportunities that are eagerly awaited by astronomers ready to take advantage of the unique set of circumstances that reveal otherwise invisible details and mechanisms of the Sun's activities.

While a solar eclipse is the gradual encroachment of the lunar silhouette across the disk of the Sun, any eclipse is an example of syzygy—the linear conjunction of three celestial bodies. In a solar eclipse, the Moon passes between the Sun and the Earth, concealing part or all of the Sun from Earthbound viewers. The illustration at right shows a solar eclipse. During a solar eclipse, several types of shadows are cast. The umbra (Latin for "shadow") is the central and darkest shadow. Viewers over whom the umbra passes see the Sun completely covered by the Moon. This is referred to as a total eclipse, which can last as long as 7 minutes, 32 seconds, and can be viewed along a track up to 150 miles wide.

Solar Eclipse Diagram

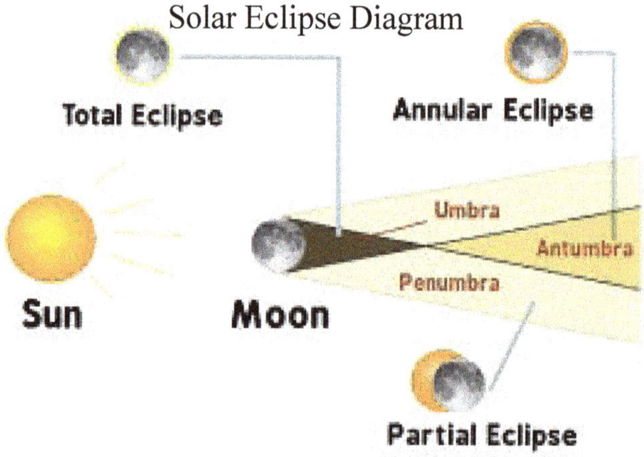

As the umbra races across the Earth's surface at more than 1300 miles per hour, temperatures can drop as much as 35° F, and stars become visible. Animals start to exhibit their typical nighttime behavior: Crickets begin to chirp, chickens retreat to their coops, and all manner of diurnal beasts prepare for sleep. Solar power plant engineers must prepare for a sudden loss of wattage, as even a partial eclipse affecting northern California can result in the temporary loss of enough solar energy to power a million homes. The word "umbra" gives rise to an alternate term for an eclipse-chaser, the kind of person who travels the world to view, photograph, and study eclipses. The term is "umbraphile," which can be loosely interpreted to mean "shadow-lover." The shadowed region outside the umbra is referred to as the penumbra. A viewer in the penumbra (Latin for "almost shadow") sees the Sun only partly obscured. The penumbra is not as dark as the umbra but is much larger. In the photograph at right, taken from the International Space Station, both the umbra and the penumbra can be seen on the curved surface of the Earth.

Umbra as seen from the ISS

By an astonishingly improbable coincidence, possibly unique in our galaxy, the size of our Moon in the sky appears to an Earth-based observer to be almost exactly the same size as the Sun. This is why the Moon can completely cover the Sun even though the Sun is over 400 times larger. It works because the Sun is also 400 times further away. But this will not last forever. The Sun is gradually growing bigger, and the Moon is slowly drifting away from the Earth at a speed of about 1.5 inches per year. This means that in 600 million years or so, the Moon will be smaller in the sky than the Sun and will no longer be able to conceal the Sun completely.

The Moon's distance from the Earth also varies slightly with each orbit. Therefore, it can sometimes happen that a "total" eclipse does not entirely block out the Sun's light. This is the case when the Moon is slightly farther from the Earth than during a normal eclipse. When this happens, a remarkable effect is seen in which a halo of sunlight is visible around the Moon. Such an eclipse is called an "annular" eclipse, the so-called "ring of fire." ("Annular" means "ring-shaped.") Also, since the surface of the Moon is jagged and irregular, sometimes a final few fleeting rays of sunlight can shoot through craggy gaps in its mountain ranges and crater walls and strike the Earth before the eclipse becomes total. If only one ray trickles through, then we see the resplendent "diamond ring" effect (right top). When more than one ray slips through, the phenomenon is called "Baily's Beads" (right middle) after Francis Baily, who observed them during an annular eclipse in 1836. Baily was a successful English businessman who retired to devote himself to astronomy.

Diamond Ring

Baily's Beads

The picture at right bottom shows the Sun's corona. This is a part of the Sun's atmosphere that is not usually visible but can be seen during an eclipse because the bright disk of the Sun is hidden and therefore does not overpower the corona. The corona (Latin for "crown") is a halo of energy a million miles thick and very tenuous but remarkably hot.

Solar Eclipse Showing Corona

Investigating Eclipses

Eclipses have fascinated, horrified, and mystified human beings since before the dawn of recorded history. Once humans learned to write, they immediately began keeping records of important events, and prominent among the events they tracked were eclipses. A carved stone in Ireland appears to record an eclipse that occurred on November 30, 3340 BCE, and an ancient clay tablet found in the Middle East makes mention of a solar eclipse visible from Syria on March 5, 1223 BCE.

One of the great talents of the human species is the ability to recognize patterns. Naturally, then, patterns in the occurrence of eclipses were quickly discerned. One example is the saros cycle, which was discovered by ancient Chaldeans around the eighth century, BCE. In this cycle, solar eclipses occur every 6,585.3 days, which is 18 years plus about 10.3 days. At any given time, there are a number of these cycles in progress.

According to the ancient Greek historian Herodotus, the philosopher-sage Thales accurately predicted the solar eclipse of May 28, 585 BCE. Thales is regarded as the first thinker to attempt to objectively determine the nature of physical reality without recourse to supernatural causes; he is also credited with having devised the first recorded rigorous mathematical proof.

Hipparchus in the second century BCE examined ancient records of eclipses and was able to discern that the Earth wobbles on its axis. This wobbling is referred to as precession of the equinoxes, and Hipparchus's discovery is just one example of how scientists seeking knowledge invariably owe a debt of gratitude to those who came before. Newton expressed this idea by famously stating that if he saw further than others, it was only because he "stood on the shoulders of giants."

The Greek astronomer Aristarchus, noted for promoting a heliocentric worldview more than 1700 years before Copernicus, also developed the first known practical procedure for computing the distances from the Earth to the Sun and Moon, and also the sizes of the Sun and Moon. Around 270 BCE, he geometrically calculated from observations taken during an eclipse that the Moon's distance from the Earth was 20 Earth radii. His measurement was inaccurate, but his method was sound; others using the same procedure and more careful observations produced estimates that were much closer to the correct distance. Aristarchus's method enabled him to use the radius of the Earth as his yardstick even though he did not know how large the Earth was. However, another giant of Greek astronomy, Eratosthenes of Alexandria, calculated the radius of the Earth a few years later.

The ancient Greeks became highly skilled at predicting eclipses. In 1901, a puzzling mechanical device made of bronze was discovered in a Roman shipwreck in the Mediterranean Sea. After years of study, scientists arrived at the conclusion that this device, known as the Antikythera mechanism after the location where the shipwreck was found, was a hand-cranked mechanical device, a sort of calculator for predicting eclipses, planetary positions, the phases of the Moon, and even the date of the next Olympic games. The complexity and cleverness of this machine are impressive even by today's standards, although it was built around 2100 years ago. The saros cycle was built into the mechanism. In addition, it is evident that the builders of this bronze device placed the Sun at the center of the Solar System, anticipating the modern view. The illustration on the next page, top left, shows the device as it survives today and the illustration on the next page, top right, shows a partial reconstruction by the science historian Derek de Solla Price). Today, of course, we are even better at predicting eclipses. The Babylonians were able to predict eclipses within four hours. Edmond Halley[2], who independently rediscovered the saros cycle after

[2] The word "saros" was adopted by Halley from an ancient Greek lexicon. The Greek word (σάρος) means "repeat" and was derived from the Babylonian word *šaru*, which likely referred to the number 3600. In modern usage, "saros" refers to the 18-year eclipse cycle.

Antikythera Device Actual

Antikythera Device Reconstructed

knowledge of it had been lost for centuries, was able to predict the time of an eclipse in England on May 3, 1715, to within four minutes. Today we are even more precise in our predictions and can be accurate to the nearest second.

Arthur Eddington

An eclipse contributed to the verification of Einstein's theory of relativity early in the 20th century. Einstein's theory predicted that light from distant stars should be affected by gravity, a finding which sharply contradicted the conventional wisdom of the time. Sir Arthur Eddington, while observing a total solar eclipse on the island of Príncipe off the coast of West Africa in 1919, saw that stars close to the Sun during the eclipse seemed to be slightly out of position. They appeared out of position because their light was following a curved path under the influence of the Sun's formidable gravity, just as predicted by Einstein. Such observations can be made only during an eclipse because the Sun's overwhelming brightness must be blocked out before the much dimmer nearby stars can be seen. In fact, some of the stars observed by Eddington were located behind the Sun and would not have been visible at all without the powerful gravitational effects produced by the Sun.

The First World War had ended just one year before Eddington made his observations. It had been a woefully bloody and destructive struggle that claimed approximately eight and a half million lives, and a large proportion of the dead were British and German. The fact that a British scientist would make a point of dedicating himself to the verification of an abstruse theory being promulgated by a German scientist was repugnant to some in that time of lingering animosity. This willingness to transcend the recent enmities of the war years was seen by many as an indication that the sharing of scientific knowledge for the greater good of humankind could be a route to reconciliation and enduring peace.

Fear of Solar Eclipses

Of all the natural phenomena that terrified and bewildered ancient peoples, an eclipse was something they could experience as it happened. During a solar eclipse, the Moon appears to slowly and methodically consume the Sun, provoking dread. To our ancestors, it seemed that the whole world order was ending before their startled eyes. Day turned into night, temperatures plummeted, the behavior of animals changed, strange shadows emerged, and the altered appearance of the surroundings generated an otherworldly sensation. At best, this disquieting sight would have been considered a kind of celestial magic or a cosmic coincidence; at its worst, to the ancient mind, it would have been a clear sign of divine displeasure. Eclipses were thus assumed to be omens of doom, harbingers of death, and portents of disaster. Kings died, dynasties were overthrown, and battles halted because of the fear these omens instilled in our stupefied primitive ancestors.

Ordinary people desperate for a comforting explanation for the alarming metamorphoses in the normally unchanging celestial order concocted a variety of creative interpretations. In various lands and at various times, the destruction of the Sun during an eclipse was ascribed to wolves, a giant frog, dogs, a bear, or a dragon.

The inhabitants of ancient Mesopotamia believed that solar eclipses heralded the death of a king. They believed it so strongly that they *made* it come true, in tragic and gruesome fashion. They picked a substitute king, and after the eclipse had run its course, they executed him and his "queen" (another substitute, of course). The real king's life was thus preserved against the deadly powers of the eclipse. Similar beliefs held sway in China, where the eclipse of 1059 BCE marked the end of the Shang dynasty.

The Greek sage Anaxagoras in the 5th century BCE was the first to suggest the correct explanation for eclipses. His Athenian compatriots were offended by his proposal, which they interpreted as an attack on their religious beliefs, and they voted to exile him from Athens for the rest of his life.

Images of a Partial and a Total Solar Eclipse

During the Peloponnesian War in 413 BCE, an eclipse accompanied the debacle that brought the Athenian invasion of Sicily to a disastrous finish. The disaster was partly due to the fact that the Athenian naval commander interpreted the eclipse as a bad omen and was paralyzed into inaction.

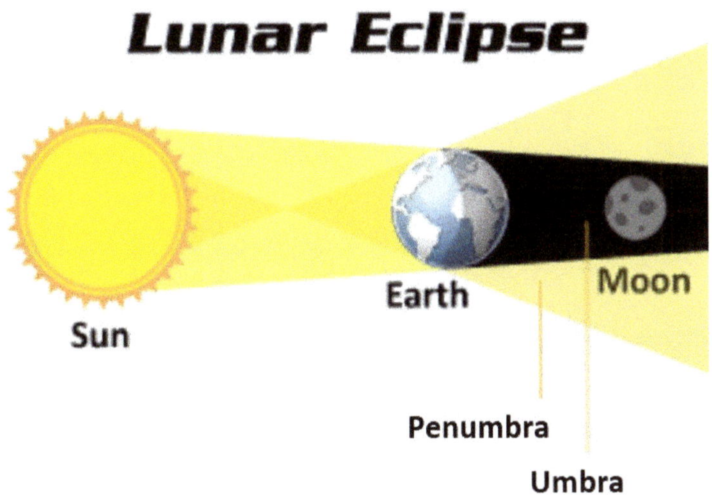

The image at left shows the relative positions of the Sun, Earth, and Moon during a lunar eclipse. As in a solar eclipse, all three must align so that a shadow is cast, this time by the Earth. Another similarity with solar eclipses is that both the umbra and penumbra are present. If the Moon is entirely within the umbra, a total lunar eclipse occurs, while if the Moon is in the penumbra, a partial lunar eclipse is the result.

Like their solar counterparts, lunar eclipses have also had an impact on history. The total lunar eclipse on June 21, 168 BCE occurred shortly before an anticipated battle between the Macedonians and the Romans at Pydna. The contemporary historian Polybius stated that "the report gained popular credence that it portended the eclipse of a king." While this lent courage to the Romans, who did not have a king, it discouraged the Macedonians, who did. The Romans achieved a decisive victory.

Columbus

An eclipse reportedly saved the lives of the famed Genoese sailor Christopher Columbus and his crew. On the famous explorer's fourth and final voyage, in July 1503, he found himself stranded on the north shore of Jamaica because his ships were being destroyed by an infestation of timber-devouring worms.

The European castaways, unable to feed themselves in that alien environment, were dependent upon the local Arawak people for provisions. Understandably, as the months passed, the Arawaks lost their enthusiasm for supporting the haughty, freeloading interlopers. It did not help that those ignoble emissaries of European civilization were murdering islanders, plundering villages, and committing other depredations among the native people.

Columbus decided he had to devise a stratagem that would return the Arawaks to their former friendly demeanor. He had in his possession a book of astronomical tables authored by the German astronomer and mathematician Johann Müller, who wrote under the pen name of Regiomontanus. This hefty tome was in widespread use at the time, and every leading explorer and navigator possessed a copy. Included in the book was a table of lunar eclipses. Columbus noted with interest that a lunar eclipse was predicted to appear on the evening of February 29, 1504.

On February 26, he summoned the Arawak chiefs into his presence. He announced to them that his god was angry with them for refusing to help Columbus and his men. Furthermore, he added, his god would show his displeasure by destroying the Moon and making it "inflamed with wrath." The Arawak leaders scoffed at Columbus's threat.

On the appointed night, the Moon began to slowly disappear as if being devoured, and the Arawaks became terrified. They came running, laden with food and begging for forgiveness. Columbus pretended to confer with his god and then announced that mercy would be shown. The Moon duly began to reappear. Columbus and his men remained on Jamaica for another six months, until they were rescued by a Spanish vessel sent from Hispaniola, and the natives kept them fully supplied the whole time.

The reason that Columbus and the native inhabitants saw the Moon as a red disk "inflamed with wrath" during the lunar eclipse was that the Earth's atmosphere bends, or refracts, red light more than blue light. The only light that reaches the Moon during a lunar eclipse is light that has passed through the narrow fringe of Earth's atmosphere and been refracted. In fact, it has been doubly refracted—once as it enters the fringe of the Earth's atmosphere and again as it leaves. Refraction bends the light rays so that a small amount of light can illuminate the surface of the Moon. That reddened light is then reflected back to the eyes of humans observing the eclipse.

The same refraction of light also explains resplendent reddish-orange sunsets, magnificently arched rainbows, and the superb spectrum that spills from a prism. An astronomy professor once shared with his class an epiphany he experienced on his honeymoon. He was strolling hand-in-hand with his new bride upon the moonlit deck of a cruise ship sailing towards an eclipse when he was suddenly possessed by a profound insight and announced, "Ah! That's why the Moon is red!" For dedicated umbraphiles, even romance can take a back seat to astronomy.

The Great Eclipse of 2017

The human sense of sight utilizes specialized cells called rods and cones. These sophisticated and delicate cells capture photons of light and then, via nerve fibers, send the information to our brains for interpretation. Vision plays a central role in many human activities, so preservation of the ability to see is of paramount importance, but while extraordinarily useful, our sense of sight is extremely fragile. Viewing the Sun directly for even very short periods can destroy forever one's ability to see. Our rods and cones simply perish before the furious onslaught of solar radiation and the overwhelming brilliance of the Sun. Viewers of a solar eclipse must therefore be very cautious and avoid injury to their eyes. Looking through welder's glass of sufficient strength is one safe way. It is also safe to use a pinhole projector to view the Sun. In this method, light from the eclipse passes through a tiny hole and an image of the Sun is projected onto a surface beyond the pinhole. On the other hand, regular sunglasses do not provide the necessary protection. In 2017, millions of Americans had the opportunity to safely view their first solar eclipse.

The Great American Eclipse of August 21, 2017, will long be remembered by those who witnessed this historic event. Many of these eager solar spectators gathered with their families and made the event a celebration, reveling in the excitement generated by this exceptional celestial event.

Numerous families and individuals traveled hundreds of miles to set up camp inside the long, narrow strip of land over which the astronomical marvel would appear most dramatically.

This eclipse was unusual in various ways. It was the first eclipse since 1776 to be visible only in the area of the 48 continental United States. It was the first eclipse to travel from coast to coast of North America since 1918. This total solar eclipse started over the Pacific Ocean and arrived at the west coast of the United States at Government Point, Oregon, at 10:15:50 AM PST. It then traveled obliquely through 14 states and reached the Atlantic Ocean near Charleston, South Carolina, at 2:49:07 PM EST. When the shadow of the Moon first touched the Earth, it was traveling faster than 8,000 miles per hour relative to the ground. As it progressed across the country, it slowed. In central Tennessee, it was traveling a little over 1300 miles per hour. It then began to pick up speed, and by the time it reached the South Carolina coast its speed was approximately 1354 miles per hour. The path of totality crossed the continental USA in 93 minutes and 17 seconds, covering a total distance of 2,496 miles.

Scientists marshaled an elaborate assortment of telescopes, cameras, and other equipment mounted on balloons, jet fighters, and the ground to study the eclipse in greater detail than any previous eclipse. After some six years of testing and preparation, NASA scientists, university professors, and student volunteers eagerly anticipated 93 minutes of frantic data-gathering. The focus of the data-gathering session was on the Sun's corona, a beautiful and mysterious region of the Sun that can only be observed and studied during an eclipse.

The corona is just one part, or layer, of the magnificent Sun which lies at the heart of our Solar System and provides us with the energy needed for life. At the core of the Sun, where nuclear fusion occurs, the temperature is 27 million degrees. From the center outward, the temperature steadily cools to approximately 10,000 degrees at the outer surface. Amazingly, from the surface outward, the increasingly tenuous solar atmosphere, the corona, skyrockets in temperature to a mind-boggling maximum of 3.5 million degrees. This coronal temperature inversion is a mystery that no one has been able to satisfactorily explain.

NASA dispatched two specially equipped WB-57 fighter jets to chase the umbra and capture images of the corona during the eclipse. These large and speedy aircraft flew 50 miles apart at an altitude of 50,000 feet, their nose-cones packed with instruments to view the eclipse in visible light and infrared, a lower-energy form of electromagnetic radiation used to detect heat signatures. By flying fast, the planes extended their time inside the umbra from two to about four minutes per plane. NASA wished to study the waves and ripples of energy that flow through this mysterious region of the Sun to determine why these temperatures rise so dramatically.

The bizarre temperature profile of the corona is undoubtedly related to the solar wind. The solar wind consists of hot, charged particles that continuously stream out of the Sun's corona into space. At the surface of the Sun they have zero velocity, but they somehow pick up momentum until, eventually, they are streaking away from the Sun at 720,000 miles per hour like subatomic super-bullets. Solar scientists took their cue from Edmond Halley, who in 1715 enlisted amateur astronomers in England to view the upcoming eclipse. Working together, they were able to gather more data and improve eclipse prediction. Thus NASA assembled hundreds of volunteers to observe the Sun through specialized equipment as part of CATE, the Continental American Telescopic Experiment. Sixty-eight teams of volunteers equipped with identical telescopes, software, and computers were stationed along the eclipse's path participated in an experiment designed to investigate the origin of the solar wind. The question of what hitherto unelucidated force of nature accelerates the solar-wind particles from a virtual standstill to 720,000 miles per hour is a baffling conundrum.

The Sun is a spherical ocean of plasma surrounded by the superheated corona, and this system has weather just like any other celestial body with an atmosphere. However, the storms that arise from the surface of the Sun's ocean of plasma have power far beyond anything seen on a planet. Shooting up from the Sun's surface are streamers, loops, and prominences of matter and energy that are larger than the Earth itself. The storms that the Sun generates can have catastrophic consequences for the planets that orbit it. Therefore a small fleet of satellites has been stationed around the Sun and tasked with monitoring the solar weather. Solar weather forecasters on Earth receive the satellites' data and pore through it looking for warning signs of impending disaster.

Solar storms can generate tremendously destructive events called coronal mass ejections, which can damage or destroy electrical circuits on Earth. Coronal mass ejections (CMEs) occur when the Sun "burps" and flings billions of tons of plasma from its surface. These bursts of superheated ionized particles can travel through space at millions of miles per hour. If such a burst of energy were to directly strike the Earth, it could easily penetrate the protective veil of our planet's magnetic field, and the result for our society would be utterly disastrous. Virtually all human activities today rely in some way upon surprisingly fragile electrical circuits. The destructive lash of a massive CME would destroy computers, make almost all vehicles inoperative, inactivate power stations, knock out satellites, ignite fires, and threaten the very continuation of human civilization. The auroras would be fantastic, but beautiful auroras would be meager compensation for the destruction of the electrical infrastructure of our society. Improved prediction of these phenomena is a necessity.

A continent-spanning armada of 70 rugged, high-altitude balloons soared to 100,000 feet in the upper atmosphere to obtain clear views of the eclipse. Images from all the balloons were merged to construct a continuous record of the eclipse from start to finish. Each balloon's four circular arrays of cameras were set to capture images of Baily's Beads. By focusing on these fleeting rays, they were able to look more closely at the Sun's surface than ever before. When the balloons finished their task, they released their precious pictures and data, which parachuted safely back to Earth. Scientists intend to use this data to improve solar weather prediction. If they learn how solar storms develop, then they believe they can better predict future storms. This would give earthbound forecasters more time to warn the population about an imminent stellar onslaught so that protective measures could be implemented.

In Lovelock, Nevada, the partial eclipse started around 9:09 AM and was over at 11:47 AM. The peak darkness occurred at 10:24 with about 83% of the Sun obscured and lasted some two minutes. Observers reported that the ambient light took on a surreal, other-worldly quality. The amount of light substantially dimmed and temperatures fell noticeably, probably 5 degrees or more. Shadows appeared sharper than normal (the quality of light was subtly changed), and the local bird population went mute. Farther north, in the path of totality, a spectacular daytime view of stars and planets was made possible by the eclipse. Jupiter, Venus, Mars, Mercury, and a host of the sky's brightest stars, including Regulus, Sirius, Arcturus, and many others, burst into visibility.

The next solar eclipse in the United States will occur on April 8, 2024, and will run through Texas to the New England states. The next coast-to-coast eclipse is expected during 2045 and will pass over Reno, Nevada. Undoubtedly, many of the 50,000 Reno schoolchildren who saw the eclipse of 2017 will be there to watch.

Saros 136

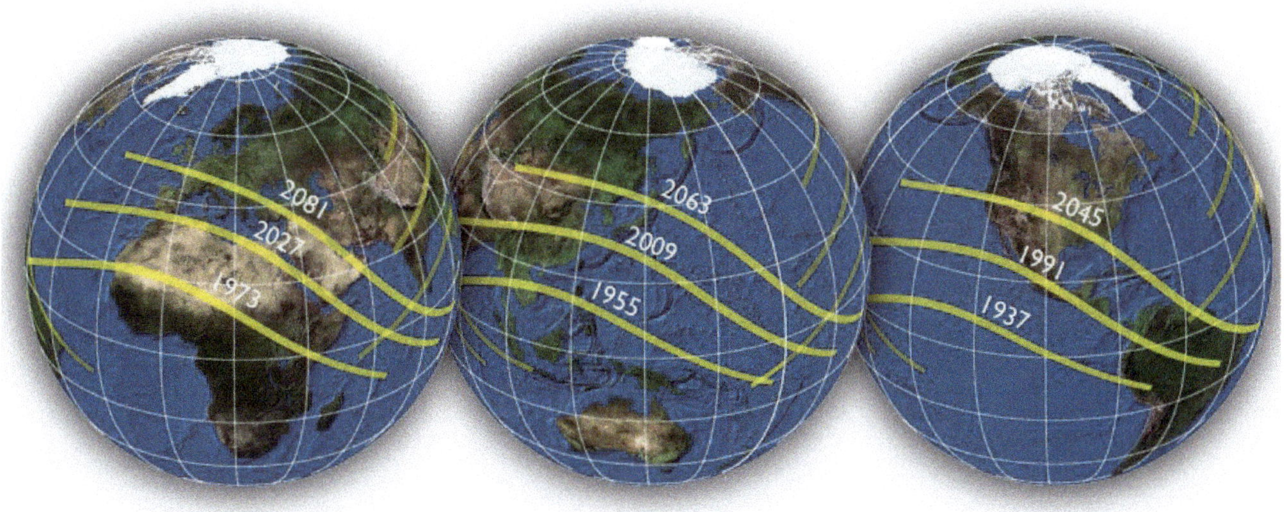

13
Mercury: Planet of Extremes

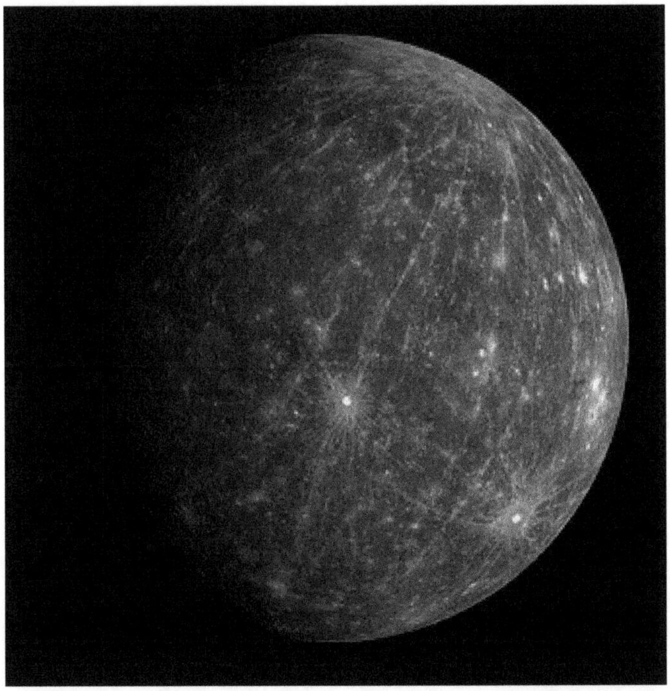

Yet in this, my stars were not Mercury as morning star in the angle of the seventh house, in quartile with Mars, but they were Copernicus, they were Tycho Brahe, without whose books of observations everything which has now been brought by me into the brightest daylight would lie buried in darkness.
Johannes Kepler

I would rather be Mercury, the smallest among seven revolving round the sun, than the first among five revolving round Saturn.
Johann Wolfgang von Goethe

The merriest of the gods, this elusive mischief-maker was as fast as the wind, lived by his wits, and served as emissary and herald of the gods. An esteemed member of the Roman pantheon of deities, he dashed forth from the radiance of Mount Olympus, always completing his assigned errands with phenomenal quickness. The son of Jupiter and the grandson of the Titan Atlas, this dashing youth seized any opportunity to rush into adventure. Of quicksilver tongue, this eloquent god was dubbed Lord of Language. Proudly plumed with wings on his cap and heels, he wielded in his left hand a striking, snake-twined staff called the caduceus, which remains today the symbol of the healing professions. God of boxers, thieves, merchants, and travelers, he was the husband of Larunda, a naiad nymph, daughter of the River Almo. His mother Maia, one of the seven Pleiades, was the eldest daughter of Atlas and a granddaughter of the Titan Oceanus. Mercury was tasked to be the "psychopomp," charged with expeditiously escorting the souls of the dead to Hades. In ancient Rome, his worshipers celebrated and honored him on the ides of May. Travelers set up roadside cairns in his honor. Mercury, the winged messenger of the Olympian gods, was the swiftest entity of ancient times and remains so today.

The planet Mercury, a cratered rock that today keeps alive the name of that ancient god, is also notable for speediness. But speed is not its only claim to fame. In addition to darting across the sky from night to night more quickly than the other sky-wanderers, it boasts a number of additional and equally impressive, unique characteristics.

Mercury is the most cratered, smallest, and innermost of the eight planets in our Solar System. This diminutive world orbits only one-third as far from the Sun as the Earth does. To stare at the Sun from Mercury's surface would be devastating to Earthly eyes since the Sun would appear nine times bigger and brighter. With the Sun so close, the dayside of Mercury absorbs vast quantities of solar energy. Here, the Sun's fierce heat produces a furnace-like environment, and the planet's surface becomes smelting hot—hot enough to melt lead and zinc. If a liquid drop of the element mercury were present on the surface of this iron-cored planet, it would boil away. The same would happen even to an element like sulfur. Under these conditions, Mercury possesses virtually no liquids or gases. It would not be accurate, however, to say that Mercury has no atmosphere at all. Its tenuous, ephemeral atmosphere consists of hydrogen, helium, and water vapor, as well as atomic sodium, potassium, and calcium. These fleeting wisps of gas are present from time to time due to micrometeorite strikes or outgassing from volcanic vents but are quickly dissipated by the solar wind. The dissipating gases create a tenuous tail similar to a comet's tail but much less showy.

A consequence of Mercury's almost complete lack of atmosphere is that the planet does not experience the effects of weathering or erosion. Thus, Mercury retains the scars of every meteor that has struck it over the last four billion years. It is a desolate, airless world dominated by craters and closely resembles the far side of Earth's Moon. Such a thin atmosphere does little to hold and disperse the Sun's heat, so the day-night temperature fluctuations are greater than anywhere else in our Solar System, ranging from −290°F to 810°F. The Mercurian poles remain permanently frigid since they receive virtually no direct sunlight. In the absence of a thick atmosphere to retain heat and distribute it around the planet, Mercury's night side radiates its previously captured heat energy directly into space. This results in plummeting surface temperatures and long periods of bitter cold. If there were any carbon dioxide or methane present at these times, it would rapidly freeze solid, only to vaporize and be blown away by the solar wind at the next sunrise.

From Earth, Mercury appears as a late morning or early evening star, while from Mercury, the Earth and Moon would appear as a pulsing and eclipsing binary star system. An astronaut looking up into the Mercurian night sky would see, in addition to the striking binary Earth-Moon system, perpetual blackness dotted with sharply defined stars. The stars would never twinkle as they do when viewed from Earth, since such twinkling is caused by atmospheric effects. No atmospheric oxygen also means no ozone layer. On Mercury, there is not even a modicum of protection from cosmic rays or from the Sun's intense X-rays and ultraviolet radiation.

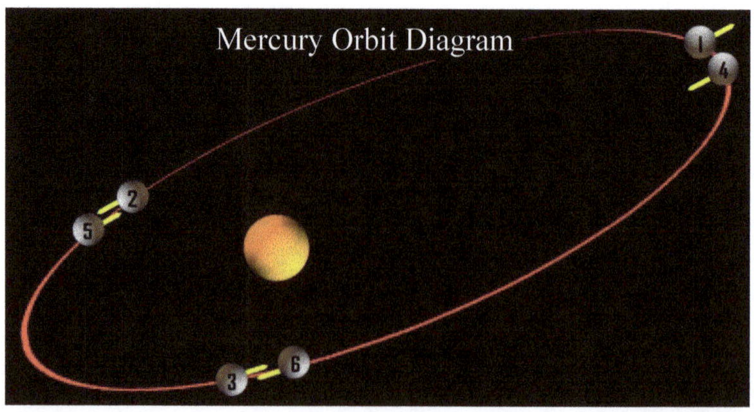

Mercury has a highly unusual rotation. It is so difficult to observe Mercury's rotation from Earth that scientists were not able to puzzle out the details until 1965, when the story was finally pieced together from data gathered by the gigantic radio telescope at Arecibo, Puerto Rico. While the Earth spins rapidly, completing over 365 rotations for each time it orbits the Sun, Mercury only executes one and a half rotations for each orbit it completes. This is called "3 to 2 resonance" because of the ratio of three rotations to every two orbits. It takes Mercury only about 88 Earth days to complete one orbit of the Sun, so a stationary observer on the surface of Mercury would see the Sun rise only once every 176 Earth days. Thus, at a particular spot on Mercury, the first rays of morning sunlight will be followed by a period of time equivalent to 88 Earth days of continuous light. Then comes dusk, followed by the equivalent of 88 Earth days of darkness before the dawn returns. Therefore, we can say that one day on Mercury is equal to 176 Earth days. The illustration on the preceding page shows how Mercury's rotation relates to its orbit. The small yellow line represents an arbitrarily chosen surface feature on the equator of the planet. In proceeding from any point on the ellipse to the next numbered point, Mercury executes half a rotation while traveling one-third of its way around the Sun.

Mercury's orbit is also unlike the orbit of any other planet. It is more elliptical, for one thing. Every planet orbits the Sun in an ellipse—Kepler's first law of planetary motion describes this—but for most planets, the ellipse of their orbit is almost a perfect circle. Not Mercury. Its orbit is so elliptical that at its farthest distance from the Sun (aphelion) it is 150% farther from the Sun than it is at its nearest point (perihelion). An orbit that deviates from perfect circularity is called "eccentric."

Mercury's orbit is very eccentric now, but computer studies have shown that it varies chaotically, at times almost circular and at other times even more elliptical than it is now. These variations are caused by minute gravitational tugs by the Solar System's other planets over millions of years. Interactions with the gravity of other planets may eventually cause Mercury's orbit to become so eccentric that, according to some calculations, there exists a tiny but non-zero probability that Mercury will collide with Venus within the next five billion years.

A consequence of Mercury's unusually elliptical orbit is that Mercury's rate of travel around the Sun varies more than that of any other planet in our Solar System. As it moves towards perihelion, its velocity increases to 36 miles per second. Moving towards aphelion, it slows to 24 miles per second. This is a natural consequence of the law of gravity: Orbits are elliptical, which means that any orbiting body, such as Mercury, is sometimes approaching the Sun and sometimes traveling away from it. When a planet approaches the Sun, the planet's velocity increases because the Sun's gravity is pulling on it, and when it travels away from the Sun, its speed decreases for the same reason. This did not become widely known until Kepler published his second law of planetary motion in 1609.

Kepler's second law accounts for one of the most startling effects that can be observed in our Solar System. When Mercury is advancing towards its perihelion, its rapidly increasing velocity approaches, matches, and then exceeds its own rotational speed, so that an observer on Mercury would get the impression that the Sun slows down in its progression through the sky, stops, and then reverses direction. Likewise, as Mercury departs from perihelion, its orbital velocity around the Sun slows until it matches and then falls below its rotational velocity, producing the opposite impression as the Sun resumes its normal course. This means that certain points on Mercury's surface experience a double sunrise. An observer at such a point would be able to watch the Sun rise in the east until it was clearly visible above the horizon. The Sun would then appear to stand still in the sky before reversing direction and descending back toward the eastern horizon. Before fully setting, however, the Sun would once again reverse direction and climb away from the horizon toward the zenith. This entire process of rising, setting, and rising again requires about eight Earth days.

There are two points on Mercury's equator, on opposite sides of the planet, which hold the distinction of being the planet's two hottest places. Referred to as "hot poles," these are the points where the retrograde motion of the Sun occurs at Mercurian noon. The Sun would appear to stand virtually still overhead and be at its most brilliant because Mercury is at perihelion.

Advance of perihelion

A planet in an elliptical orbit has a perihelion at which it is closest to the Sun and an aphelion at which it is farther away. These points do not stay fixed, but advance slowly with each orbit, so that in effect the entire orbit is turning in a wide, slow rotation that will bring it back to its starting point in millions of years, or sometimes billions. In the case of Mercury, this takes 300 million years. The curve of the orbit traces out a striking rose-petal pattern as the ellipse turns. The rate at which the aphelion and perihelion change their position, called the rate of precession, can be predicted from Newton's law of gravitation.

All planets precess, but Mercury, because it is closest to the Sun, feels the strongest gravitational force from the Sun's immense mass, and its orbit precesses more than other planets. Therefore, Mercury's precession is the easiest to observe, and astronomers very early focused their attention on this phenomenon. Unfortunately, what they found did not make sense. In 1859, the French astronomer Le Verrier announced that the rate of precession of Mercury's orbit was different from the rate predicted by Newton's law. It was off by 43 arc seconds per century, a vanishingly small amount but maddeningly inexplicable.

Many possible explanations for the discrepancy were proposed, but it was not until the advent of Albert Einstein's general theory of relativity that the problem was fully solved. Einstein's theory accounted for the missing 43 arc seconds and explained the observed precession; calculations based on his work matched the observed deviation almost perfectly. This provided much-needed evidence in favor of Einstein's theory, which because of its surprising and counterintuitive approach was slow to be accepted by scientists. Mercury occupies a singular place in the history of science for having provided this vital support at a crucial time for Einstein's theory. This badge of honor is just one more of Mercury's signs of uniqueness among the planets.

Mercury is the tiniest planet in our Solar System. It has borne this title since the controversial demotion of Pluto from among the ranks of the planets. Mercury is the size of a large moon and has no moon of its own. The illustration at right shows the size of Mercury compared to some of the Solar System's largest moons. Clockwise from upper left, they are Mercury, Jupiter's moon Ganymede, Saturn's moon Titan, and the Earth's Moon. Even though Mercury is smaller than the moons Ganymede and Titan, Mercury still qualifies as a full-fledged planet because it satisfies the three criteria for planethood as established by the International Astronomical Union: It is roughly spherical; its orbit is clear of other bodies; and, most importantly, it orbits the Sun, which is something that Ganymede and Titan do not do.

Mercury, Moon, Titan, Ganymede

Mercury's axis has the least amount of tilt of any of the Solar System's planets, only about $\frac{1}{30}$ of a degree. That is to say, it is the most vertical of all the planets. One consequence of this is that Mercury has no seasons. Another effect is that if you stood at one of Mercury's poles, you would always be able to see the Sun peeking about halfway over the horizon, no matter what time of year it was—unless you happened to be standing at the bottom of a crater. On Earth, by contrast, the Sun disappears from the view of observers at the poles for six months out of the year, due to the Earth's axial tilt.

Mercury reflects light poorly because of its dark color and the unevenness of its cratered, rocky surface. Despite its low albedo, Mercury is the sixth brightest object in our night sky, surpassed only by the Moon, the bright planets Venus and Jupiter, the giant star Sirius, and the planet Mars.

Not everything about Mercury is strange. It is a rocky planet just like Earth, Venus, and Mars. Rocky planets are referred to as "terrestrial," which means "Earth-like." Interestingly, the Solar System's four innermost planets are all terrestrial, and their densities are similar. Mercury and Earth are the two densest. In the competition for the title of "Solar System's Densest Planet," the preponderance of experts cite Earth over Mercury, but the issue remains unresolved since various methods are used for determining density.

Mercury's great density is most likely due to the presence of large amounts of iron in its core. Mercury may consist of as much as 60% iron. It is similar in size and appearance to our Moon and is pockmarked with craters, indicating that during its existence it has experienced many meteor strikes. The craters remain intact because there is no wind or rain to erode them. Furthermore, the presence of craters indicates that Mercury has been geologically inactive for billions of years. If it were active, those old craters would have been covered up or filled in by lava flows. Hence, the Mercurian surface presents a relatively pristine, primeval portrait of a planet that has persevered since the Late Heavy Bombardment approximately 3.8 to 4 billion years ago.

Exploration

Being so close to the Sun makes Mercury challenging to study. With the naked eye, an Earth-bound observer can only see it early in the morning or late in the evening because the bulk of the Earth blocks it from our view. Our night side never faces Mercury, and in the daytime the Sun's light overwhelms the reflected light of Mercury. In fact, the Hubble space telescope cannot observe Mercury because the planet is too close to the Sun. The delicate optics of the telescope, just like the delicate optics of our eyes, would be damaged if they looked too directly toward the Sun.

One of the unique problems of trying to study a planet like Mercury is that a probe launched from Earth towards Mercury will experience the gravitational pull of the Sun during the entire journey. The resulting acceleration will make it very difficult for the probe to slow down sufficiently to go into orbit when it reaches its destination. To prevent a probe from sailing uselessly past the planet, copious amounts of fuel must be burned, and this is often impractical. In fact, a trip to Mercury requires more rocket fuel than would be needed to leave the Solar System. As of 2016, only two probes from Earth have

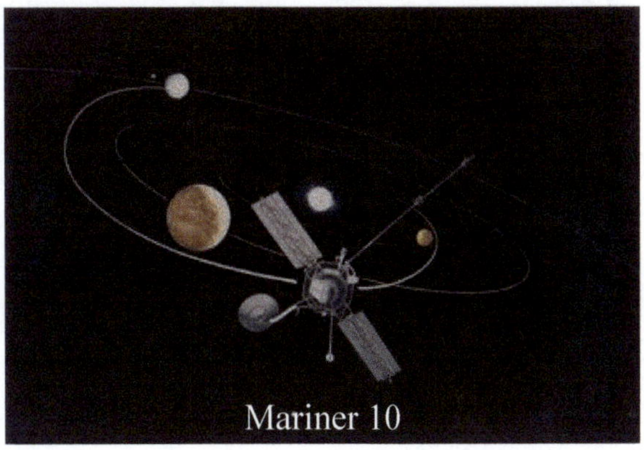

Mariner 10

visited Mercury. They were Mariner 10 and MESSENGER (MErcury Surface, Space ENvironment, GEochemistry, and Ranging).

Mariner 10 (previous page, in an image showing its trajectory) traveled to Mercury with an assist from the gravity of Venus. It was the first spacecraft ever to use this gravitational "slingshot" effect. Although it never orbited Mercury, it was placed in an orbit around the Sun that enabled it to have several close encounters with the planet. Mariner 10 initially flew by Mercury on March 29, 1974. It then circumnavigated the Sun before returning to Mercury on September 21, 1974. After executing another orbit of the Sun, it flew by Mercury for the last time on March 15, 1975, passing almost over its north pole. Unfortunately, it was able to photograph only 45% of the planet's surface. When Mariner 10's batteries ran out of power, its systems were shut down, and the spacecraft was left to continue following the same orbit. Presumably, Mariner 10 is still adrift in the space between Mercury and the Sun.

MESSENGER

Nonetheless, Mariner provided the first close-up images of Mercury, revealing the planet's heavily cratered surface and a diverse array of geological features, including giant scarps that scientists suspect were created by the slow shrinking of the planet due to the cooling of its core. These scarps are up to 300 miles long, and their height of up to 8,000 feet above the surrounding plains suggests that Mercury has shrunk by as much as 0.1% of its diameter. As a result, the terrain in some places is wrinkled, fractured, and sunken like the skin of a drying prune.

After Mariner 10, no probes were sent to Mercury for more than 30 years. Then the MESSENGER spacecraft (right) was launched in 2004. It flew by Mercury in 2008 and 2009, and in March 2011 it began to orbit Mercury. Thanks to MESSENGER, we know more about Mercury than ever before. We have detailed images of its surface and reams of data from which scientists can draw conclusions about Mercury's structure, composition, and geologic history. MESSENGER orbited Mercury over 4,000 times in four years before exhausting its fuel and crashing into the planet's surface in 2015.

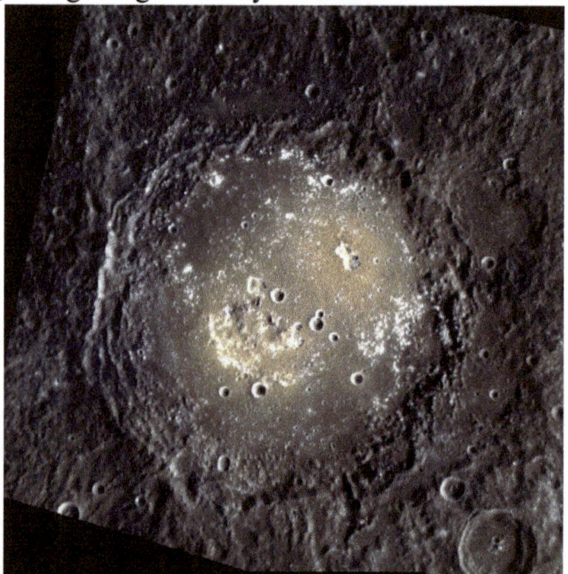
"Hollows" on Mercury

One of MESSENGER's discoveries was that Mercury is freckled with mysterious, shallow, rimless, irregularly shaped pits (right). They are unlike any other landform yet seen in the Solar System. High-resolution photographs from the probe revealed these holes, which resemble the holes in Swiss cheese, on the floors and walls of impact craters all over Mercury.

The features have been dubbed "hollows" and can be as wide as a football field or as large as a small town. They are superficially similar to structures that form on Mars when frozen carbon dioxide sublimates during seasonal temperature changes; therefore, planetary scientists suspect that a similar process is happening on Mercury. However, the sublimation process on Mercury must involve solid rock, while the same process on Mars involves frozen water and carbon dioxide. It has also been theorized that the hollows could form when volatile materials such as sulfur on the surface are exposed to the solar wind, which is a continuous stream of charged particles from the Sun. Since the tiny planet has no permanent atmosphere, these particles can hit the surface directly, vaporizing volatile substances. The intense heat from the nearby Sun might also "boil" away these volatiles.

On October 20, 2018, the BepiColombo mission was launched as a joint mission by the European Space Agency and Japan. The BepiColombo spacecraft is expected to reach Mercury in 2024 when it will split into two parts. Both parts, or probes, are designed to operate for at least one Earth year. One probe, in a circular orbit, will map the planet's surface using infrared, ultraviolet, X-ray, and gamma ray technology. The second probe will measure the planet's magnetosphere from an elliptical orbit, hopefully solving the mystery of Mercury's curious magnetic field.

Mercury's magnetic field is weak, less than one percent as strong as Earth's, yet this field is stronger than the magnetic fields of the Moon, Mars, and Venus. Planetary magnetic fields are thought to be generated by flows in the hot, liquid iron cores of the planets. Because Mercury is so small, most scientists expected its core to have cooled and solidified long ago, in which case it should not have a magnetic field. Many speculated that the magnetic field detected today may have been "frozen" into the planet when the core cooled, which is similar to how household magnets are manufactured. Today, however, there is evidence that Mercury's core is still molten. Possibly heat generated by the uneven pushing and pulling of the Sun's gravity keeps the core at high enough temperatures for it to remain molten so that circulating iron can create the planet's magnetic field. Alternatively, heat could be coming from the slow but steady decay of radioactive elements in Mercury's core.

Nevertheless, if Mercury's core is molten, then its magnetic field should be about as strong and reliable as Earth's. One theory to explain how Mercury could have a molten core and still have such a weak magnetic field is that the highly energetic charged particles of the solar wind counteract the dynamo effect of the hypothetical liquid iron core. This would result in a magnetic field that appears anomalously weak. Another theory is that, even though molten, the core generates less of a dynamo effect because Mercury's rotation is so slow. Scientists, however, remain uncertain.

Mercurian Symbol

Another mystery emerged from observations by MESSENGER. Mercury has an ion tail like a comet, which is believed to be caused by the interaction of the planet's magnetic field with the solar wind. Bizarre findings such as this remind us that there is much left to learn about Mercury.

Name

Mercury's name comes from Roman mythology, according to which the god Mercury was the most fleet-footed of the gods and served as their messenger. He is usually depicted with wings on his shoes and helmet (see illustration at right). The name assigned to the planet by the ancients reflected their observation that

Mercury traveled through the sky from night to night more swiftly than the other planets. Indeed, scientists today have confirmed that Mercury travels faster in its orbit than any other planet: up to 36 miles per second—an impressive velocity for a planet or a god. Not coincidentally, the ancient Babylonians, who preceded the Greeks, also identified this planet with their messenger-god, Nabu, and for the same reasons. There is evidence in the form of clay tablets dating to the 14th century BCE that Assyrian astronomers knew about the planet Mercury as well.

The ancient Greeks had two names for Mercury. When they saw it in the evening sky, they referred to it as Hermes, and when they saw it in the morning sky, they called it Apollo. It was not until the fourth century BCE that the Greeks came to the realization that Hermes and Apollo were the same planet, the one known today as Mercury.

In ancient China, Mercury was referred to using the characters 辰星, which mean "Hour Star" and are pronounced *Chén Xīng*. East Asian cultures today refer to Mercury as the "water star" (水星, *Shuǐ Xīng*), water being one of the five elements in traditional East Asian cosmology. Ancient Indian astronomers referred to Mercury as Buddha and this god was associated with Wednesday. The Norse god Odin, or Woden, was also linked with the planet Mercury and Wednesday. In Romance languages such as French, Spanish, Romanian, and others, the name for Wednesday is derived from the name "Mercury," as in French *mercredi* and Spanish *miércoles*.

Viewing Mercury

Mercury is never visible from Earth in the middle of the night because of its proximity to the Sun. It can only be seen near sunrise or sunset. Because Mercury is closer to the Sun than the Earth, like Venus, it exhibits phases as it changes its position relative to the Earth and the Sun. When Mercury's phases were discovered by Giovanni Zupus in 1639, it was a blow to the geocentric theory of the universe because the simplest, and almost incontrovertible, explanation for the phases was that Mercury was orbiting the Sun. Any body closer to the Sun will display a full range of phases to a more distant body; for example, a Martian could view phases of Earth.

Galileo, early in the 17th century, was the first to turn a telescope on Mercury. However, his telescope was not powerful enough to see the phases. Mercury was the first planet ever observed to cross the face of the Sun, an event known as a "transit." Johannes Kepler predicted that there would be a transit of Mercury, and in 1631 the event was duly observed by the French savant Pierre Gassendi. Such transits occur about once every seven years.

Surface Features

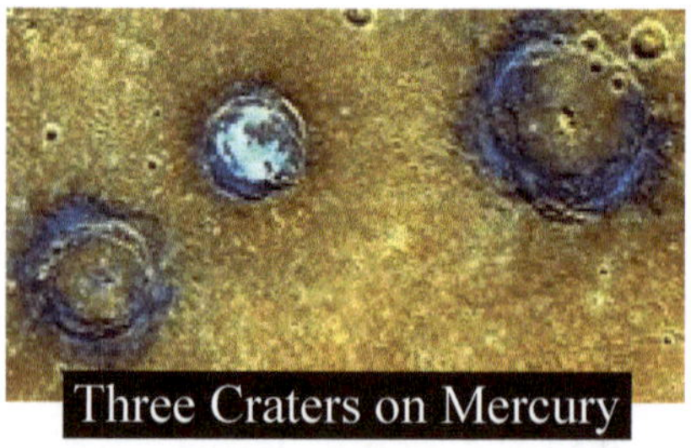
Three Craters on Mercury

Mercury has bowl-shaped craters, some of which are hundreds of miles across. From their appearance, it is evident that some are recent and some are ancient. The International Astronomical Union has set guidelines for the naming of craters on every planet in the Solar System. Craters on Mercury must be named after artists who have been dead for at least three years and who meet a 50-year standard of notoriety. The enhanced color image at right shows the craters Munch, Sander, and Poe. Munch was a Norwegian painter, Sander a German photographer, and Poe an American writer.

The largest known crater on Mercury is the Caloris Basin (left, in a false-color view with the basin shown in orange). Its diameter of 1,000 miles makes it larger than the state of Texas. On Earth, this crater would stretch nearly from New York City to Chicago. It was formed about 3.85 billion years ago when an object approximately one hundred miles across struck Mercury. The impact was so powerful that it caused lava eruptions and left a set of encircling mountain ranges, some over 1.2 miles high, surrounding the resulting crater. At the antipode of the Caloris Basin, that is, the point exactly opposite it on the other side of Mercury, there is a sizable region of unusual, hilly terrain known as the "Weird Terrain" (left). Possibly the shock waves from the Caloris impact circled the entire planet and converged on the far side, and where they met, the surface fractured. Another possibility is that this terrain was formed by material that was hurled from the impact site and then traveled all the way around the planet, where it converged, creating the weird terrain.

The MESSENGER probe clearly revealed that craters at the Mercurian north pole contain water ice (right). Since Mercury's poles lie in permanent darkness and the bottoms of some craters do not receive any direct sunlight, the ice could persist there indefinitely. Blanketing the ice is a dark substance as black as coal. The leading idea is that the black material is a complex organic substance, which could have been deposited by comets at the same time they left behind their water. Taking into consideration Mercury's dramatic temperature extremes, the presence of water there conjures inevitable mental images of "a snowball's chance in hell"—which is evidently not such a long shot after all, at least not in planetary science.

The Future

Mercury may never see a human visitor. It is exceptionally unwelcoming—blisteringly hot or bitterly cold and lacking utterly in the air that humans require. Furthermore, orbital mechanics make entering orbit around Mercury costly in energy terms.

Given these difficulties and the high cost, what could lure people to Mercury? There may be valuable mineral resources there, as Mercury's abnormally high density implies, and wonders waiting to be plucked from the baked regolith. There may be exciting and impressive scientific discoveries that can be made only in Mercury's unceasing shower of solar radiation. There may be ways for future populations of humans to live comfortably and thrive in Mercury's seemingly hostile environment. But these benefits will be subsidiary to the true benefit, the prime motivation that has impelled humankind across oceans, into the mouths of volcanos, to the Earth's far poles, and even to the Moon and beyond. That motivation is the pure urge to discover. Mercury, the planet of extremes, could one day represent the most extreme expression of humankind's urge to explore. In the meantime, it will maintain its sedate flower-petal of orbits, eccentrically weaving in and out of the Sun's harsh glare, turning slowly as it waits for human visitors to tread upon its surface.

Mercury, Venus, Earth, and Mars

Mercury

14
Venus: Queen of the Solar System

Bluish Venus

Now the bright morning-star,
Day's harbinger,
Comes dancing from the East...
John Milton

She walks in beauty, like the night
Of cloudless climes and starry skies
Lord Byron

Farewell, Morning Star, herald of dawn, and quickly come as the Evening Star, bringing again in secret her whom thou takest away.
Meleager

Venus, the second world from the Sun, is named after the Roman goddess of love and beauty. This confers upon the planet the distinction of being the only planet in the Solar System to be named after a female deity. Aside from the Moon, Venus is the brightest object in the night sky; it is no less than ten times more luminous than Sirius, our brightest star. Occasionally, Venus is so bright that it can be seen with the naked eye in the daytime. On cloudless nights when the Moon is absent, Venus becomes the dominant source of light in the starry sky and occasionally even casts shadows on Earth. These rare, delicate, ephemeral shadows are referred to, appropriately enough, as "Venusian shadows." All that brightness is perfectly logical. Venus is the Earth's next-door neighbor, but aside from simple proximity, Venus is bright because it is resplendently robed in highly reflective clouds. The sight of Venus has dazzled civilizations ancient and modern, and many have marveled at the exquisite splendor of this luminous wandering divine entity.

Earth and Venus

Venus is often called Earth's twin or sister planet because its mass, volume, density, composition, and surface gravity are remarkably similar to Earth's. (See size comparison at right.) Earth and Venus are both terrestrial planets that possess a central core of iron, a mantle of molten rock, and a crust composed of silicates, or sand-forming compounds such as quartz, feldspar, mica, and many others. When melted, silicates form glass.

A noteworthy feature of Venus' orbit around the Sun is its lack of eccentricity. Although technically an ellipse, this orbit is closer to circular perfection than the orbit of any other planet in the Solar System. Like Mercury, Venus' axial tilt is so small that it has no seasons. Also like Mercury, Venus has no moons, and, like Earth, Venus has no rings.

While Earth and Venus as siblings share many similarities, there exist important differences. If one were to gaze down upon the Solar System from the direction of the North Star, Polaris, all the planets would be observed to orbit the Sun in a counterclockwise direction and also spin on their axis in a counterclockwise direction—except Venus and Uranus. While Venus rotates around the Sun counterclockwise, its axial rotation is retrograde or clockwise. The reason for this is unknown, but scientists have advanced a number of theories. One theory is that Venus' original counterclockwise rotation could have been altered to its current retrograde motion by a collision with a Moon-sized asteroid that turned Venus completely upside down and significantly slowed its rotation. The rotation of Uranus is "backward" too, but this is more or less moot since Uranus' axis of spin is not vertical but is effectively horizontal.

For centuries, it was impossible to determine Venus' rate of rotation since the planet was entirely enveloped in dense, heat-absorbing, opaque clouds. However, in 1965, astronomers were able to beam pulses of radio waves from the Arecibo Radio Astronomy Observatory in Puerto Rico and bounce them off the planet, which allowed them to accurately calculate Venus' spin. They found that Venus' rate of rotation is the slowest of any planet in the Solar System, requiring approximately 243 Earth days for a single rotation with respect to the distant stars, or a "sidereal day." The time from one sunrise to the next as viewed from the surface of Venus would be 117 Earth days, or a "solar day." The difference is caused by the fact that Venus alters its position with respect to the Sun much more quickly than its position with respect to the stars. Thus, from sunrise to sunrise of a sidereal day, Venus has completed over half an orbit around the Sun. By contrast, Earth only completes 1/365 of an orbit in the time from one sunrise to the next. The difference between Earth's sidereal day and its solar day is only about four minutes. Venus' sidereal day is longer than its year, as Venus takes approximately 225 Earth days to orbit the Sun.

Over the years, astronomers have determined the conjunctions of Earth and Venus, which occur when the two planets are nearest each other. The two planets' minimum distance is 25.5 million miles. At that distance, one-way communication from Venus would take about two minutes and 17 seconds to reach Earth. The response time from Earth to Venus would require an additional two minutes and 17 seconds, so the overall transmission time would be approximately four and a half minutes.

From 1610, when Galileo first turned a telescope towards Venus, until 1962, astronomers' telescopes revealed only a globe veiled in an impenetrable layer of swirling clouds. Venus was seen from Earth as breathtakingly brilliant and blazingly beautiful, but astronomers could see only Venus' outer beauty; the same clouds that made Venus so picturesque masked whatever lay beneath. Throughout the centuries, a few scientists and science fiction authors envisioned a world slightly warmer than Earth and speculated that this intriguing planet might have ample amounts of liquid water. They wondered if Venus could be a tropical paradise covered in rainforests, jungles, and coral reefs and asked: Does Earth have a twin sister of even greater loveliness? For centuries, astronomers and the general public wanted to know what lay beneath that outer layer of clouds that reflected light so brilliantly. People asked, "Was Venus' beauty only skin-deep?"

Humanity began to discover what Venus was really like when Mariner 2 flew by on December 14, 1962, and made several provocative discoveries about the planet. Twenty-eight years later, NASA's Magellan spacecraft orbited Venus for two years, zipping from pole to pole and mapping

98 percent of the planet's surface using radar in the microwave spectrum. As of 2016, 50 different missions have been dispatched to Venus, of which 38 have been at least partially successful. During the past half-century, we have learned much about Venus.

Dante Alighieri, an Italian poet of the 14th century CE, wrote *The Divine Comedy*. In this epic poem, the poet describes a harrowing descent into Hell. Dante's Hell is a vividly depicted collection of concentric rings, each of which is characterized by a different kind of torture, depending on the sins of the souls enclosed within it, and the torments become more severe as one nears the center. Dante named the first part of his epic *Inferno*, which is the Italian word for Hell. Venus, sizzling hot, deadly toxic, and destructively volcanic, is a high-pressure world that warps, melts, and tortures metals, swiftly destroying any human-made object. Venus is a veritable inferno.

Let us descend towards the surface of Venus as if aboard an earthly lander capable of transmitting all that it detects. The first circle of hell is Venus's thick bank of clouds. This turbid layer absorbs half of the solar energy that hits it and reflects 60% of the visible light it receives back into space, imparting to the planet a glowing appearance. Few would suspect that this glistening brightness is the first circle of Hell. For these lovely clouds are not the fluffy, cottony, water-vapor clouds of Earth. These clouds consist of corrosive sulfuric acid, commonly found in battery acid, and are laced with pungent, noxious sulfur dioxide along with reeking hydrogen sulfide. This visually impenetrable global blanket of concentrated sulfuric acid droplets is stained with sulfur, and thus the initial color of Venus as seen by descending Earthly probes is a pale, creamy, lemon-yellow hue. The combination of these chemicals creates a witch's brew of caustic substances that consume, corrode, and dissolve most metals. No probe should expect to slip through unscathed; this deadly cloud layer is twenty miles thick, twice as thick as Earth's total atmosphere.

The European Space Agency's Venus Express, which crashed into Venus in 2014 after a very successful period of operation, detected "flux ropes," tubes of concentrated magnetic energy along which charged ions flow, high in the atmosphere above the poles. Other planets with stronger magnetic fields also have flux ropes; however, on Venus, these ropes are rare and represent the only known form of magnetism created by the planet. At the same time, an abundance of titanic lightning bolts suffuses these clouds with light, though the bolts never reach the ground. Among other strange features of this truly alien atmosphere are an ozone layer, a super-cold stratum that may host dry ice, and an unpredictable and turbulent polar vortex at each pole.

The polar vortices are the only breaks in this all-encompassing layer of clouds. The gaping maws of these 1,800-mile-wide megacyclones have been photographed as the tornadoes swirl, twist, and continuously morph into new shapes. They occasionally split in two or twine into figure-eight formations as they recirculate the superheated, superthick Venusian atmosphere.

The upper clouds of Venus travel at furious speeds of 250 miles per hour, pushed by winds blowing as fast as the mightiest hurricane or tornado on Earth. Propelled by such powerful winds, the upper clouds circumnavigate the planet in just four days. The inner layer of clouds forms a jet stream where winds are even more powerful, exceeding 400 miles per hour.

As clouds do on Earth, Venus's clouds release the chemicals with which they are laden in the form of precipitation that falls toward the surface. On Earth, of course, this chemical burden is life-giving water, which showers land and sea, human and beast, with the precious fluid that sustains all living things. On Venus, however, the rain is sulfuric acid. Fortunately for Earthly probes, none of the metal-dissolving sulfuric acid reaches the surface because the acid evaporates and circulates back upward before it gets anywhere near the ground.

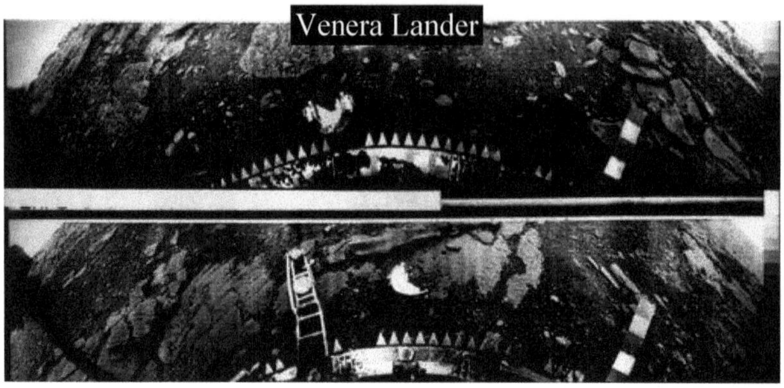

Now our descending probes approach an ever more malevolent milieu. Below the 20-mile-thick outer layer of acid clouds lies the second circle of Hell. Here is a dense, light-scattering layer of almost pure carbon dioxide gas 30 miles thick. This suffocating, heavy gas scatters light waves and casts a reddish-orange glow over the landscape. The eerie photographs above were taken by the Venera 13 probe in 1982. (The saw-toothed ring is part of the craft's landing mechanism.)

As Earthly probes near the planet's surface, they move through a Venusian atmosphere that has become superdense and stagnant. It flows like molasses, but despite its slow speed, its density gives it the power to push small rocks around as if they were caught in an ocean current. Without running water or fast-moving low-altitude winds, there is little erosion of surface features on Venus. Nevertheless, the slow and inexorable waves of highly pressurized carbon dioxide rub the surface of Venus and gradually wear it down. There is undoubtedly also chemical weathering from the corrosive atmosphere.

Venus's atmosphere consists of 96% carbon dioxide and 3.5% nitrogen, with the remaining half percent split among sulfur dioxide, argon, carbon monoxide, and even a tiny amount of water vapor. Venus's atmosphere is over five times as deep as Earth's. This volume of atmospheric gas makes the pressure at the planet's surface 92 times greater than Earth's pressure at sea level. The Venusian pressure at its surface is so immense that it is equivalent to being 3,000 feet beneath the Earth's ocean surface. At these depths, the hull of a WWII-era submarine would be crushed like a soda can under the heel of an elephant.

The temperature at the surface is condemned to remain at a metal-melting 865° F. Night and day, pole to pole, the dense carbon dioxide holds the heat, and, in combination with the ongoing greenhouse effect, refuses to let it dissipate. Meanwhile, the Sun, the ultimate source of all this light and heat, remains hidden by the thick cloud cover except for a diffuse, discolored illumination that leaks through. From the surface of Venus, the sky would resemble Earth's sky on a very bleak, overcast day.

The mechanism by which Venus accumulates and retains its sweltering heat is similar to what happens in Earth's atmosphere (but, thankfully, to a lesser extent), as well as in a greenhouse. Simply put, heat energy is allowed to infiltrate but is not permitted to leave. In a greenhouse, infrared radiation can enter through the glass, but once the heat is absorbed by objects inside the greenhouse, including even air molecules, very little of the heat escapes from the sealed building. In the same way, infrared radiation penetrates the clouds of Venus and, once inside, remains captured. The carbon dioxide in Venus' atmosphere is a strong "greenhouse gas." Some planetary scientists fear that if the composition of Earth's atmosphere changes sufficiently, it could become enough like Venus that we could eventually find ourselves experiencing 865°F temperatures and skies filled with sulfuric acid. Indeed, careful measurements indicate that the amount of carbon dioxide in our atmosphere is increasing; however, Earth has a long way to go to get to where Venus is today.

Today, Venus is the hottest planet in our Solar System. Venus is even hotter than Mercury, even though Mercury is closer to the Sun. This is because of Venus' atmospheric greenhouse gases. Billions of years of scorching heat have evaporated Venus' oceans of water, leaving a dehydrated rock of a planet. Today, Venus has become a seething cauldron of caustic gases and boiling lava.

Landforms

The third circle of Hell is the baked, desiccated surface of Venus itself: a parched, fractured, cratered wilderness. Due to the thick clouds cloaking Venus, the only practical way to map the surface is with radar. The first radar mappings of the Venusian surface were completed in 1978 by a Pioneer Venus spacecraft. The resulting maps revealed a surface consisting primarily of plains formed by ancient lava flows, with only two highland regions, or "continents," called Ishtar Terra and Aphrodite Terra.

Magellan's radar images revealed a volcanic wasteland. The surface of Venus shows about 20% lowland plains, 70% gently rolling uplands and lava flows, and 10% highlands. Overall, Venus is relatively flat compared to Earth but possesses a complex surface consisting of mountains, plains, high plateaus, canyons, volcanoes, ridges, impact craters, chasms, and rivers of lava. Curiously, Venus does not experience the movements of tectonic plates as happens on Earth. Consequently, on Venus the ground does not quake as it does on Earth. However, the planet is dominated by volcanism and experiences ferocious geological forces that cause faulting and folding, and wherever the ground fractures, rivers of lava burst forth.

The mountaintops and steep slopes of Venus sparkle with a type of snow, but this is no ordinary snow—it consists of flakes of metal. The surface of Venus is so hot that metallic compounds melt and vaporize. The resulting vapors then waft upward until they cool, condense, and float back down to rest on dormant volcanic mountaintops in the form of a bizarre "snow" of lead sulfide and other compounds.

In 1990, the Magellan spacecraft arrived at Venus and began a much more detailed radar mapping. Magellan found approximately 940 impact craters ranging in size from one mile to 170 miles. All were greater than 5000 feet in diameter. This suggests that meteoroids small enough to create a crater having a diameter less than 5000 feet would have broken apart and disintegrated due to friction with the dense Venusian atmosphere before reaching the surface.

Interestingly enough, there is also an absence of large craters. This suggests that the Venusian surface formed after the period of the Late Heavy Bombardment, a span from 3.8 to 4.5 billion years ago when a large number of impact craters were created on the inner planets. Thus, geologically speaking, the present Venusian surface is relatively young.

The surface's most prominent features are those produced by the planet's volcanic activity. This is not surprising, since Venus has more volcanoes than any other planet or moon in the Solar System. Magellan found more than 1500 major volcanic areas and as many as a million lesser ones. Due to the venting of sulfur dioxide gas, some of these volcanoes must be semi-active.

Over 40 major rivers of lava flow across the planet. Many meander, create oxbows, and then dump their fluid into vast delta-like plains. The lava conduits vary from a third of a mile to a mile in

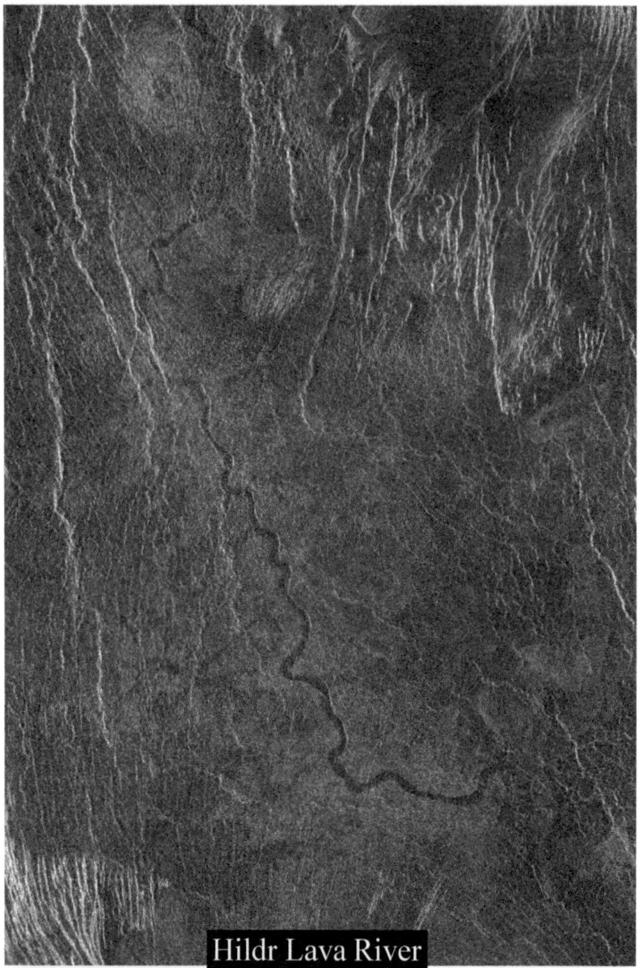

Hildr Lava River

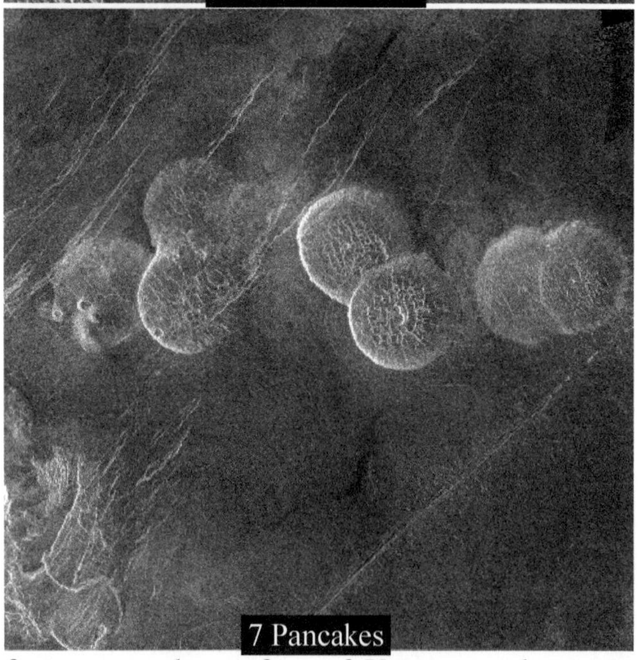

7 Pancakes

width and can be hundreds of miles in length. The longest one, Hildr (left top) is longer than the longest river on Earth. Hildr was named after a Valkyrie of Norse mythology who possessed the power to revive slain heroes. The Hildr lava canal extends from Fortuna Tessera in the north down to the eastern Sedna Planitia in the south. This channel is nearly a mile wide and has many branches and islands along its length. Sedna Planitia is a large lowland area south of Ishtar Terra and is similar to a lunar mare.

The surface of Venus has vast, smooth plains that resulted from numerous lava flows, including over 50 major fields of step-like hill formations similar to the Siberian Traps of Russia. The volcanoes on Venus come in all shapes and sizes but are not randomly distributed. Most are found in a zone called the "uplands" of Venus, a region between the highlands and vast plains of the planet. There are broad, flattened shield volcanoes of the type found on Earth. These give rise to numerous "pancake" structures, up to 40 miles across. These uniform, disc-like formations (left bottom), result from an eruption where the volcano's magma is ejected through a single opening in the planet's crust. As the lava flows, it spreads in a flat circular manner.

Coronae, unusual features found only on Venus and Uranus's moon Miranda, are high-rimmed oval features with diameters typically greater than a hundred miles. It is theorized that a local hot spot in the mantle created an upwelling of magma that forced the surface of the planet to expand and bulge above the surrounding plain. Later the magma dome collapsed over the magma chamber, leaving a ring of mountains surrounding a depression in the surface.

Among the most unusual and fascinating features on the surface of Venus are the mysterious arachnoids. An arachnoid consists of concentric rings surrounded by a network of fractures that bears a likeness to a spider's web. They

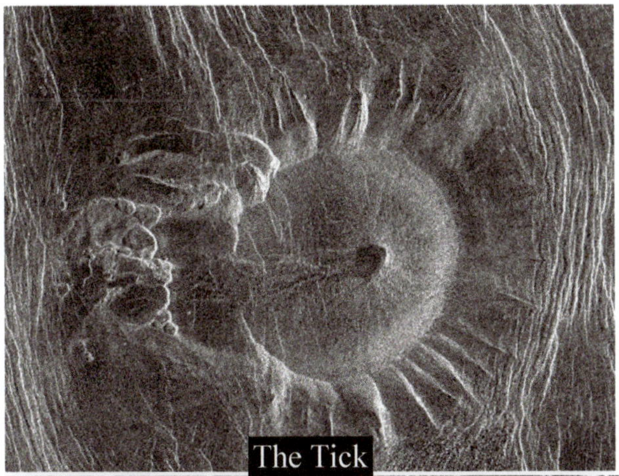

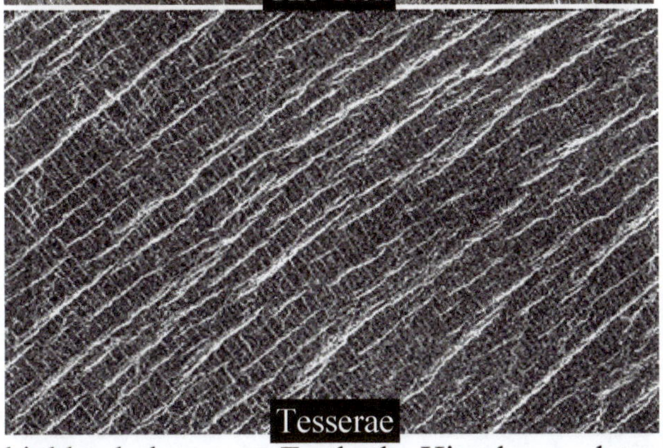

are generally smaller than coronae and vary in diameter from 30 to 150 miles. The structure in the picture at left top has been dubbed "The Tick."

Another odd surface feature, unique to Venus, is the mysterious tesserae. The word "tesserae" originally referred to the tiny colored pieces used to make mosaics. Venusian tesserae are tile-like structures in which many ridges and valleys have formed in different directions. This terrain of complex ridges forms when the planet's crust is forced together, which causes the surface to fracture, fold, and fault (left bottom).

Thrusting up from the Venusian lowlands are three "continents," the two that Magellan found and an additional one near the planet's South Pole. On Earth, continents are surrounded by water, but on Venus, the continents are raised landmasses that stand like vast plateaus above the lowlands. Ishtar Terra is located in Venus' northern hemisphere and is about the size of Australia. It is more extensive than the largest highland plateau on Earth, the Himalayan plateau. The interior of this continent is called Lakshmi Planum.

The only feature of Venus that is named after a male is the mountain range Maxwell Montes on Ishtar Terra. The mountains of Maxwell Montes are higher in elevation than any other feature on Venus, up to 39,000 feet—10,000 feet higher than Mount Everest. The mountain range is named after James Clerk Maxwell (1831-1879), a Scottish physicist. The reason for the deviation from the naming convention dictated under the hegemonic authority of the International Astronomical Union is that these mountains were assigned their name before the IAU decreed that solely female names would be used.

The second continent, Aphrodite Terra, is situated near Venus' equator and is comparable in size to South America. The recently discovered landmass near Venus' South Pole has been named Lada Terra. This region is lower-lying than the other two continents.

The shield volcano Maat Mons reaches 5 miles into the thick Venusian atmosphere. It is the tallest volcano on Venus.

The table below gathers a few of the names and origins of the various goddesses and other figures (some were historical figures) after which features of Venus are named.

NAME	CULTURE	GODDESS OF	FEATURE
Venus	Roman	Love	Planet
Aphrodite	Greek	Love	Terra
Ishtar	Akkadian	Love	Terra
Lada	Slavic	Love	Terra
Lakshmi	Hindu	Love	Plain
Hildr	Norse	(Valkyrie)	Lava Channel
Maat	Egyptian	Justice	Volcano
Sif	Norse	Earth (Wife of Thor)	Volcano
Inanna	Sumerian	Love	Planet (ancient)
Astarte	Western Semitic	Love	Tholi (small domed mountains)
Sedna	Inuit	Sea	Planita-lowlands
Fortuna	Roman	Fortune or Luck	Tessera
Gula	Neo-Assyrian Neo-Babylonian (3000 BCE)	Fertility	Volcanic mountain
Trotula*	Italian	Doctor, not a goddess	Arachnoid
Cleopatra	Egyptian	Queen, not a goddess	Crater
Laima (Lakshmi)	Baltic	Fate-childbirth, marriage and death-pregnant women	Tessera
Diana	Roman	The hunt and moon	Chasm
Miralaidji	Australian Aborigine	Fertility	Corona

*During the twelfth century in southern Italy there dwelt a female physician named Trota of Salerno who wrote a useful textbook on gynecology. Later, additional medical texts came to be associated with her, and these texts were collectively identified as “Trotula.” These texts were known throughout Europe from Poland to Spain and all the way to Ireland. Over time, “Trotula" came to be considered a real person.

The Morning and Evening Star

Ancient humans knew well which star of the night sky was the brightest. Furthermore, they knew that this star appeared only near dawn. It was the harbinger of morning, the bringer of light. Brightness is synonymous with beauty, so when ancient storytellers populated the sky with heroes, gods, and goddesses, it must have been perfectly natural to personify the brightest of the sky's lights as a divine female, the very goddess of beauty and love.

Ancient Sumer, the southern region of the land between the Tigris and Euphrates Rivers in what is now Iraq, is notable for having been the location of the very first civilization, around 3300 BCE. In Sumer, the goddess of beauty, love, and war was known as Inanna, the Queen of Heaven (𒀭𒈹 in cuneiform). In the Akkadian civilization that followed some thousand years later, the goddess of beauty, identified with the evening star, was called Ishtar, and a little west of Akkad in Sidon and Tyre she was known as Astarte. By 1500 BCE, the Babylonians were calling Venus the "bright Queen of the sky." When the Greeks rose to prominence, their goddess of beauty and love was known as Aphrodite. The Romans called the brightest planet Venus, and we still do today.

Venus is the morning star because it is so close to the Sun. Because it orbits closer to the Sun than the Earth, it can never be seen in the middle of the night, but only close to sunrise or sunset. When Venus is seen close to dawn, it is the morning star, and when it is seen close to sunset, it is the evening star. Primitive humans did not suspect that the morning star and the evening star were the same "goddess." Even the knowledgeable Greeks and Romans retained separate names for these two manifestations of Venus. In Greece, the evening star was Hesperus, and the morning star was Phosphorus. In Rome, the evening star was Vesper, and the morning star was Lucifer. This "Lucifer" has nothing to do with Satan. In fact, it is a literal translation of the Greek "Phosphorus," meaning "light-bearer." Despite the pairs of names, it is very likely that the Greeks and Romans understood by then that they were seeing one planet. In fact, there is evidence that the Babylonians, long known as skilled astronomers and mathematicians, were aware of this in 1581 BCE. We know it was exactly 1581 BCE because the Babylonians recorded their astronomical observations so carefully that modern-day astronomers using computers are able to work backward and reconstruct the exact positions of the stars on that date and compare it to the Babylonians' clay-tablet records.

By the time of the ancient Romans, it was understood that Venus was one of five planets. As the brightest and most visible of these planets, Venus was named by the Romans after their goddess of love and beauty. The planet has been associated with love, femininity, and romance throughout history. Morningstar as a symbol of fertility and womanhood remains entrenched, best demonstrated by the fact that the astronomical symbol for Venus happens to be the same as the one used in biology for the female sex: a circle with a small cross beneath. ♀

Venus, to our ancestors, was a goddess of the eternal heavens, but today we know that she rules a planet that rapidly disables Earthly probes by roasting, corroding, and melting them. There is no doubt that this regal empress of the firmament is worthy of being called the "bright queen of the sky"—but her domain is a glowing land of unquenchable destruction.

Astronomical History of Venus

In the geocentric theory that was commonly accepted before the sixteenth century, all the planets were believed to orbit the Earth, attached to crystal spheres. This may seem bizarre today, when we know that the planets orbit the Sun, but even though ancient astronomers were handicapped by belief in their misguided theory, they still achieved impressive insights. For example, Claudius Ptolemy placed Mercury and Venus between the Earth and the Sun in his astronomical treatise Almagest. The 11th-century Persian astronomer Avicenna claimed to have observed the transit of Venus, an event in which Venus passes in front of the Sun as viewed from Earth. This event was thought by later astronomers to confirm Ptolemy's theory. In the 12th century, the Andalusian astronomer Ibn Bajjah observed "two planets as black spots on the face of the Sun," which were later identified as the transits of Venus and Mercury by another Arab astronomer, although it is equally likely that Ibn Bajjah was seeing sunspots since there were no transits of Venus during his lifetime.

On December 11, 1610, in Florence, Galileo discovered that Venus goes through phases just as the Moon does. The first telescopic observations of Venus occurred in December of 1610 when Galileo studiously observed the phases of Venus for months and created detailed drawings. The fact that Galileo could see phases of Venus from Earth convinced him that Venus orbits the Sun. Galileo, although confident in his knowledge that the Earth orbited the Sun as Copernicus had taught a century before, recanted when he was hauled before the Inquisition and threatened with torture. Galileo's conversion to Copernicanism was a pivotal moment in the modern scientific revolution that eventually shunted the dogmatic geocentric model of the universe into the dustbin of history's mistaken ideas.

By 1639, it was realized that both Mercury and Venus sometimes pass in front of the Sun. Such a passage is referred to as a transit. In October of 1639 English amateur astronomer Jeremiah Horrocks determined that transits of Venus come in pairs and that each pair is separated by 8 years. (1631 and 1639—the next observable transit from Earth was not until 1761, 122 years later.) We now know that Venus crosses in front of the Sun roughly twice per century about eight years apart. Thus, there were transits in 1761 and 1769, 1874 and 1882, and June 2004, with the last observed transit on June 5, 2012. The next transit will occur in 2117. It might seem that because Venus is closer to the Sun than the Earth, Venus should transit the face of the Sun every time it treks past. However, the slight inclination of Venus' orbit means that it more often passes above or below the Sun as viewed from Earth.

It was realized early on that transits of Venus could be utilized to measure the size of the Earth's orbit. The method was to observe the transit of Venus from as many places around the world, and as widely separated as possible. From each location, Venus appeared to cross the Sun in a slightly different position. Observers in the northern hemisphere would see Venus crossing the Sun's face lower down, while observers in the southern hemisphere would see it farther up. Likewise, viewers farther east would start to see the transit slightly earlier than observers farther west would, but the difference would be very small, just a few seconds. The observations of the 1761 and 1769 transits of Venus were massive international scientific endeavors. Even Captain Cook in the South Pacific made an observation. Despite numerous difficulties, including bad weather and limited equipment, the distance to the Sun was measured to be 93,726,900 miles, which differs from the correct value by less than one percent. The distance was further refined by observations of the 1874 and 1882 transits. By the time of the 2004 and 2012 transits, the size of Earth's orbit was already known to far greater precision from radar measurements, but as a matter of general interest, the observations were repeated. This time, the results were within about 7000 miles of the correct value, or about one-thousandth of one percent.

Russia directed a series of 16 space probes named "Venera" to Venus between 1961 and 1984. Ten of the Venera probes landed successfully on the surface of Venus. Despite the exceptionally harsh and hostile environment, Venera 7 in 1970 became the first probe to transmit data from the planet's surface. It was only able to broadcast for 23 minutes. Russia sent an additional nine probes, the longest lasting of which was Venera 13. Venera 13 transmitted data for 127 minutes before its signal terminated.

Conclusion

Venus, that stunningly radiant, wandering beauty in the firmament, has enchanted humankind with its intense luminosity since before the dawn of history. The advent of the telescope in 1609 only added to the allure. Today, humanity remains bewitched by the exquisite elegance of this exceptionally enigmatic planet that is perpetually arrayed in reddish-orange. Our intrepid probes have penetrated the roiling, glistening exterior of this spheroid and yielded insights into Venus' complex, rocky terrain, its volcanic soul that oozes rivers of lava, and its corrosive, crushing atmosphere. Yet, these are only glimmers of the planet's essence, for we desire to plumb its depths and unlock the treasure chest of secrets that remain hidden in its heart. Scientists are eager to equip and dispatch cutting-edge spacecraft to explore further this fascinating orb. As engineers develop new technologies, and science advances in knowledge and understanding, we may yet discover the answers to the mysteries held deep in the interior of this phenomenal incarnation of celestial rock. In our Solar System, Venus is the epitome of light, heat, and mystery; and to our civilization, it is the embodiment of feminine mystique, romance, and beauty—a true Queen of the Solar System.

Whitish Venus

15

The Challenges of Interplanetary Exploration

The universe is probably littered with the one-planet graves of cultures which made the sensible economic decision that there's no good reason to go into space – each discovered, studied, and remembered by the ones who made the irrational decision.
Randall Munroe

It is difficult to say what is impossible, for the dream of yesterday is the hope of today and the reality of tomorrow.
Robert H. Goddard

Sail fast, sail fast,
Ark of my hopes, Ark of my dreams;
Sweep lordly o'er the drowned Past,
Fly glittering through the sun's strange beams;
Sail fast, sail fast...
Sidney Lanier

Spacecraft (Futuristic)

Throughout humankind's 200,000-plus years of existence, there have been critical moments when irrevocable decisions were made that forever altered the course of human history. Today, another crucial juncture is at hand for *Homo sapiens*, and the choice made will have profound implications for the future of our species. We will either send humans to other planets and spread to nearby stars, or we will remain passively on Earth awaiting our eventual, yet certain, demise. The sudden onslaught of a giant asteroid or comet may plunge the Earth without warning into a perpetual winter; or a nearby hypernova could blast away the Earth's atmosphere, meanwhile devastating the rest of the Solar System; or the gradual enlargement of our Sun over eons could result in the slow roasting of the Earth and eventually transform it into a rippling spheroid of molten rock. The choice is clear: either we choose our own destiny, or the universe will choose it for us.

Our species arose here on Earth, and Earth is the only planet we have ever called "home," though it need not forever be so. Except for twelve men who walked on the Moon, all of human history has occurred on this planet. Now we must ask: Will we leave the treasured soil from which we sprang? The Russian space scientist Konstantin Tsiolkovsky (1857-1935) noted, "Earth is the cradle of humanity, but one cannot remain in the cradle forever." What heritage will we bequeath to our progeny? Will our species' epitaph read, "They chose to remain on Earth for the rest of their days," or will our descendants in the far future exuberantly express their elation at our boldness in having chosen to "slip the surly bonds of Earth"? Daring to burst the gravitational shackles that restrain us, daring to escape the physical and social constraints of Earthly civilization, will humans of the next and following generations send colonizing vessels on far-flung voyages carrying the seeds and hopes of humanity to distant planets? Will those children of ours "sail fast" as they "fly glittering through the sun's strange beams"?

The human brain not only exists; it is aware of its own existence. It has transcended the mundane and entered the realm of consciousness. Humanity is earthy matter that has progressed beyond mere assemblage and has become more than the sum of its parts. Wondrous manifestations of life, emanating from the union of atoms, chemistry, and time, we are creatures of the cosmos grown sentient. Aggregated atomic dust from the dawn of time has learned to dream, reason, and create. Visualizing the possible, humanity aspires to leap beyond its limitations and reach for the sublime. Ancient hunter-gatherers spreading outward to find new and fertile ground were not so different in spirit and nature from those who today peer hopefully into huge telescopes, shoot mighty rockets into space, and long to journey between the lonely stars. Our species has always had its starry-eyed wanderers. We were born as wanderers and wanderers we remain. It is in our DNA.

If we value life, thought, and our own existence, we will continue to express our true nature by taking the steps necessary to preserve ourselves. We are living, breathing, thinking organisms that have struggled mightily to survive by maintaining and adding to the hard-won knowledge and wisdom of countless generations, and the struggle will continue. If we seek the perpetuation and spread—in essence, the immortality—of the human species, we will inevitably dispatch pioneers to distant worlds. We will have to, because our planet will not last forever. Even if human beings were suddenly to embrace peace, universal compassion, and civility, there would be no guarantee of survival. Making our way through the next few decades, let alone centuries, without obliterating ourselves by war, overpopulation, or destruction of the environment could prove difficult.

An unexpected cataclysm could destroy the Earth. An unstoppable giant meteor could ram our planet. The Sun could hurl a colossal coronal mass ejection at Earth. A nearby star could supernova, firing massive amounts of deadly radiation at Earth from deep space. These types of events are rare; however, over a billion years, the probabilities add up to a near certainty. No less than five different mass extinctions are recorded in the fossil record. Among these are the Great Dying of 250 million years ago, which was caused by a massive outpouring of magma for hundreds of thousands of years, and the Chicxulub meteor that hit the Yucatan peninsula 65 million years ago and exterminated the dinosaurs. Both of these events radically altered the climate. We may rest assured that in the future another extinction-level event will occur on Earth, forever altering our planet's ability to support life. Stephen Hawking states that humankind's "only chance of long-term survival is to spread out into space."

Ultimately, even if we manage to avoid the dangers listed above, we will eventually face the exhaustion of the Sun. A billion years or so from now, the Sun will have expanded and will be producing ten percent more light and radiation than it does today. A star holds together because there is an equilibrium between the outward push from the high-energy fusion reactions in its interior and the inward crush of gravity. As the Sun's supply of hydrogen is consumed, there will be less outward push of energy, and thus the equilibrium will be disturbed and the Sun will contract. This contraction will increase the pressure in the core, causing it to heat up. Once the core gets hot enough, helium will start to burn, and the energy outflow from this reaction will once again cause the Sun to grow in size. It is this growth that will eventually engulf Earth and the other inner planets.

The increased heat will shift the "Goldilocks," or "just right," zone outward to Mars, and the Earth will become an uninhabitable, floating cinder engulfed in the roiling flames of the Sun's gaseous atmosphere. Although the exact details are hard to predict, the Earth may have its matter reduced to plasma and cease utterly to exist as a planet.

A few billion years later, the Sun will have exhausted its fuel completely. If anything is left of the Earth at all, it will exist in a perpetual oblivion of cold and darkness. By then it will be far too late for humans to decide that they must venture forth into the Solar System, the galaxy, and the universe. This is the time for determining our destiny, and the path we choose will make all the difference. If it so happens that we choose the road that leads to the stars, then the planet Mars is an obvious—and some would say necessary—first step.

Mars

Mars is the most natural destination for twenty-first-century human interplanetary explorers. It is nearby, it has some gravity, and it is not as bitterly hot or cold as some of the other candidates. Furthermore, the Red Planet occupies a special place in our dreams and imaginations, as shown by the constant fascination with which popular culture has embraced it. If we go to Mars—which we must, if for no other reason than to advance the goal of ensuring our species' survival—it will be a momentous turning point. We will be opening a new frontier of discovery. Mars will become a steppingstone and a way station as we progress toward interstellar travel.

Robotic exploration of Mars, while it has been generally successful and has contributed immensely to our knowledge of the red planet, can never be considered the primary objective. Robots do not eat, drink, sleep, breathe, or experience psychological problems. While they are cheaper to hurl through space, at present they are deficient in analytical and problem-solving skills. A robot faced with a complex challenge it has not been preprogrammed to overcome will most likely fail. Computers have long been highly skilled at activities like chess-playing, in which the rules are well defined and the goal is simple, but something as easy for humans as walking across an uneven landscape could defeat an autonomous robot. When a computer program receives paradoxical data, it can crash or freeze in situations where a human could cut through the Gordian knot of stimuli and seize upon the singular strand leading to a solution. Furthermore, no number of robotic missions can save humanity from the existential threats it faces as a one-planet species. Only the presence of humans can confirm the viability and survivability of a distant planet's unfamiliar topography.

Therefore, although robots may prepare the way, the ultimate goal must be the dispatching of human beings and the establishment of a permanent human colony upon our nearest practical neighbor. While infrared machine eyes can peer through storms and dust clouds to see the surface of a planet or moon, only a human mind will feel the wonder of joyous realization when the cloudy veil is lifted from a future Martian atmosphere and its rich transplanted life is at last revealed.

Will humankind ever be willing to spend the necessary hundreds of billions of dollars and centuries of time for the settlement of another planet? Quite often, the benefits of exploration are not initially evident. When Thomas Jefferson negotiated the acquisition of the vast expanse of the Louisiana Purchase in 1803, he did so more or less sight-unseen. The explorers sent to reconnoiter the area returned two long years later with accounts that did not fully capture the full extent of the vast wealth in land and resources that the United States had obtained for a paltry fifteen million dollars, amounting to about three cents per acre. When the U.S. bought Alaska from Russia in 1867, the price of that apparently barren wasteland was $7.2 million, about two cents per acre. No one recognized at that time the true worth of that frozen wilderness, known as "Seward's Folly," that was to become a cornucopia of natural and mineral wealth.

Interestingly, Mars and the Earth have approximately the same amount of dry land, 35.5 billion acres. If journeying to Mars were to cost $355 billion, the equivalent of 100% of NASA's budget for nearly two decades, then the price would be $10 an acre. This is not so different from the price of Louisiana or Alaska once inflation is factored in, but the surface of Mars will be more like Antarctica without air than like the lush, fertile hinterlands of the Mississippi and Missouri rivers.

The attempt to travel to and live on other planets will inevitably give rise to explosive growth in new technologies. There will be—there must be—thousands of new discoveries in rocketry, space science, agriscience, communications, power generation, storage, and even psychology and human relations, before a successful crewed interplanetary mission can come to fruition. The positive effects from these innovations will reach far into the future, transforming the lives of all people for the better. The advances that were made during the Apollo program of the 1960s have had similar far-reaching beneficial effects on human existence, and are now so ubiquitous that they are taken for granted—advances in telecommunications, data processing, rocketry, and even nutrition, all due to the spurt of energy and creativity that sprang from a galvanized nation's self-confident and ambitious attempt to reach the Moon before the decade was out.

The Mars Society meets annually in Washington, D.C., for a conference on traveling to Mars. Members of this organization understand that pushing out into the Solar System—not just to visit, but to stay—and becoming a two-planet species is necessary if we are to remain a viable civilization over truly long time-scales. Already more than 200,000 volunteers have signed up to undertake the fateful journey to Mars. Many believe that there are people already born who will be the first to stand upon Mars and gaze with pride and wistfulness across space to the planet of their birth, the planet of their ancestors.

Setting foot on Mars is one thing—remaining is another. It will be relatively easy to fling humans, their paraphernalia, and their dreams across the gulf separating us today from Mars, but making Mars a long-term home will be a much greater challenge. The humans who settle on Mars will spawn a second genesis of life.

Emboldened by a vision of a better tomorrow, humans will alter Mars and make it a comfortable home. However, to make this happen, Martian pioneers will have to be cutting-edge, inventive visionaries with skill and daring. They will have to be gifted, consummate dreamers—beings endowed with a fertile, vivid imagination and the ability to create. They will need to be able to fabricate materials and make real the ideas, designs, and solutions conceptualized in their own minds and the minds of others. They will need a strong sense of what is realizable, based in reality and practicality. They must be capable of resourcefully transforming raw materials into useful devices under harsh and unforgiving conditions. These courageous trailblazers must possess an eclectic range of foundational knowledge and a deep understanding encompassing the full breadth and interconnectedness of scientific and technological disciplines. They must be keen observers of nature, disciplined researchers, and originators of new ideas. These pioneers must be infused with courage to carry them through, around, and beyond risk, driven by the dream of developing a new world. They must be endowed with spirit, determination, indomitable will, and a belief in the successful culmination of their exploratory endeavors. They must be team players, willing to lead—and follow, when necessary. They must own an inner reserve of confidence, optimism, and tenacity that will tide them through the tough times. They will have to demonstrate willingness to endure hardship, face daunting obstacles, and manufacture success no matter how unfavorable the odds. They must be willing to venture into and settle unknown territory. Only born risk-takers, those willing to stray from the safe path, need apply.

Astronauts must be physically robust, mentally sharp, and technically skilled. They must be able to resist boredom for months on end and yet be ready at a moment's notice to spring into action making critical life-and-death decisions. These astronauts will be traveling the dark depths of space for at least five months at four miles per second, and it will seem to them like they are not moving at all. The explorers will face tremendous physical challenges. On the flight to Mars, they will be forced to live a confined life inside their tiny vessel. During the voyage, which could take as long as ten months, they will be experiencing microgravity and will have to exercise a minimum of two hours per day to retain their strength, unless a passive system, yet to be invented, is put in place to counteract the effects of microgravity and help the astronauts maintain their physical condition. Once they reach Mars, they will have to be ready to launch themselves into vigorous action. The transition from relative inactivity inside a cramped capsule to explosive, mission-critical exertion could prove to be quite problematic. The explorers may have to be selected mainly on the basis of their ability to make this rapid adjustment. Early crews will build a subterranean maze of habitats or erect a complex of dwellings on the surface, and this will require hard labor. It might be possible for robots to be sent first to complete some of these more arduous tasks, but in any case, the work will have to be done. They will be charged with maintaining their vessel and creating domiciles, habitats, and greenhouses on Mars. They must learn how to extract water from a frigid and forbidding desert environment. Growing crops will be a necessity, and that will require oxygen, nitrogen, carbon dioxide, water, light, and temperature control inside Martian greenhouses. Food plants need copious amounts of light, but the sunlight that reaches Mars is only half the intensity of the sunlight that reaches Earth. Plants also need to be protected from harmful radiation just like humans. Specialized plant-breeding experiments will need to be conducted to develop varieties more suitable for a Martian environment. There will be no insect pollinators, so food plants will need to be selected on the basis of their ability to self-pollinate or be wind-pollinated; while hand-pollination is possible, it is an arduous and time-consuming task. Wastewater will be recycled. The soil will initially be far from ideal for growing crops. On Earth, soils are packed with living organisms and decaying organic matter which plants can use as nutrients. None of that will be present on Mars, at least not at first. Eventually machinery and buildings will have to be constructed of local Martian materials. A method is being developed of using the Martian regolith and a few simple chemicals to produce sturdy construction materials that will never rust. Self-sufficiency will be the ultimate goal in order to cut the umbilical cord linking Mars with Mother Earth.

Those first pioneers must be motivated by the long-term good of humanity, confident that their individual endeavors will make a difference and enhance our species' long-term prospects. These utterly unselfish, intrepid individuals must survive, exist, act, explore, and create with the conviction that the initial seeds they sow will ultimately blossom for the benefit of future generations. The next chapter in human history will be written in their blood, toil, tears, and sweat. They will be perpetual learners capable of turning crisis into opportunity. They will need flexibility; upon reaching a dead end, they must be willing to reverse course and try a new direction. Their own spirit of exploration will be the wind constantly at their backs. They will be measured by their character; Martian pioneers will need all the grit, and ingenuity that characterize humanity at its best. They will also require emotional aplomb and a practical approach to problem-solving. Problems will range from the mundane, such as a repetitive diet and lack of meat, to the dramatic and unprecedented, which will be nigh impossible to anticipate. As in most successful enterprises, prior training enhances the effectiveness of future efforts. The problems will appear insurmountable, but solutions will inevitably be found. The nearly impossible will become possible, and the nearly unachievable will eventually be achieved.

Mars can be the next small step and giant leap for humankind. The adventure will be ambitious, risky, and expensive. The force of imagination filled ancient sails and propelled bold explorers across unknown seas; today we need new technology—engines, fuels, and vessels—to take us to the next-best planet in our Solar System and eventually to the stars. Imagination continues to fuel our dreams; more powerful than any rocket fuel, it impels us toward the attainment of the grand goal, the vision of a new home on another world. Humankind's most momentous achievements are awaiting us in the future, and the trip to Mars will be the next great leap in human exploration to add to our astounding history of discovery.

Practical Difficulties to Be Overcome

The idea of sending humans to another planet is exhilarating and generates an abundance of enthusiasm, but the trip will be exceptionally hazardous. Humans must be protected from the unrelenting four furies of space: vacuum, cold, radiation, and weightlessness. These potentially lethal risks must be mitigated. Mission planners would prefer to send astronauts on a trajectory that would minimize the astronauts' time in space. However, this is not possible because the amount of fuel required would be prohibitive. As of this writing, the plan is to follow the most fuel-efficient path through space.

The period during which Mars and Earth are in the right locations for efficient travel between the two is called a "launch window." There is a launch window every 26 months, and if for some reason the window is missed, there will be no option but to wait for the next window. Once launched, the spacecraft will follow a curved path. It is not aimed directly toward Mars; instead, it is directed toward the point in space where Mars will be when the craft arrives. This is analogous to how a quarterback throws a forward pass. He anticipates where the receiver will be when the ball gets there. But regardless of the trajectory chosen or how long the journey takes, the merciless four furies must be dealt with. Evolution has designed the human body to thrive under the force of gravity that is normal near Earth's surface, under one atmosphere of pressure, in a mild radiation environment, and in warm Earth-normal temperatures. None of these conditions is found either in space or on Mars.

The incapacitating effects of weightlessness are readily apparent every time a space-worker returns from the International Space Station after a six-month stay. These astronauts return to Earth visibly weakened by their extended sojourn in microgravity. They have trouble readjusting and must be physically helped out of their return capsules. For safety, they are barred for a time from driving a car or lifting heavy objects. Since they no longer have to constantly resist the pull of gravity, muscles atrophy and bones fail to continuously rebuild themselves as they normally do on Earth. The body's muscle mass (including the heart muscle) and bone density steadily diminish by about 1% per month. The inner ear is also affected, reducing the body's ability to maintain its balance. In space, sensory data supplied to the brain by the inner ear can conflict with the input from the eyes, and the brain is often unable to resolve the contradiction. The result can be nausea and vomiting. Physiologists have found that astronauts returning from long periods in weightlessness may have the agility of a toddler and require many weeks or months to return to normal. In addition, weightlessness can cause fluids around the brain to shift, distorting the shape of the eyeball, which can detrimentally affect vision.

To shed more light on this vital issue, NASA performed a 25-month study of two identical twins. One of the twins lived on the International Space Station for 340 days while the other remained on Earth as a control. Besides the expected loss of muscle and bone mass and the shifting of fluids in the eye, the orbiting twin experienced a number of unexpected and alarming physiological, biochemical, and cognitive changes, including changes in telomere length (related to cell growth and aging) and decreases in mental abilities. Fortunately, within six months of returning to Earth, most of the detrimental changes returned to pre-flight levels.

The Twins Study generated a large amount of data that scientists will be studying for a long time. However, an expedition to Mars will require much longer than one year, with the current plan involving a six-month flight to Mars, 14 months in the reduced gravity of Mars, and another six-month flight home, a total of about 26 months in reduced gravity. It is impossible to know in advance how astronauts' bodies will respond to these unprecedentedly long periods without Earth-normal gravity.

A serious challenge that astronauts will face in microgravity is that every action must be carefully thought through in advance until an astronaut develops the necessary intuition to function successfully in this alien environment. For example, new astronauts attempting simple tasks like tightening a screw are likely to find that their body will turn more than the screw does, since the friction of the screw-thread would overcome the minute resistance offered by a nearly weightless astronaut's body. Surfaces within the International Space Station are generously equipped with perforations and grooves in which workers can anchor their feet for such tasks. Once having adapted to microgravity, astronauts arriving at Mars would then have to readapt. This is a chancy prospect indeed, especially considering that piloting a spacecraft toward landing in unfamiliar gravity could require lightning-fast responses based on an instinctual feeling that is no longer present, having deteriorated during eight months in space. Furthermore, a spacecraft arriving at Mars on the Hohmann transfer orbit will be going fast enough to vaporize itself if it crashes into the Martian surface.

One solution would be for a rocket ship transporting astronauts to Mars to create its own artificial gravity. This can be done by creating a rotating vessel shaped like a wheel. The centripetal force generated by the rotation would create an artificial gravity much like a spinning wheel at a circus. The people on the ride are spun in rapidly moving circles, which forces their bodies against the outer rim of the spinning device, creating artificial gravity. In the 1950s, Wernher von Braun and Willy Ley envisioned a rotating wheel with a diameter of 250 feet that would revolve at three revolutions per minute and create an artificial gravity equal to 40% of the gravity on Earth. Some scientists contend that spinning faster than one revolution per minute is harmful to humans over the long run, but in any case, NASA's plans as of this writing do not include any artificial gravity. Spacecraft dimensions are also a concern. The Orion spacecraft is currently designed to carry a crew of between two and six astronauts into orbit in a habitable volume of 316 cubic feet, which is larger than previous crew-carrying modules. A much larger craft to make the journey to Mars could be constructed in orbit and could be of almost any size and configuration, limited only by the cost of lifting structural components into orbit piece by piece and module by module. Planners have set a benchmark of 65,000 cubic feet for the craft that will carry astronauts to Mars. This is twice the size of the International Space Station. NASA planners hope to reach the capability, with the Space Launch System currently under development, to lift 143 tons into orbit. The current limit is 40 tons.

Humans need oxygen to breathe, but, just as importantly, they need to be surrounded by gases exerting pressure on every square inch of their bodies. In deep space, the lack of oxygen would result in almost immediate asphyxiation; but even if there were oxygen to breathe, death would come quickly if we were to find ourselves in a vacuum. As if that were not enough, humans also have very picky temperature requirements, requiring balmy temperatures near the low-end temperature of liquid water; forty degrees above the freezing point of water would be quite comfortable, assuming the pressure is right.

Astronauts will need a dependable supply of breathable air at a suitable temperature and pressure. Any structure that will serve as a habitat for humans must be sufficiently sturdy and impermeable to hold atmospheric gases securely inside. A high-speed micrometeorite can punch a hole through three inches of aluminum. Such an event in space or on Mars could disastrously deflate a habitat, cubicle, spacesuit, or spaceship. The ships and dwellings used by the explorers will have to be constructed like today's ocean-going vessels, which are compartmentalized so that damaged sections can be closed off until repaired. The Earth is struck by between 5 and 300 tons of meteorites per day, most of which are burned up by Earth's atmosphere. The number on Mars will be comparable except that the meteors will be less likely to burn up in the thinner atmosphere.

Extensive studies have been made of how the body reacts to very low air pressures like those found on Mars. The Martian atmosphere is about 1% as dense as Earth's and consists almost entirely of carbon dioxide. In one experiment, a volunteer subject performed a free fall from an altitude of 100,000 feet. Unknown to experimenters and to the subject himself, there was a small hole in one of his gloves that let air escape from his life suit. This tiny, localized loss of pressure was enough to cause his hand to expand to twice its normal size. Fortunately, in this case, the effects were not permanent. One subject who experienced an unexpected loss of atmospheric pressure related that he noticed a fizzing sensation on his tongue; this was his saliva boiling away. Astronauts in chambers that mimic being at an elevation of 25,000 feet can lose their ability to function mentally within ten seconds and pass out in 15 seconds. This means that in the case of a catastrophic malfunction of a spacesuit, as would happen, for example, if a micrometeorite were to punch a hole through it, the wearer could find himself mentally incapacitated that quickly. Then after 60 seconds, the dissolved gases would start to leave the blood. This is excruciating and often fatal. Deep-sea divers who experience this call it "the bends." If the punctured spacesuit is repaired within 90 seconds, the astronaut may survive without any ill effects; otherwise, death is certain.

Weightlessness is debilitating, and lack of air is quickly fatal; however, perhaps the most terrifying danger associated with long space journeys is the unique danger posed by radiation. There are three main types of radiation that space explorers must face: solar wind, cosmic rays, and solar flares. The damage they cause to the human body can be swift and deadly, or cumulative and remorseless. The solar wind is the outflowing of charged particles from the Sun's surface, and cosmic rays are the inflowing of similar solar radiation from distant stars. Neither the solar wind nor cosmic radiation is overwhelmingly powerful, yet their detrimental effects are deadly over time. A solar flare is something else entirely. A solar flare is a massive eruption of matter and radiation that the Sun violently discharges into space at irregular intervals and at speeds reaching a million miles per hour. Such events are comparable to the explosion of thousands of nuclear bombs detonated simultaneously with all their fury hurled into space in a particular direction. If a spacecraft ferrying astronauts were in that line of fire, the vessel's protective measures could be overwhelmed in a matter of minutes, and the inhabitants would perish almost immediately.

Space travelers are assailed by both the highly energetic particles of the solar wind and by cosmic radiation emanating from the unimaginably vast reaches of deep space. Radiation exposure is measured in millisieverts. We on Earth receive about 150 millisieverts worth of radiation throughout our lifetime—about 2.4 millisieverts per year. However, in space where we are not protected by Earth's friendly atmosphere and magnetic field, we would be exposed in 6 months to 150 millisieverts, and the deleterious effects would not become apparent until it was too late. These types of radiation attack at the microscopic or cellular level, killing cells and causing mutations. The impinging radiation weakens DNA and tears it apart. Our skin and hair would appear to age at the rate of a decade per year. Bone marrow would make fewer red blood cells, and the ones it did produce would tend to be defective. The digestive system would cease to function as the lining of the gastrointestinal tract accumulated more and more damage. Even nerve cells would shrivel and die.

Studies are underway to solve this problem. One approach is to have a system of moving magnets to create a magnetic field that would shield astronauts inside the ship from the insidious and deadly effects of penetrating radiation, just as Earth's magnetic field protects humans on its surface from radiation. New plastics are being developed that are proving effective as shielding. Furthermore, surrounding the capsule with the fuel or water supply could protect against much of the incoming radiation. Colonists on Mars would face the same threat, but living underground would mitigate the danger. One of the solutions proposed is that the initial pioneering astronauts could choose to live underground, perhaps in the hypothesized lava tubes under Olympus Mons or one of its sister volcanoes, such as Pavonis Mons.

The health hazards for a journey to Mars are both physical and mental. While physical problems could include improper blood circulation, space sickness, weak immune systems, backaches, muscle loss, and radiation-induced cancer, challenges posed by psychological issues could be just as severe. Such problems can arise easily in cramped work areas and living spaces where there is nowhere to go to escape from an irritant. Interpersonal conflicts, depression, anxiety, insomnia, psychosis, and even suicidal behavior can result. As astronaut Mike Massimino noted, mental isolation is not good for mission success.

The Mars 500 project, conducted in Moscow, Russia, was a full simulation of a voyage to Mars. One of its primary goals was to study the psychological effects of a long space mission. Between June of 2010 and November of 2011, the participants resided in the experiment's four hermetically-sealed habitat modules while simulating every phase of a Mars mission, including takeoff, the months-long journey across the gulf between Earth and Mars, landing on the Martian surface, residence on Mars for over a year, and the return flight. During this time, the test subjects endured all the rigors of space travel, except weightlessness. They lived and worked together in cramped isolation, and had to overcome resource rationing, communication delays, and hazardous bacterial biofilms that accumulated on life-support equipment. The participants were able to maintain relative harmony, though it was noted that they sorely missed family and friends. One finding of this type of research is that just because an astronaut works well on one team does not mean that that astronaut will be a good fit on another. The crew for a Mars mission will be selected as a complete team, not as a set of individuals.

More Hurdles

An interplanetary spacecraft could be assembled in space and then stocked with the necessary consumables and fuel. The main difficulty and cost associated with this will be the lifting of materials and supplies out of the Earth's gravity well. There is no way to alleviate or obviate this requirement. Gravity is an inexorable and unavoidable tyrant, and the Earth's mass will exact its gravitational tax on all space missions.

Many technical hurdles must be overcome before it will be possible to land a crewed vessel on the surface of another planet. There will be docking and undocking in orbit above the planet, and the mode of descent to the surface has not yet been decided. Some Mars rovers have landed on elaborate "space cranes," in which a battery of retrorockets fires to stop the descent of a parachuting vehicle and then a cable unspools to gently lower a rover to the ground. Other probes intentionally crash-landed with giant airbags inflated around them; the probe and airbags then bounced and rolled across the rocky Martian surface until they came to rest. SpaceX is working on reusable rockets that will land on Mars, or anywhere else, and take off again.

Propelling a rocket to Mars is no simple matter. Until recently, all rockets worked by burning chemical fuel with an oxidizer. The fuel basically explodes like a bomb, and if the rocket is designed correctly, the force of the explosion is controlled and directed to propel the ship forward. This approach has limitations, however. For one thing, there is a constant risk that the explosion will surge out of control, obliterating the vessel and its occupants. Also, voluminous amounts of bulky, heavy fuel and oxidizer are required. With chemical rockets, because of the limited amount of fuel they can carry, the trajectory for the trip to Mars must be selected wholly on the basis of which one will use the least energy, and this trajectory, the Hohmann transfer orbit (named after the German engineer who first proposed it in the 1920s), requires eight months travel time from Earth to Mars. However, new engine technology is under development, such as plasma ion engines, which could cut the travel time to less than two months. If such technology proves feasible, the difficulties faced by astronauts and mission planners would be greatly diminished.

Once the first Martian pioneers have landed, they will start to work immediately to make Mars habitable for humanity. Hazards specific to Mars include planet-girdling dust storms and 250°F temperature swings, and if something goes wrong, the nearest help is at least 34 million miles away. The necessary protective equipment, such as helmets and space suits, is in development. Helmets will have a full faceplate, and suits for journeying on the surface of Mars must carry an oxygen supply and an electrical system for heat and air purification. The suits will never be worn in living areas; Mars-dwellers intending a walkabout will step through a portal into their suit, which will then be disconnected from the habitat and free to roam. When they return, they will reattach the suit to the habitat and step back inside. This procedure will be necessary to keep dust and contaminants out of the astronauts' living areas.

Explorers will want a safe and maneuverable vehicle for traversing the Martian surface. The lunar rover, which was used on the surface of the Moon by the Apollo 15 astronauts, could travel at 10 miles per hour and had a 20-foot turning radius. A twelve-wheeled Mars rover currently under development will carry its own pressurized cabin, will turn on a dime, and will even be able to move sideways, like a crab.

Mars is believed to have plenty of water. Much of it is frozen at the poles, but there is clear evidence that liquid water sporadically flows across the Martian surface. A settler might just bring in chunks of ice harvested from an icecap and let it melt in a bowl—a simple procedure indeed, although the excess CO_2 in the ice would turn gaseous and need to be vented away before it could build up to high enough levels to cause narcosis. However, even with abundant Martian water, all the astronauts' water would, initially at least, have to be hauled to Mars on their spacecraft, and water is very heavy. The question of food is at least as serious. Planners estimate that each Martian astronaut will consume about two tons of food during a three-year mission. Any crewed mission will need to transport 20 to 30 times as much mass as NASA now routinely sends to Mars for a typical robotic-rover mission. This is the difference between sending a car and a jetliner. Sending a rover is like sending a car, while sending people, which will require a habitat, atmosphere, water, food, and myriad other necessities, would be like sending a jetliner.

Martian pioneers must eventually terraform the planet, a difficult and time-consuming process that, if it worked, would change Mars to make it more nurturing of human life. This would mean creating, somehow, an oxygenated atmosphere on Mars, increasing, somehow, the temperature, and arranging for liquid water to flow in rivers, lakes, and even perhaps briny oceans, or at least into cisterns. Until that happens, humans will need to live in airtight structures, possibly domes, and will have to carefully maintain their limited supplies of water for drinking and irrigation. Enclosed habitats will also need areas for growing food, which would likely consist of super-productive, multi-tasking strains of algae that can yield tons of protein, fuel, and sugar per acre, and even medicines. There will have to be a source of energy for heat and power, and the structures in which humans live would need to be shielded from deadly radiation. For safety, the airtight habitats would need to be compartmentalized so that a zone can be sealed off from the other zones in case of a breach.

Finally, a three-year space trip requires backups of all critical mechanisms, and backups of the backups. If there is a mechanical failure comparable to what happened to Apollo 13, the outcome could quickly turn tragic.

One of the greatest fears of NASA supervisors before and during a mission is failure to anticipate the unforeseen calamity. Preparedness for any eventuality is the hallmark of a successful NASA enterprise. As John Wooden pithily stated, "Failing to prepare is preparing to fail." But the next interplanetary venture will be the first one ever, and the unexpected will occur, resulting in losses. It is far safer and cheaper to practice in Antarctica or even on the Moon, where at least an accident may not be so immediately fatal. NASA learned from accidents in the past, in some of which astronauts lost their lives, and the learning process is ongoing. It is a part of exploration.

Conclusion

When humans begin to go to Mars, at first the flow will be a trickle, but as the trip becomes safer and cheaper, the spigot will open and the flow will turn into a gushing torrent of people. They will go to Mars in search of freedom, adventure, and opportunity, wishing to find their own ideal life on a strange, red paradise where clouds of dust fill the skies and humankind can live self-sufficiently for the first time untethered to the Earth or a space station. Adversity and danger will not deter them, and eventually there could be millions, a nascent nation, a new community of extraterrestrial humans.

The dangers of space exploration are clear; the rewards are, in the eyes of many, insubstantial and hypothetical. Will we summon the collective will to cross this Rubicon and sail to other planets and moons? It is earnestly to be hoped....The human race desperately wants to go, and, if given the chance, some will inevitably go.

We can break the bonds that bind us to Earth and sail to the far heavens. Many will emulate the myths of ancient heroes and boldly set off on their own epic journey, taking wing on improbable flights in ways our ancestors could not have dreamed. On Earth, we have been able to create, expand, and dominate, and the same drive will carry us to Mars or to the moons of Jupiter and Saturn. The naked apes have swung out of their treetops and may land with their feet on the surface of another world.

The journey outward into space has been written into the core of our being since the first humans raised their eyes to observe the revolving patterns of the star-studded nighttime sky. And dwelling inside this core is a magnificent shining light that rules our passions, guides us along our path, and illuminates our future. It is a craving that permeates our lives and souls. Incredible discoveries await our species as we prepare to sail glistening argosies across the ancient, vast, and austere sea of deep space.

To explore, to grow, to learn, indeed to survive—these are what it means to be human, and this is what interplanetary exploration symbolizes. And when we achieve this aim, when we succeed for the first time in living and thriving on an alien world, we will inevitably begin looking outward and upward toward the next awe-inspiring adventure, forever seeking to broaden our horizons, always in quest of the next world to explore.

This will be the dawning of a new era in human history: the age of the multi-planet species. Never again will the human race labor under the dread specter of imminent destruction by any of the dozen perils listed above. We will have begun the wonderful and necessary process of spreading out into the universe. Crossing the gulf that separates the Earth from other planets is the next step, in many ways the first step, on the path to the achievement of humankind's destiny, which is someday to span the universe and endure as long as time.

16

Mars: Our Captivating Red Neighbor

It is human destiny to expand from Earth, to the Moon, to Mars, to the stars.
Barney Oliver

An eye like Mars, to threaten and command,
Shakespeare

Mars has captivated the human imagination since our ancestors first noted that there was a reddish point of light in the heavens which wandered from night to night at variance with the motion of the "fixed" stars. Mars's red hue, which is detectable by any sharp-eyed human on a clear night, inspired inhabitants of ancient Asia to dub it 火星, *huǒ xīng,* the "fire star." Meanwhile, early Western civilizations saw the red color as blood and personified the planet as a bloodthirsty god of war. Babylonian astronomers called the planet Nergal for their god of death, scorched earth, war, and pestilence. The ancient Egyptians referred to the planet as "Horus of the Horizon." Their god Horus was famous for the bloody vengeance he exacted upon the murderer of his father Osiris and was portrayed as a human being with the head of a fierce bird of prey. The modern name of Mars comes from the ancient Romans. Their militaristic and violent nature catapulted their empire to a position of dominance in the Mediterranean region, and they naturally placed their god of war, Mars, high in the divine order.

These ancient civilizations knew nothing about Mars except its color and its odd path through the sky. Little did they know that our neighbor planet indeed bears huge scars as if from battle—including a gaping laceration 2,500 miles long as if the planet had been slashed by a heavenly halberd. An entire hemisphere of Mars has been gouged by an unimaginable celestial cataclysm, which created a depression the size of the United States and Canada combined and left the planet looking like an enemy's crushed skull. Nor did those ancients have any idea that the Martian terrain and climate are as harsh and murderous as some ferocious god of death or war, with bitter temperatures, suffocating atmosphere, and abrasive dust storms.

Astronomical History

Since the seventeenth century, numerous astronomers have noted Mars's unique features. The first person ever to point a telescope at Mars was Galileo in 1608. Christiaan Huygens is credited with the first realistic drawing of the planet's surface in 1659, and he also discovered the planet's rotation. Giovanni Cassini in 1666 noted its polar caps and calculated its day at 24 hours 40 minutes, an error of only three minutes. In 1877, the Italian astronomer Giovanni Schiaparelli reported seeing straight lines on the surface of Mars, and he used the Italian word *canali*, which means "channels," to refer to them. *Canali* was later translated into English as "canals," helping to foster the erroneous impression that these lines were something other than naturally occurring features.

Percival Lowell, born to a wealthy Boston family, was a brilliant student who graduated with distinction from Harvard University with a degree in mathematics. This American businessman and astronomer built his own observatory in Flagstaff, Arizona, in 1894. Lowell later wrote extensively about the "non-natural features," or canals, he saw on Mars. Lowell believed that the canals he observed were the creation of a dying civilization that was transferring water from the planet's water-rich poles to its arid equatorial regions for irrigation purposes. His prose excited the populace, and a cottage industry of science fiction sprang up about life on Mars. Perhaps the most famous exemplar of this genre was Edgar Rice Burroughs' series about John Carter, a veteran of the American Civil War who was astrally transported to Mars, where he found adventure, romance, and alien races at war. When better telescopes were eventually built, it became evident that Lowell's straight lines were not straight at all, but naturally occurring features of the cratered Martian surface. Later, the Mariner missions of the 1960s and 1970s proved conclusively that there are no canals on Mars. Even up to that time, questionable images had led some to believe that there were features on the Martian surface that resembled a human face, or pyramids, but these too were later shown to be perfectly ordinary rocks and shadows in the Martian landscape.

The moons of Mars were discovered by the American astronomer Asaph Hall in 1877. They are two tiny, irregular chunks of rock, mere moonlets in comparison to the stately, planet-sized Galilean moons of Jupiter or Earth's mighty companion. Mars's minuscule, tumbling moons were named Phobos (Fear) and Deimos (Terror) after two sons of Ares and Aphrodite, whose respective Roman counterparts were the gods Mars and Venus.

Phobos, the inner and larger moon (right), revolves around Mars in 7 hours and 39 minutes. It orbits so near Mars—nearer to its primary than any moon in the Solar System—that the planet's gravitational pull slows the moon. Thus, in the near future (less than 100 million years), Phobos will either be torn apart by tidal forces or crash into the planet. Deimos' fate will be different. It moves in a more distant orbit, and tidal forces are causing it to

Phobos

recede from the planet like our Moon. If left alone, it will inevitably drift away from the planet.

Physical Description of Mars

Mars is the planet of the highest mountains and deepest valleys ever discovered. Less massive than Earth, it possesses only 37% of Earth's gravity. Because Mars is smaller than Earth, it cooled much more quickly, resulting in a solid interior. Although Mars appears to be geologically dead, the land occasionally buckles due to cooling and shrinkage, and this causes Marsquakes. Because the interior is solid, there is no circulating liquid core to create a magnetic field and no flowing mantle to produce plate tectonics. Because the planet has no protective magnetic field, the solar wind stripped away most of its atmosphere billions of years ago.

With virtually no atmosphere, Mars's liquid water boiled away and left the planet a barren, desiccated desert. The atmosphere that does remain is a diffuse layer of carbon dioxide, at pressures less than one-tenth of a pound per square inch (0.1 psi) as compared to Earth's 14.7 psi at sea level. Since there is almost no water vapor in the atmosphere, there is no precipitation to shower the Martian landscape. A land where it never rains is a land of dust, dunes, and rocks. Mars is covered in dust the color of rusty iron—hydrated, oxidized iron ($Fe_2O_3 \cdot xH_2O$). This is the same rust one might see on the body of an abandoned jalopy, and it tints almost the entirety of the Martian surface.

Because there is virtually no atmosphere to absorb, transport, or retain heat, Mars's surface temperature varies widely: plunging to -284°F at the poles, which is cold enough to freeze carbon dioxide, and rising to 86°F at the equator, which would be shirtsleeve weather on Earth. These are extreme temperatures, but the planet-wide average temperature has been calculated to be -67°F, which is a little colder than Antarctica. Since the Martian atmosphere is so thin and has no ozone layer, the planet's surface is constantly bombarded by solar and cosmic radiation—250 times the radiation that strikes Earth—that over time would be deadly to humans. The ceaseless barrage of lethal radiation, coupled with the lack of liquid water on Mars's surface, and the discovery of toxic perchlorate salts make it no surprise that NASA's rovers have so far failed to detect any signs of life there.

Mars orbits roughly 140 million miles from the Sun, so it receives less than half as much light and heat as the Earth receives. The Martian year is approximately equivalent to two Earth years. In some ways, Mars is remarkably similar to Earth, in that Mars rotates on its axis once every 24 hours and 37 minutes and has an axial tilt of 25 degrees. However, unlike Earth, which is stabilized in its spinning by its hefty companion, the Moon, Mars's axial tilt gyrates wildly between 10 and 80 degrees over a 120,000-year cycle. This implies that Mars's normal-appearing tilt is just a coincidence: we happen to have arrived fortuitously on the scene at a moment when its tilt closely matches ours.

Mars experiences the same four seasons as Earth, except that the seasons on Mars are almost twice as long. In winter, the temperatures plummet to below -110°F and carbon dioxide freezes. In spring and summer, temperatures rise, and the frozen carbon dioxide sublimates to create a slightly thicker atmosphere of carbon dioxide gas. The temperature gradients increase, and the marginally thicker atmosphere begins to more vigorously transfer the heat around the planet. So, just as on Earth, the

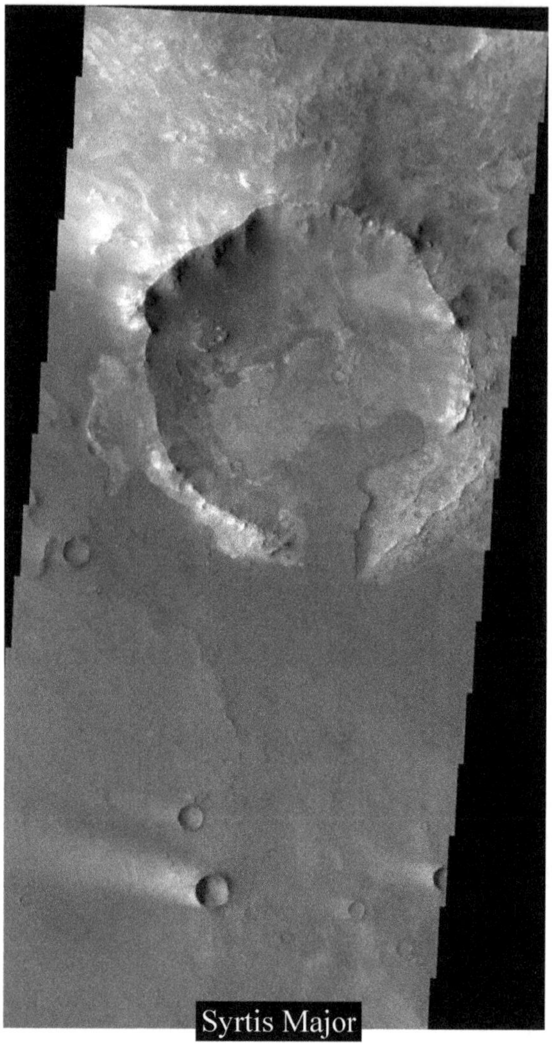

Syrtis Major

spring and summer bring Mars to life. However, this "life" is no annual rebirth of green foliage and frolicsome newborn animals. Instead, this planet transforms into a springtime of whirling dust devils up to half a mile high and raging tornadoes that can scour the surface of wandering, shifting dunes and create the impression of terrain that has been scored by gargantuan, sidewinding sandworms. Mars has the largest dust storms in the Solar System, which at times can encompass the entire globe and block our view of the planet's surface. Despite their size, these storms do not pack a powerful punch. Since the atmosphere of Mars is so thin, the power of their seemingly impressive winds is equivalent to that of a mild breeze on Earth. The image at left shows clear wind streaks spreading leftward from several small craters. The large crater at the top of the picture, bearing an impressive lava flow, is Syrtis Major. This crater, situated near the equator of Mars, is so prominent that Christiaan Huygens in the seventeenth century was able to use observations of it to measure the length of the Martian day, accurate to within seven minutes.

Water on Mars

There is little liquid water on Mars, but there is evidently plenty of H_2O in solid form. For many decades after telescopic observation of the Martian surface began, Mars was believed to be absolutely dry and desolate. However, we now know that glacial swirls dot the Martian landscape like bright swatches on an abstract artwork. In September 2015, NASA's Mars Reconnaissance Orbiter (MRO) detected evidence that occasionally, during the Martian summer, liquid brine can briefly break through the planet's sandy soil and create rivulets that streak the surface for hundreds of feet. The MRO also discovered deposits of ice covered by layers of dark-colored dust. If melted, the water in these deposits could cover the entire Red Planet in a sea three feet deep. That ought to be enough to meet the hydration needs of human explorers. The poles of Mars maintain vast sheets of ice covered in a layer of frozen dry ice (carbon dioxide) and dust. (Martian North Pole shown at right). Including the water from the icecaps would increase the depth of the hypothetical Martian meltwater-ocean to 60 feet.

Polar Region Showing Ice Cap

The presence of water in such large amounts makes it likelier that there could be microbial life. However, regardless of the presence or absence of microbes, the availability of a plentiful water supply would be of paramount importance to human residents on Mars and would significantly improve the prospects for successful settlement of the planet. NASA projects that each human settler on Mars will need a minimum of 100 tons of water for a 15-year stay.

Mars in its past was a warmer, wetter world than it is today. Rocks on the Martian surface contain hydrated minerals, which could only have been formed in a water-rich environment. Also in evidence are eroded landforms, salt deposits, and water-carved canyons and valleys, as well as structures resembling rivulets and deltas. The striking Gale Crater, visited by the Curiosity rover, is the site of an ancient lake that has lost its water, and Terra Sirenum, found by the Mars Reconnaissance Orbiter, has thick salt deposits left behind by evaporation of a broad sea.

The Surface of Mars

Mars has two distinct faces. Half of its surface is pockmarked with craters (100,000 of them), mountains, valleys, and highlands. The other half is covered with smooth plains, the result of a cataclysmic ancient asteroid strike and subsequent lava flows. In Mars's South Polar Region, seasonal geysers of sublimating carbon dioxide have created strange, spider-like features called aranea. These mysterious shapes, discovered by the MRO, can cover hundreds of miles and take thousands of years to form. They are oddly reminiscent of the arachnoids seen on the surface of Venus, a planet one thousand degrees hotter with an atmosphere 25,000 times thicker.

Meanwhile, near Mars's equator lies an immense feature called the Tharsis Bulge. The Tharsis Bulge covers one-fifth of the planet's surface and contains an inactive volcanic mountain chain that features a dozen very distinctive volcanoes that stretch 16 miles above their surrounding plains. The biggest and tallest is Olympus Mons (below). Three times as tall as Mount Everest and much larger than Earth's tallest volcano (Mauna Kea), Olympus Mons covers an area the size of New Mexico and is 88,500 feet tall and 374 miles wide. Its central caldera is almost as wide as the mountain is tall. Although volcanism is common on planets and moons throughout the Solar System, no other known volcano comes close in size to this mind-boggling juggernaut.

Olympus Mons, like the rest of the Tharsis Bulge, is believed to rest atop a labyrinth of lava tubes carved by flowing magma millions or billions of years ago. If this is true, these tubes could be hundreds of miles in length and may be suitable for human habitation. Living in these deep, underground tubes, humans would be protected from dangerous micrometeorites and deadly radiation as effectively as if they were living on Earth. Scientists estimate that five yards of Martian regolith overhead would be sufficient to provide this protection. The Olympus Mons lava tubes might also provide access to precious metals and minerals, although these would be of secondary importance compared to safety.

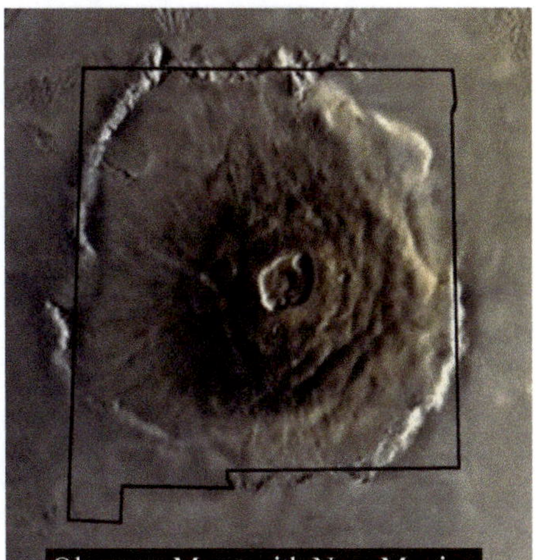

Olympus Mons with New Mexico

Another unique Martian landmark, Valles Marineris is an immense canyon system that is as long as the United States is wide and is easily seen from space. It is the largest known canyon in the Solar System. It consists of interconnected canyons, several of which merge to form a deep depression 375 miles across and as much as 5.6 miles deep. Valles Marineris is twelve times wider, ten times longer, and five times deeper than Earth's Grand Canyon and stretches across 20% of Mars's circumference—about 2400 miles. This immense scar upon the surface of Mars is a visual reminder that vast rivers of water once gouged the surface of this planet. Its imposing size can be seen in the photograph of Mars at the opening of this chapter. Valles Marineris was named after the Mariner 9 probe.

Astrobiologists consider Valles Marineris to be the best possible landing site on Mars because it is flat, not too dusty, not too rocky, has relatively high atmospheric pressure, and is warmer than anywhere else on Mars. There may even be some water there. Such a gaping, open scar would give geologists outstanding opportunities to probe deeply for Martian rocks with a minimum of difficulty. Astonishingly, this deep cleft in the Martian surface lies near the immense Olympus Mons, creating a breathtaking study in contrasts.

Finally, the North Polar Basin, or Borealis basin, is elliptical in shape and covers 40 percent of Mars's surface. This basin is a smooth area that is 6,600 miles long by 5,300 miles wide and most likely formed from a single enormous impact. There is evidence that it has a secondary rim as well, which is additional evidence that it was created by an impact event. If this was created by a meteor, then it is the largest known crater in the Solar System. It is four times larger than the next largest crater, Utopia Planitia, which is located within the North Polar Basin. (The next two largest in the Solar System are the Aitken basin at the south pole of the Moon; and Hellas Planitia in Mars's southern hemisphere.) Another huge Martian crater is Orcus Patera, with an odd shape resembling a whale or a teardrop. It is about 250 miles long and 50-150 miles across, with mile-high walls. This unusual crater was probably formed by a giant asteroid striking at a shallow angle, as low as 15 degrees.

Exploration of Mars

More than 50 exploratory missions have been sent to Mars. Some of the probes flew by the planet and took pictures; others orbited the planet and studied it for long periods. Some dropped to the surface and took their places as stationary data-gathering sentinels. The rovers which have traversed the Martian surface have shown themselves to be the most impressive and capable of humanity's robotic planetary explorers. These versatile vehicles, miniature dune buggies studded with prehensile arms, tools, instruments, and sensors, have revolutionized our understanding of Mars.

A total of four American wheeled rovers have operated on the planet's surface. These robotic heralds of humanity methodically rambled over the Martian terrain searching for signs of water, microbial life, and anything earthbound scientists believed might be of importance. The rover named Opportunity, a golf-cart-sized rolling science laboratory, continued to function well past its thirteenth birthday, making it older than the average car on Earth, before it went silent during a planet-wide dust storm in 2018. The rover called Spirit quit communicating with Earth shortly after becoming stuck in soft sand. The first rover, Sojourner, is shown at right just after rolling down a ramp from the Pathfinder probe. Sojourner, whose name means traveler, was the size of a microwave oven and roved about 100 meters over its three-month lifetime. Rovers have quite literally come a long way since then.

Sojourner

Curiosity (below) is one of NASA's most successful Mars rovers. Curiosity is a billion-dollar, nuclear-powered, mobile, insect-like vehicle with a weight on Mars of over 700 pounds, and is about the size of a large car. It has six appendages attached to its lower body that end in wheels for rolling locomotion. Its mechanical arm is capable of drilling into the crust of Mars and scooping up material to be chemically analyzed in a sophisticated onboard laboratory. The rover's upper body bears antennae for receiving and sending data, three camera systems that act as eyes, and two computers for brains. A radioisotope thermoelectric generator provides dependable power no matter the season. This device generates heat from the decay of radioactive plutonium dioxide. Some of the resulting heat is used to warm the rover, and the rest is converted into about 125 watts of electricity to power the probe's systems. The nuclear generator is more reliable than solar panels since solar panels on Mars inevitably get covered with dust, eventually rendering them useless. Unused electrical energy from the nuclear generator is stored in two lithium-ion batteries to supply an extra boost of power if the rover ever requires it.

To transmit news of its discoveries to scientists on Earth, the rover uses only ten watts of power. A large radio station on Earth, by comparison, broadcasts using 50,000 watts. Automobile travelers on Earth know how quickly a radio signal can get lost in just a 100-mile trip; Curiosity has to send its signal millions of miles: Depending on where Mars and Earth are in their respective orbits, the signals must travel anywhere from 34 million miles to 249 million miles, which means a time lapse in communication of between 4 and 22 minutes.

To receive Curiosity's faintly beamed messages, scientists employ the largest receiver to which NASA has access. This receiver, the DSS-14, is a 230-foot-diameter dish located in the Mojave Desert near Barstow, California. In addition to Curiosity, the DSS-14 sends instructions and receives data from other rovers and orbiters. Since Mars and Earth rotate at the same speed, Curiosity is often on the wrong side of Mars, and there is no direct line of sight with the Earth. A partial solution has been to bounce signals off one of two Martian orbiters, the MRO or another one known as the Mars Odyssey orbiter.

Curiosity

Curiosity's resemblance to an automobile or a scurrying insect can give a false impression of speed. In fact, Curiosity trundles over the Martian surface at the cautious and deliberate pace of around 7 miles per year, which is about three times as fast as a snail. Part of the reason for Curiosity's dawdling pace is that scientists insist on planning the rover's route well in advance to avoid obstacles and other hazards that might damage or incapacitate it. The path the rover is to follow is meticulously plotted and beamed as a set of instructions once per day to the rover.

Curiosity's primary mission is to detect the presence of water and microbial life. To date, it has found neither. One of Curiosity's most intriguing discoveries so far is a dry, ancient streambed bearing pebbles in conglomerates of clay, and clays should form only in the presence of water. As for the quest for Martian life, none of the Martian rovers have discovered any life on Mars. However, the absence of evidence is not evidence of absence. As of this writing, Curiosity is still gathering and transmitting data, and the future may hold amazing surprises.

Three Rovers in the Desert

The picture at left shows prototypes of Opportunity (left), Sojourner (bottom), and Curiosity (right), on a desert proving ground with two researchers.

The Mars InSight lander (right), which arrived on Mars near the end of 2018, is not a rover. It remains stationary on the Martian surface to carry out a set of thermal, gravitational, and seismological observations. An 800-pound device with round solar panels seven feet across, a powerful lithium-ion battery, and a robot arm, it will investigate Mars's interior by taking readings of seismic events caused by meteorite strikes or Marsquakes. This will generate valuable data about the inner structure of the planet, including its presumably solid core.

Insight Lander

Conclusion

Mars—this striking, airless stone floating 140 million miles from the Sun—is a world of possibilities waiting to become reality. Its arid terrain may soon be traversed by human explorers. Since being envisioned by earliest humans as the home of a bellicose god, rubicund Mars has captivated the imagination of ancient sky-watchers, classical fabulists, and contemporary astronomers. It has swirled in the thoughts of dreamers and doers for millennia and today looms on the horizon of human aspiration as engineers, entrepreneurs, and political leaders set their sights upon it as the next destination for space pioneers, those valorous visionaries yearning to capture a new beginning.

Tantalizingly nearby, and alluringly similar to Earth, Mars holds many wonders already seen and many more waiting to be discovered. There is no way to know what will be revealed by further exploration in addition to the wandering sand dunes, towering dust devils, titanic glaciers, and impressionistic landscapes already known to adorn its ruddy surface, not to mention the tallest volcanoes, the largest chasm, and the broadest expanse of sandy ground ever discovered anywhere in our Solar System.

Mars's contrasts and dichotomies are incongruous yet delightful: surprisingly smooth regions next to expanses of jagged craters, rusty red ground juxtaposed with mosaics of white, the highest volcanoes of the Solar System adjoining the deepest trenches. This is a desiccated world holding enormous blocks of ice, a world that appears calmly benign yet is intensely hostile to life, a world marked by the tracks of robotic probes though not a single earthly visitor has left a footprint upon its soil, a world that beckons humanity to visit but would kill an unprotected human in a matter of a few heartbeats. To pursue our destiny in space, humankind must establish a beachhead on Mars and build a staging area from which to launch forays to the outer Solar System and then an eventual giant leap to the stars. Soon, the shrieking wail of human-crewed vehicles descending for the first time through the thin Martian atmosphere at 20 times the speed of sound may pierce the unpopulated calm of slumbering Mars.

Topographic Map of Mars

The surface of Mars from 70° south latitude to 70° north latitude. Clearly visibly on this false-color topographic map are the broad low elevations (blue) and the huge extinct volcanoes (at left, capped in white). Note the marked difference between the cratered southern hemisphere and the smoother northern hemisphere.

17
Asteroids: Blessing and Danger

We have the capability – physically, technically – to protect the Earth from asteroid impacts. We are now able to very slightly and subtly reshape the solar system in order to enhance human survival.
Rusty Schweickart

This rock shall fly.
Sir Walter Scott

Asteroids are cryptic, enigmatic, obscure chunks of rock and metal careening around and through our Solar System. Dark, small, and difficult to see, they appear through a telescope as tiny points of light like dim stars; hence the name asteroid, which means "star-like object." Billions of these cosmic conundrums dart among the planets. Many are mere grains of dust or pebbles, while others are as large as cars, and the giants among them are mini-worlds hundreds of miles across. A few asteroids are so massive that they are spherical and even possess their own moonlets. The foundational substance of these random rocks was likely the original embryonic material from which the Solar System formed. They arose by the same process and from the same motley array of materials that built the Earth, Jupiter, and the rest of the Solar System's planetary siblings; however, for unknown reasons, the asteroids failed to aggregate to become full-sized planets.

Most of the asteroids in the Solar System, billions of them, reside in a 150-million-mile-wide swath of space between the orbits of Mars and Jupiter. Some asteroids are rich in carbon while others are stony or metallic, but regardless of the type, asteroids almost always contain some water. There are gaps in the Asteroid Belt, called "Kirkwood gaps," that are caused by Jupiter's gravitational influence. What we call the Asteroid Belt, then, is really several distinct belts, similar in some ways to Saturn's rings.

The intriguing Trojan and Greek asteroids occupy the same orbit as Jupiter. They occur in two groupings, or swarms, one of which (the Greeks) flies ahead of Jupiter while the other (the Trojans) trails behind. Within each swarm, the asteroids oscillate around their Lagrangian point. The Lagrangian points are the locations where the gravity of the Sun balances the gravity of Jupiter. Astronomers estimate that the Trojan and Greek asteroids are as numerous as all the asteroids in the Asteroid Belt.

The Hildas are a group of asteroids numbering about 4,000 that orbit beyond the Asteroid Belt but inside the orbit of Jupiter. They orbit the Sun three times for every two orbits of Jupiter, and

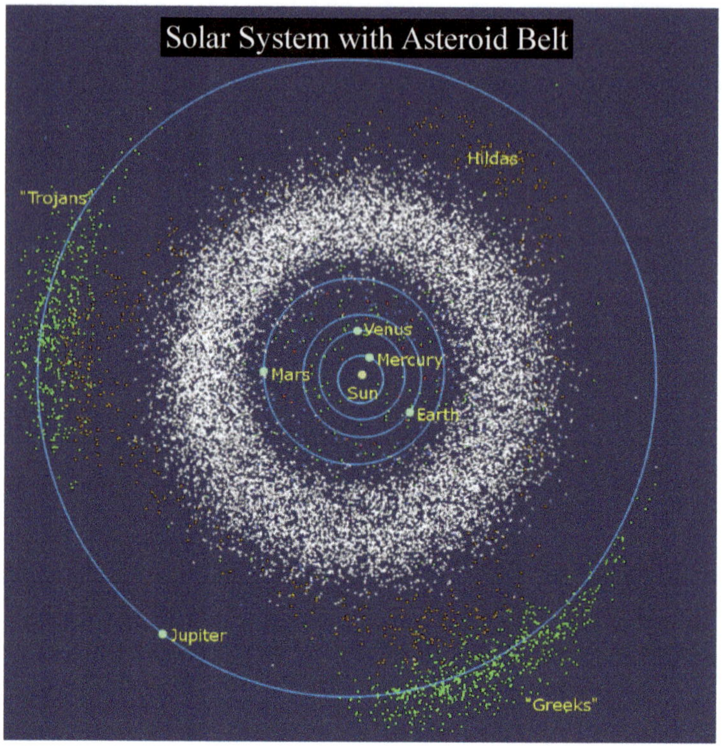

this resonance allows them to maintain their anomalous position outside the main belt. The main asteroid in this group, Hilda, was named by astronomer Theodor von Oppolzer after one of his daughters.

As we continue to amass evidence about asteroids and comets, it is becoming apparent that asteroids and comets are not very different from each other. They are closely related cousins, maybe even siblings. A comet puts on a more dramatic display than an asteroid, but this may be due merely to differences in the object's orbit or history. A comet approaches the Sun closely enough that the Sun's heat boils off some of its water and volatile gases, creating a tail, while an asteroid either stays in an orbit far from the Sun so that sublimation cannot occur, or else all of its volatiles have long since boiled away. Comets and asteroids, in this view, are not different at all. They are the same type of space object, just at two different stages in their life span. The most compelling evidence for this hypothesis is the asteroid Chiron. Chiron follows a 51-year orbit beyond Saturn, and sometime after its discovery it was observed sprouting a tail like a comet.

Origin

Four and a half billion years ago: The Solar System is an immense, swirling, cosmic whirlpool consisting of a tremendous cloud of microscopic dust grains and gas particles. As time passes, the grains of dust begin to stick together by electrostatic attraction and create small clumps and strands that over time grow into grit and pebbles. Eventually, the pebbles coalesce to form large rocks through a process called accretion. These rocks are nowhere near the size of planets; trillions of them whirl in elliptical orbits around the newly born Sun. Some of these rocks gather to form floating piles of rubble. In all different sizes and shapes, they continue to aggregate and spin, developing into planetesimals. Planetesimals are the tiny precursors of the planets, the building blocks of the planet on which we live and all the other planets that will so picturesquely populate our Solar System.

Scientists have been able to examine some of the actual ancient material that formed these planetesimals. The Allende meteorite, which struck Peru in 1969, was discovered to consist of material older than any rock ever found on Earth. It is a carbonaceous chondrite containing light-colored inclusions that are thought to be the earliest-formed solid material in the Solar System. These inclusions are dated at 4.57 billion years old, and are probably older than the Earth itself.

This meteorite with its primordial minerals must hold many secrets just waiting to be revealed, and scientists continue to study it.

To form a planet requires billions of small asteroids to clump together. However, asteroids race at thousands of miles an hour, and when they collide, they do not always stick together. The asteroids in the early Solar System are like racecars speeding around an oval racetrack. Eventually, asteroids smash into each other. However, to join, they must collide and stick. When they do, we have proto-planets in the making. Asteroids gather like cars in a multi-car pileup and lock together. Mixed in among the stony and metallic constituents of an asteroid is water which acts like a super adhesive. The water was there in the primordial dust cloud with the other gases and particles. When asteroids collide, they generate heat that melts the frozen water. This temporarily liquid water flows into the cracks of the newly created rock and refreezes, thus cementing the smaller pieces together to form a larger asteroid.

As more and more asteroids stick together, the rock piles grow. If this were the only process, then planets would require billions of years to form. Yet, planets emerge in just a few million years. How? Gravity becomes the dominant factor when the rubble piles are several miles across, and this dramatically speeds up the process. The early Solar System is full of these growing asteroids. The larger an asteroid becomes, the more rock it pulls in, and the faster it grows. As soon as it has enough mass, the asteroid will crush itself into a roughly spherical shape under the force of its own gravity. It attracts even more rocks to itself until there are no more left in the vicinity. Then the asteroid is no longer an asteroid; once it has cleared its orbit, it is a planet. Mercury, Venus, Earth, and Mars are all former asteroids that kept on accumulating additional material and increasing in size until they became what they are today: full-sized rocky planets.

However, out beyond the orbit of Mars, something goes awry. An immense cosmological force disrupts the planet-building process. There, in the Asteroid Belt, the process comes to a halt before any rocky planet can form. According to present models, this happens because of the influence of Jupiter. Four and half billion years ago, Jupiter already gravitationally dominates the space around it, and its immense gravity is a source of chaos, altering asteroids' courses and flinging them willy-nilly in every direction. Giant boulders, tossed onto unpredictable paths, no longer follow each other in the same direction. The inevitable result is a large number of violent collisions as speeding missiles of rock smash into each other from every angle.

A collision, according to physicists, can be either elastic or inelastic. In an elastic collision, the colliding objects bounce apart. In an inelastic collision, the colliding objects stick together. Naturally, in any collision in the real world, a little of both happens, as with cars that collide inelastically and travel together a distance from the point of impact, yet parts of them are ejected and can be found a good distance away. This is how rocks from Mars have sometimes found their way to Earth. A partly elastic collision occurs when an object strikes Mars and part of it (or part of Mars) is flung into space and eventually hits Earth. This might seem like a longshot, but, according to experts, over 132 meteorites that originated on Mars have been found on Earth, including the famous Nakhla meteor, which landed in Egypt in 1911. Scientists are confident that these meteorites originated on Mars because the isotopes and elements in the meteorites match rocks analyzed on Mars by the Curiosity rover, even up to the atmospheric gas molecules trapped in the meteorite material.

The rocks in the early Asteroid Belt collide and shatter, and these violent collisions are still rocking the Asteroid Belt today. As recently as 2010, scientists were able to examine images captured by the Hubble Space Telescope of the aftermath of a high-speed asteroid collision. Scientists believe that such collisions are the reason there are billions of pebbles, rocks, and boulders in a location in the Solar System where a planet could have formed. When asteroids slowly accumulate material, a planet gradually takes shape, but haphazard, violent collisions result in an asteroid belt.

Boon and Bane

Asteroids can be viewed as either a boon or a bane. They hold within themselves the clues to how the Earth was formed, but they also have the potential to be killers on a terrifying scale. They represent a rich repository of valuable resources, but is humankind prepared to meet the threat of Armageddon from above? Asteroids may even serve one day as steppingstones to the stars, but if a large one were to strike the Earth, it would signal the end of a cosmic epoch and the demise of our civilization. Clearly, asteroids are a double-edged sword, simultaneously good and bad. The ancient Greeks, with their gift for exquisitely clear language, had a term for referring to this type of situation: agathokakological. The agathokakological nature of asteroids is fascinating.

Boon

Asteroids have been a beneficial and almost miraculous presence in our Solar System. As noted above, the Earth owes its very existence to asteroids, since they were the building blocks of its construction. One of the chief benefits from asteroids, then, is that they are ancient repositories of knowledge, pristine pieces of the past, and may one day allow humans to decipher the mysteries of the origin of the Solar System. Another, and presumably more vital, benefit from asteroids is that they brought to Earth its supply of life-giving water. Also, as humans reach out to achieve their destiny of space exploration, asteroids might even serve as terminals, fuel depots, or even the very starships that carry the seeds of humanity to the nearest exoplanet or the farthest reaches of our galaxy and beyond.

Asteroids could prove invaluable as sources of metals. Their valuable mineral resources are just ripe for the taking: Asteroids could be literally mines in the sky. They are not just gold mines, but copper, nickel, zinc, aluminum, platinum, and iridium mines. One asteroid of average size could contain minerals worth trillions of dollars. Ironically, the easiest asteroids to reach are the same ones that threaten the Earth. An asteroid closely approaching Earth could be easily reached by a spaceship that could dispatch personnel or equipment to extract some of the asteroid's mineral wealth. Yet, even when asteroids come so close, the idea of mining them may be too good to be true. The reason we are not mining the Asteroid Belt today and letting the space program pay for itself is cost. It would take an enormous amount of rocket fuel and expertise to bring back even a part of an asteroid from outer space. But perhaps we do not need to bring the minerals back to Earth. Perhaps we could utilize the resources of asteroids to build settlements out in space. We do not have to build a space station—it is already there. There would be no need for shielding from deadly radiation because rock and ice for shielding against radiation and micrometeorite strikes would be readily at hand. Low gravity makes it easy to come and go.

Space-Wheel Habitat

There is plenty of water, and the asteroid contains all the construction materials the colonists could ever need.

The largest asteroid in the Asteroid Belt is Ceres. When Ceres was discovered in 1801, it was considered a planet. It is 590 miles across, or about as big as Texas. Since Ceres has been observed to spout plumes of water into space, scientists are confident that it holds a vast reservoir of water. Ceres might consist of as much as 25% water, which would be more than all the fresh water on Earth. This water could be the key to deep space exploration because it could be used for drinking and for rocket fuel. Today NASA is building the Orion rocket, the most powerful rocket yet developed. Its fuel is liquid hydrogen and liquid oxygen, and its byproduct is water. Simple electrolysis of water produces hydrogen and oxygen gases, which could then be readily condensed to liquid form and stored easily and inexpensively at asteroid temperatures. For hydrogen and oxygen fuel to power rockets throughout and beyond the Solar System, Ceres and its sister asteroids could serve as refueling depots.

Someday there may be human colonies and cities in the Asteroid Belt, like the artist's conception above. Asteroids contain large quantities of readily available minerals, metals, and water, which could provide the raw materials for a manufacturing zone that could not exist on Earth. Factories, mines, and chemical processing plants on the asteroids themselves would be able to create everything needed for a comfortable life a half billion miles from the Sun. The fabrication of mini spaceships for local travel and gargantuan galleons for making the circuit around the belt would form the basis for an emerging economy that would anchor humanity for its next foray into space. Thus asteroids, which may have brought the material for life to early Earth, could serve as platforms from which the human race will eventually launch itself to the outer planets and beyond, far into interstellar space.

Bane

Asteroids, the original building blocks of Earth, are rightly called the bringers of life. But someday (and it may be soon), an asteroid could become the ultimate bringer of death. Death, the most dreaded of the Four Horsemen of the Apocalypse, will come for humanity if we do not take these ominous rocks from space seriously. Otherwise, one will indubitably strike the Earth and leave nothing but detritus in its wake. Huge rocks hurtling at speeds over 50,000 miles per hour have collided with Earth before, leaving scars, gashes, and craters in the Earth's crust. The air and water flowing over Earth's surface gradually smooth out almost any wound or deformation, but all one has to do is view the Moon's 80,000 craters to recognize the devastation that similar missiles from space have scattered across Earth's landscape. Within recent memory, Earth has been walloped with meteor strikes near the city of Chelyabinsk in Russia, and Nakhla in Egypt. A crater three-quarters of a mile wide, formed by a meteor impact tens of thousands of years ago, still stands out in contrast to the surrounding Arizona flatland. We have almost absolute proof that a planet-buster the size of Mount Everest struck the Yucatán peninsula 65 million years ago, and the result of this moderately large asteroid's impact was that half the life on Earth was wiped out, including the dinosaurs. And yet all these various impactors pale in comparison to some of the gargantuan rocks hurtling through space at this very moment. To be struck by a rock the size of a small moon would fracture the Earth, shatter the continents, and cause innumerable earthquakes and volcanic eruptions around the globe. The devastation would be complete. The Earth would be transformed into a molten sphere of lava surrounded by an all-encompassing cloud of incandescent, fiery meteoroids extending to the orbit of the Moon and beyond. The resulting rain of fire would last for centuries as the Earth, now devoid of life, slowly remolded itself. This destruction of Earth and humanity could be described succinctly as "Armageddon from above."

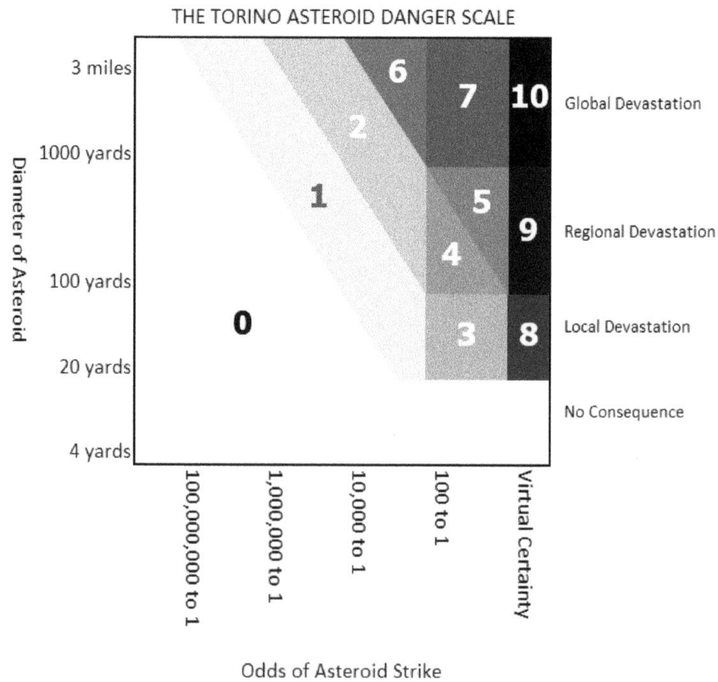

The danger posed by an asteroid is measured on a scale called the Torino scale. On this scale, 10 is the highest, representing certainty of global destruction, while 0 is the lowest, representing almost no chance of impact or no damage if impact occurs. See chart on previous page. For example, the asteroid Apophis is a 4 on the Torino scale. It is 80 feet in diameter, and if it hits Earth, its impact will be equivalent to a 500-megaton bomb. It will come menacingly close to Earth in 2029.

K-Pg (Chicxulub)

A massive asteroid strike over 65 million years ago in the Yucatán peninsula was so violent that it created in effect a nuclear winter and brought an end to the long-running reign of the dinosaurs. In addition to the initial rain of fire, there were huge shock waves and tsunamis, along with volcanic eruptions and pyroclastic flows (especially along fault lines). An immense quantity of dust and ash was blasted into the upper atmosphere where it blocked sunlight from reaching Earth for years. Many plants died, and with them the animals that depended on them for food, and the ecological disaster propagated all the way up the food chain, culminating in the extinction of even the mighty apex predators.

The extinction of the dinosaurs occurred at what paleontologists today refer to as the K-Pg boundary. (It was formerly called the K-T boundary.) There is surprising geological evidence for this devastating asteroid strike. All around the world, a thin rock layer dated to the time of the extinction event contains the rare element iridium in comparatively huge amounts. This iridium, brought by the doomsday asteroid, was spread all around the world as dust. In addition, the resulting rain of fiery remnants of the Chicxulub object resulted in a global firestorm that sent soot high into the atmosphere. The iridium-enriched soot and dust formed a gigantic high-altitude cloud which in the following years slowly settled down to Earth, resulting in a layer that covered the entire surface. The rock layers below this boundary contain dinosaur fossils, and the layers above it do not. This layer is seen around the world, the clear metallic fingerprint of an asteroid strike. In the image above, from the Raton basin in Colorado, the K-Pg boundary is visible as an abrupt change from dark-colored to light-colored rock.

K-Pg Boundary

In addition to the abnormally high levels of iridium, the K-Pg boundary also contains glass beads the thickness of a dime. These are remarkably similar to the glass nodules found at Trinity, the site of the first atomic bomb explosion. The layer also contains "shocked quartz," a type of quartz that has had its crystalline structure violently ripped apart. It is only found in places where there has been a massive release of energy like an atomic bomb blast.

From the estimated amount of iridium in the K-Pg boundary, scientists were able to calculate the size of the asteroid. They deduced that the killer asteroid was six miles wide and traveling at 70,000 miles per hour when it entered Earth's atmosphere at a shallow angle. They calculated that at the point of impact it released the energy of a billion Hiroshima bombs, creating heat that instantly vaporized the water and melted rock leaving a crater two miles deep and 112 miles across—twice the size of the island of Hawaii. But the scientists had one problem—where was the crater? A crater that size should be clearly visible, even after millions of years. Then in 1980, geologists mapping the ocean floor looking for places to drill for oil found something spectacular: a huge formation of concentric circles near the northern coast of the Yucatán peninsula in Mexico. This spectacular underwater formation proved to be the missing crater caused by the asteroid that killed the dinosaurs. Evidence from as far away as Texas, where tell-tale contemporaneous deposits of gravel and rock were found, gave strong support to the hypothesis. The crater was given the name Chicxulub (CHEEK-shu-loob) after a town in the vicinity. The name means "Tail of the Devil" in the Mayan language.

Arizona Meteor Crater

Near Winslow, Arizona, there is an impressive, well-preserved impact crater. Though much smaller than the Chicxulub crater, it may be even more worrisome because it was caused by a much more recent asteroid strike.

A mere 50,000 years ago, a meteor slammed into the Colorado Plateau in what would later become Arizona and created a crater 4,000 feet wide and 600 feet deep. The crater is officially known as Meteor Crater, and is occasionally called Barringer Meteorite Crater after a mining engineer who was the first to suggest, in 1903, that the crater was the result of the impact of an iron-rich meteorite.

Imagine you are present on that day fifty millennia ago as an immense fireball flashes through the sky at 30,000 miles per hour. In a matter of seconds, it crosses the entire sky and smashes into the ground. Heat from the fireball scorches the surface, but most of the damage happens after the initial impact. Once it strikes the ground, it only takes a few seconds to create the crater. The fireball delivers an enormous amount of energy, blasting a hole in the terrain and hurling molten material high into the sky. The asteroid is obliterated instantly, creating a shockwave. The force of the blast throws boulders the size of cars several miles, hurling debris that would crush buildings and homes if it hit today.

It does not take much imagination to grasp that if such an event were to occur today in the vicinity of a population center such as New York City or Chittagong, all the inhabitants would be killed, most of them vaporized instantaneously. And that asteroid was relatively small—a mere 150 feet across, much smaller than the Chicxulub asteroid which brought on the worldwide devastation described above. A larger asteroid, around half a mile wide, would be sufficiently destructive to end human civilization, partly through direct effects and partly by creating a nuclear winter as a result of billions of tons of dust thrown into the atmosphere. The life-sustaining light and warmth of the Sun would be blocked out, perhaps for years or decades, making it impossible to grow food until the dust settled.

Millions of such asteroids are zipping around in the vicinity of Earth. An impact by any one of them would cause worldwide carnage and hurl us back to the Stone Age. The largest asteroids have cores of iron. They are, in essence, iron cannonballs surrounded by rock and ice hundreds of miles wide. One of these galactic missiles could cause our extinction, and, if large enough, could shatter the planet into a pile of rubble. Asteroid experts estimate that an asteroid one-fifteenth the size of Earth would smash it completely to bits.

Chelyabinsk

On February 15, 2013, at 9:20:21 AM in the south Siberian metropolis of Chelyabinsk, a bright object streaked unexpectedly above the city, leaving a blazing, smoky trail across the sky and baffling the surprised inhabitants. People up to 60 miles away saw a flash as bright as the Sun, and within minutes of the blinding flash, a powerful shockwave struck the city. The shockwave broke thousands of windows, sent shards of glass flying, collapsed a factory roof, blew doors open, and flung people, furniture, and loose items across entire rooms. Residents reported a smell of gunpowder in the air.

Chelyabinsk Meteor

Chelyabinsk is a city of over one million people, so there were many witnesses. Also, due to a peculiarity of the Russian automotive insurance system, many vehicles there are equipped with dashboard cameras. Almost immediately, amateur photographers and possessors of dashboard camera footage began to post their pictures on the internet. Most of them had no idea what they had seen. Scientists around the world rushed to piece together the story of what had happened. Never before had scientists had so much data, so many pictures, and so many first-hand accounts of an extraordinary event of this magnitude.

Scientists quickly arrived at the conclusion that the bright streak in the sky marked the passing of a meteor. They converged on the rural zone five miles away from the city where fragments of the meteor had rained down. The pieces were not hard to find in the snow-covered fields: Each fragment, most about the size of a dime, left a clear hole in the snow. Two small chunks of the meteorite are shown at right. The largest single piece, weighing around 1450 pounds, punched a 20-foot-wide hole in ice two feet thick on Lake Chebarkul and was recovered from the lakebed in October, 2013.

Chunk of Chelyabinsk Meteor

Chunk of Chelyabinsk Meteor

From the evidence they gathered, scientists were able to deduce that a meteor weighing about 13,000 tons had flown in at a low angle before it exploded into shards 18 miles above the Earth's surface. The velocity of the meteor was over 40,000 miles per hour (60 times the speed of sound), and the kinetic energy of the airburst was equal to about 30 fission bombs of the type dropped on Hiroshima. The event registered on seismographs at 2.7 and was detected by 20 infrasound monitoring stations designed to detect nuclear weapons. In fact, it was detected twice. The very low-frequency sound waves generated by the blast were so strong that they circled the entire globe and then kept going until they were picked up by the same detectors a second time.

Over 1500 people were injured enough to seek medical treatment. More than 7200 buildings were damaged in six cities across the region. Nonetheless, Chelyabinsk was phenomenally lucky. If the meteor had struck just a few miles closer to the city, the destruction would have been virtually complete. It would have been almost exactly equivalent to a hydrogen bomb attack. Tens or hundreds of thousands of people could have been killed.

The Chelyabinsk asteroid was a messenger of death in the form of a missile no one expected. When it rocketed through the atmosphere and exploded into a million shards of metallic rock, it shattered the peaceful solitude of Chelyabinsk with a blast that was heard around the world—twice. Only 60 feet wide, it was of a size that NASA considers too small to look for, yet its burst produced 450 kilotons of energy. If the blast had been directly over Chelyabinsk, all of its million inhabitants could have been vaporized. Chelyabinsk was less than half a second from the center of the explosion. With the minutest of differences in path or trajectory, the planet could have gone from 1500 people seeking medical help to one million souls vanishing from the face of the Earth. In terms of human life, the Chelyabinsk event was, luckily, a non-event, but it could have been far worse, a veritable sounding of Gabriel's horn. If this was not a wake-up call, then nothing is.

Amazingly, the Chelyabinsk event occurred the same day as another asteroid made an uncomfortably close pass by the Earth. NASA had been tracking this asteroid, called 2012 DA 14, for a year and estimated that at its closest point it would be about 18,000 miles from Earth. This is well inside the orbit of the Moon. In fact, it is inside the orbits of most communication satellites. Naturally, scientists wondered if the two asteroids arriving so close together were somehow related. However, the Chelyabinsk asteroid came from the East during the daytime, while DA14 came from the south during the nighttime. Thus, their orbits were very different, and since they approached the Earth from two different directions, they could not possibly be related. Two celestial hammers aimed at Earth on the same day turned out to be pure coincidence. But given the huge number of asteroids on orbits that bring them near Earth, this will not be the last time it happens.

Hoba Meteorite

Defense

Earth's atmosphere provides reliable protection from the thousands of tons of pebble-sized and smaller asteroids that have assaulted the Earth nightly for millennia. These "small fry" flash through our air, burning out brilliantly in the night sky, and are called "shooting stars." However, boulder-sized asteroids and larger pass through the atmosphere in mere seconds, striking the Earth's surface with the force of nuclear bombs. The Hoba meteorite, a 60-ton titan of iron and nickel (previous page), struck in what is now Namibia 80,000 years ago. There have been millions of impacts like this in Earth's past, and there will be more.

Fortunately, we have sophisticated telescopes that may give us advance notice of impending impacts. However, space is vast, asteroids are dark and much smaller than stars or planets, and finding all of them is exceedingly difficult. Once an asteroid is noticed, though, it is easy to determine if it is on a collision course with Earth. Scientists use giant radio telescopes like the Arecibo Observatory in Puerto Rico to track asteroids. Arecibo has one of the largest telescopes in the world, with a radio receiver that is 1,000 feet across. It and telescopes like it are vital for defense against an asteroid onslaught, since "to be forewarned is to be forearmed." Arecibo has more than enough resolving power to track near-Earth-objects (NEOs). An asteroid can be considered a NEO even though it is millions of miles away, because the issue is how close it will come, not how far away it is at the moment. Unlike a visible-light telescope, radar can directly measure an object's distance and reveal exactly where it is.

The asteroid Apophis, mentioned above and named after the enemy of the Egyptian sun-god Ra, will pass near Earth on April 13, 2029, which happens to be Friday the thirteenth. Apophis is large enough to be a continent-crusher. Luckily, with the help of the Arecibo telescope, scientists have determined that Apophis will not strike the Earth in 2029, but the margin for error is razor thin by astronomical standards. Apophis will pass within a mere 20,000 miles of Earth.

It is only a matter of time before we find an asteroid on a collision course with Earth. When that day of dread comes, what can we possibly do to save ourselves and our beloved planet? The most challenging part of dealing with an asteroid is not to destroy it but to remove the threat it poses. While a nuclear explosion could break a giant rock into thousands of pieces, the blast would have little effect on the giant rock's center of mass, which would still be on a collision course with Earth. In essence, instead of one giant bomb hitting Earth, we would have created thousands of little missiles headed for Earth. Thus, exploding an asteroid can do more harm than good. Better to make it miss completely.

When an asteroid is heading for Earth, it means that the Earth's and the asteroid's paths cross and that they both will reach that crossing point at the same time. Imagine the Earth as a passenger train, and the asteroid as a semi-truck, both headed for a railroad crossing. If they reach that crossing at the same time, they will collide. One or the other needs to slow down, and the Earth cannot slow down. What is needed is a way to slow an asteroid down—or speed it up.

Gravity can be used to change an asteroid's speed. The simple expedient of stationing a large object like a spacecraft near an asteroid could be enough to gravitationally nudge it out of its path. This is what is known as a "gravity tractor." Care must be taken that the tractor does not collide with the asteroid—an occasional blast from rocket thrusters should do the trick.

Another way to protect the Earth from asteroids is being explored in Scotland. Scientists there are working on lasers to heat the surface of a threatening asteroid. With enough laser light, spots on the surface of the asteroid could be heated to 5,000 degrees Fahrenheit, and the resulting mini-explosion would create a new force that could alter the trajectory of the asteroid just enough to make it miss the Earth. If this concept is shown to work, then in an asteroid emergency an armada of laser-equipped spacecraft would take off from Earth and focus their beams on the intruder at point-blank range, causing it to veer away or at least slow it down enough to save humanity.

Finally, a mission planned for launch in 2020 will test the possibility of ramming an asteroid with a "kinetic impactor." The DART (Double Asteroid Redirection Test) spacecraft is equipped with roll-out solar panels and a cutting-edge ion-thrust engine. Traveling at high speed, it will journey outward from Earth on a spiral trajectory and attempt to strike the moonlet of the asteroid Didymus in 2022. The resulting crash will transfer momentum from the craft to the target, changing the asteroid's direction of travel. The moonlet, dubbed "Didymoon," is the size of the Great Pyramid of Khufu, which is large enough that it would pose a serious threat to Earth. If the test is a success, the kinetic impactor could become the preferred anti-asteroid technology.

For any of these interdiction methods to be effective, time is needed. Months or years may be required to move a massive object out of its deadly trajectory. Congress has directed NASA to track NEOs that are large enough to pose serious danger to human life, in hopes that when an asteroid strike becomes likely, we will have sufficient advance warning to take effective action.

Exploration

There are billions of asteroids, each with a story to tell, yet we have barely explored any of them. The first mission committed to the exploration of the Asteroid Belt was the Dawn probe. It was launched in September of 2007 with the purpose of studying two of the largest asteroids in the Asteroid Belt: Vesta and Ceres. Dawn was the first spacecraft to visit either of these two asteroids (and the first to orbit a dwarf planet), arriving at Ceres in March of 2015. The Dawn probe is an engineering marvel. With its wing-like solar panels fully extended, it resembles a giant bird soaring through space at speeds approaching 5,000 miles per hour.

In July 2011, Dawn arrived at Vesta and sent back numerous photographs. Named after the Roman goddess of the hearth, Vesta is the brightest asteroid of the Asteroid Belt and was the fourth such object to be discovered. It was discovered by the German astronomer and physician Heinrich Wilhelm Matthäus Olbers on March 29, 1807. One of the common misconceptions about the Asteroid Belt is that objects in it are small. Many asteroids are indeed small, but some of the largest ones are several hundred miles across and are correctly labeled minor planets. They are not considered full-fledged planets because they have not cleared their orbits of other objects, so even though Vesta has a mountain three times higher than Everest and is as wide as Arizona, it cannot be considered a planet. After leaving Vesta, Dawn proceeded toward the Asteroid Belt's heftiest denizen, icy Ceres.

Ceres

Ceres (right) was found, serendipitously, by the

Italian astronomer Giuseppe Piazzi of the Palermo Observatory on January 1, 1801. Additional observations of the object by Piazzi were cut short by illness, but Ceres was rediscovered on January 1, 1802, by the German-Hungarian astronomer Franz Von Zach, using an orbit calculated by the German mathematician Carl Friedrich Gauss. Ceres was named after the ancient Roman grain goddess and patron goddess of Sicily, thus beginning a long-standing tradition of naming main-belt asteroids after female personages from Greco-Roman mythology. Ceres, when first found, was believed to be the "missing" planet between Mars and Jupiter. A German mathematician of the eighteenth century found that the distances of the then-known planets from the Sun fell in the proportions 4, 7, 10, 16, 28, 52, 100, and 196. These numbers (after the first in the sequence) can be calculated using the expression $3 \times 2^n + 4$. Although Ceres—and the next two asteroids discovered, Pallas and Juno—are located near the distance predicted by Bode's Law for the "missing" planet between Mars and Jupiter, most asteroids and planets found subsequently are not so located, so the agreement with this "law" appears to be coincidental.

Ceres contains one-third of the mass of the entire Asteroid Belt. Almost twice as big as the next largest, Vesta, Ceres is very reflective, which leads scientists to believe that it is ice-rich. Its outer layer may consist of almost pure ice, which remains perpetually frozen so far from the Sun. If Ceres really has as much frozen water as it appears to, this could help explain why the Earth has so much water. It is believed that shortly after its birth, the young Earth had no water because it was so hot. Then, some four billion years ago, an event occurred which has been dubbed the Late Heavy Bombardment. This was a period during which millions of asteroids bombarded the Earth. At this time massive icy bodies like Ceres could have brought immense amounts of precious water to Earth. It is possible that every drop of water in Earth's seas, lakes, ponds, and streams came from asteroid sources.

OSIRIS REx and Bennu

Our knowledge of asteroids is about to take a major step forward due to a sample-return mission to the asteroid Bennu launched in September 2016. The mission is known as OSIRIS-REx (Origins, Spectral Interpretation, Resource, Identification, Security, Regolith Explorer) which is an acronym of a list of the mission's science objectives. If all goes as planned, it will be the first mission to bring back to Earth a piece of an asteroid dating from the dawn of the Solar System. The asteroid Bennu was chosen as the target of the mission because it is a "primitive" asteroid, a carbon-rich body resembling a flying rubble pile, with a composition that is not much altered from four billion years ago. There is some evidence that Bennu once contained water, and investigators are eager to see whether a sample from Bennu contains organic compounds, such as amino acids, that were the precursors of life on Earth. These compounds are delicate. In the past, our knowledge of bodies like Bennu came from meteorites that had struck Earth. However, these meteorites had invariably been fried by intense heat from entering the Earth's atmosphere. The samples brought back by OSIRIS-REx will be pristine and unaltered.

In addition to being the right type of asteroid for exploration, Bennu is near the Earth. Although it probably originated in the Asteroid Belt, it has been migrating inward for perhaps a billion years. Its orbit is now similar to Earth's, and it approaches Earth very closely every six years.

Moreover, Bennu is very large. At around a quarter mile in diameter, it is big enough that OSIRIS-REx will be able to safely approach it and collect a sample, probably from near the asteroid's equator.

When OSIRIS-REx arrived at Bennu in 2018, it began a period of observing and mapping the asteroid's surface. It will eventually move slowly toward the asteroid, deorbiting and aligning itself with the sampling site. The attempt to retrieve a sample will take place in 2020. If unsuccessful, OSIRIS-Rex has enough fuel to make two more attempts. Once a sample is obtained, a specialized return capsule will blast off and return it to Earth. Researchers are hoping for at least two ounces, but the probe could potentially grab up to four pounds. In addition to the sampler arm, OSIRIS-REx is equipped with an 8-inch telescope, a specialized mapping camera for imaging the sample site, and a camera for recording the sample-acquisition procedure. A laser altimeter will continually determine the probe's distance from the asteroid. Visual, infrared, and x-ray spectrometers will provide mineral and organic spectral maps, helping to locate the most promising sample sites.

The return capsule will take two years for its voyage back to Earth. It will enter Earth's atmosphere traveling at more than 25,000 miles per hour and will release the collected asteroid material in a pod that will parachute down to the Utah desert for retrieval. Meanwhile, the main OSIRIS REx spacecraft will place itself, with a final burn of its thrusters, into a safe "graveyard orbit" around the Sun.

The name of the probe and the name of the asteroid complement each other. The Egyptian Bennu is a mythical bird with the appearance of a heron. It corresponds to the mythical Greek phoenix, and like the phoenix, it was said to periodically burst into flames, from the ashes of which a newborn Bennu would arise. In Egyptian funerary art, Bennu was sometimes shown wearing the crown of Osiris, who was the god of rebirth, among other things. It is to be hoped that Bennu's association with fiery destruction does not turn out to be a dire prophecy. The asteroid Bennu's frequent and very near approach to Earth means that it is relatively likely to strike the Earth, eventually. Scientists anticipate that in 2135, Bennu will pass nearer to Earth than the Moon. Passing that close to the Earth will subject Bennu to the Earth's gravity, causing its trajectory to change. If at that time Bennu passes through one of several mile-sized regions in the space near Earth, dubbed "keyholes," then scientists will know with near-certainty that Bennu will impact Earth during a later orbit. However, it is impossible to be certain about Bennu's future course. One astrophysical phenomenon that complicates the question is the Yarkovsky effect. In this effect, sunlight warms one face of a small, spinning asteroid more than the other, and the reradiated heat from the warmer side creates a thrust that alters the asteroid's trajectory. It also changes the asteroid's spin rate and orientation. Since 1999, Bennu's orbit has shifted by about 100 miles, and its rate of rotation has increased. In short, its orbit and movements are not perfectly predictable.

Given the large size of the asteroid, the survival of humanity hangs in the balance. If Bennu were to strike the Earth, it would release more energy than all the nuclear bombs in the United States' arsenal. As noted above, numerous researchers are already working on ways to safely deflect any large asteroid that threatens the Earth, but their ideas and plans are speculative and almost impossible to test. Let us fervently hope that the only thing the Earth ever sees of Bennu and asteroids like it is two ounces to four pounds of sampled dirt and rocks.

Conclusion

Asteroids, envoys from the distant past, remind us that life on our planet is fragile and that we are intrinsically connected with the cosmos. Although at present these sphinxlike rocks menace us with destruction, one day they could deliver the clues we need to solve the deepest mysteries of the origins of the Solar System and the human species. Furthermore, these inscrutable boulders could one day serve us as way-stations to the stars or even vehicles to transport us to new domains. Asteroids are a wonder and a treasure, a boon and a bane, and could provide the key to humanity's future. These heralds from the remote and cold reaches of space speak to us. They bestow upon us by their mere presence not only the inclination but also the responsibility to act to preserve our future. That future depends upon our ability to act in unison as a species. Our increasing understanding of asteroids and growing power over them is evidence of the first step in humankind's transition toward becoming a people that is at home among the stars—and rocks—of space.

Asteroid Streaking Toward Earth

18
Comets: Icy Remnants of the Embryonic Solar System

For most of the history of our species we were helpless to understand how nature works. We took every storm, drought, illness and comet personally.
Ann Druyan

I conclude, therefore, that this star is not some kind of comet or a fiery meteor... but that it is a star shining in the firmament itself, one that has never previously been seen before our time, in any age since the beginning of the world.
Tycho Brahe

Ancient humans saw messages in the night sky. The constellations seemed to act out stories, familiar stories that everyone loved to hear told around the home fires. It was clear that the messages in the night sky were for their edification and survival. Moreover, when the divine order was violated by an apparition of a comet in the sky, they took it personally. Can we fault them? It must be human nature to interpret an unexpected event as a bad omen because throughout history, cultures of every time and on every continent have unfailingly associated comets with doom and disaster. The human brain recognizes patterns even when they are not there, a phenomenon called false pattern recognition. We continually search for significance and meaning. To people who saw patterns and meaning everywhere, a comet was a divine warning, and once received could not be ignored. Disruption of the otherwise orderly heavenly pattern was obviously a bad sign.

Comets are conspicuous by virtue of their unusual tails. The word "comet" comes from a Latin word that means "long-haired one." In other times and places, comets were given names like "broom star" or "star with long feathers," but regardless of the name, they were always considered omens of evil or danger. Comets, then, were deeply misunderstood objects that our ancestors feared as messengers of unfathomable evil, purveyors of doom, death, and destruction. They were omens of despair and portents of mortal destiny, announcing the imminent arrival of the four dread horsemen of the Apocalypse: Pestilence, War, Famine and Death. The very word "disaster" carries this meaning forward from ancient times. Its Latin roots literally translate as "bad star."

In a sense, the ancients were right. Comets have the ability to cause enormous destruction. If a massive comet were to collide with the Earth, it could be a planet-buster. An average-sized comet would be sufficiently large to wipe out all life as we know it. Although it is rare for a comet to be as big as, say, Connecticut, it is not so rare to find a comet the size of Mount Everest, and comets the size of a football field are not rare at all. Regardless of their size, these wandering menaces contain enough mass traveling with sufficient velocity that when they hit, they pack an enormous wallop. The amount of concentrated energy released upon impact by a large comet would be devastating on a global scale.

String of Pearls

The destructive power of a comet was amply demonstrated in 1994–when a very typical comet named Shoemaker-Levy 9 inflicted mind-boggling damage on mighty Jupiter. The original comet broke into 21 fragments that resembled pearls on an invisible string (right) headed straight toward Jupiter. The 21 comet fragments smashed into Jupiter's southern hemisphere between July 16 and July 22, 1994, at speeds of approximately 134,000 miles an hour. Some scientists felt that Jupiter could swallow such tiny projectiles without any noticeable effect on its gaseous bulk. However, each impact released more energy than all the world's nuclear arsenals combined, and even though the comet was not particularly big or dense, the impacts hurled plumes of debris thousands of miles high and scarred Jupiter's swirling atmosphere with dark, lingering wounds the size of Earth. This signal event rocked the scientific community and galvanized it into action. The immense destructive power of such an insignificant object, an object about the size of a small hill, was terrifying. It awoke astronomers and scientists to the fact that impact events happen now, and they could happen here on Earth. If such a comet just a few miles across were to hit the Earth, the result would be catastrophic. The initial heat from the collision would burn, melt, and vaporize a large portion of the surface of the planet. The subsequent shock wave, traveling faster than the speed of sound, would spread destruction in all directions. Tidal waves would drown the land for hundreds of miles inland. Molten debris would be thrown up into the atmosphere and would inevitably rain back down around the world in a series of devastating secondary impacts. It would be a true rain of fire. Life as we know it would be over. We have already had close calls.

Tunguska

One of the most powerful and mysterious explosions in history occurred on June 30, 1908, in the Tunguska region of Siberia. This rustic, sparsely populated area of pine and cedar forest is home to nomadic reindeer herders called the Evenk. An adult Evenk reports that his great uncle, who was living 45 miles from the presumed center of the blast, was sitting on the steps of his house when he was suddenly thrown several feet through the air to his fence. The great-grandson of an eyewitness states that his clan lived in the area when the explosion swept through and killed everything. A hunting expedition came to an abrupt end when, as eyewitnesses describe, "A light brighter than the Sun suddenly appears, streaks across the sky, and explodes." The locals truly believed they were witnessing the end of the world.

The massive explosion destroyed 800 hilly square miles of dense coniferous forest, an area more than twice the size of New York City, and produced shockwaves powerful enough to cause vibrations 3,500 miles away. Although only one person lost his life, hundreds of reindeer were killed, and horses were knocked to their knees by the blast. This powerful event shook the Eastern hemisphere, and seismographs all over Europe recorded the violent tremors. A thousand times more powerful than the Hiroshima atomic bomb, this shockwave and fireball could have incinerated a large city. This site sits on the same parallel as the megacities of Saint Petersburg, Helsinki, Stockholm, and Oslo. If this event had happened farther around the globe to the west, one of those cities could have been destroyed and millions of people killed.

Tunguska Trees

In 1927, when the first team of serious investigators reached the site, the only clue they found was the striking arrangement of the dead trees (left). Approximately 80 million dead trees were flattened over 500,000 acres in an area shaped roughly like a butterfly. They had been knocked down in parallel rows radiating from a central point. In fact, the only trees left standing in the blast area were at the very center. Investigators also reported that "something burned and stripped the trees." The way the trees had been burned, stripped, and forcibly blown down at Tunguska is remarkably similar to what is observed in a nuclear explosion, but nuclear weapons would not be created for another 37 years. Regardless, after World War II, Soviet scientists descended on the area looking for radioactive fallout, metals, and meteorites. They burned trees and chemically tested the ash for radioactivity. Numerous expeditions found nothing. With no Earthly explanation, scientists turned to the heavens.

For most of the twentieth century, scientists believed that a massive asteroid strike caused the Tunguska event, but they lacked crucial pieces of evidence. There was no crater at the site, and they found no rocky-metallic material in the surrounding area that could have come from an asteroid. However, scientists were able to calculate the energy of the blast. The hypocenter of the explosion (the point on the ground below the airburst, where the few trees were left standing) was readily determined from the direction of the downed trees and from the size and shape of the affected area. From the number of trees that were knocked down, scientists made an educated guess of the explosion's energy. They were even able to estimate the angle of approach, speed, mass, and altitude of the mysterious exploding visitor. In fact, modern computer simulations match up exceptionally accurately with the known detailed information of timber fall and the formation of the ground, including the trees left standing.

However, the biggest question still remained—why were there no asteroid pieces to be found? Geophysicists have spent years investigating. One surprising clue was the report by numerous witnesses of the event that glowing night-time clouds were visible for days afterward. The clouds were giving off sufficient light that "you could read a book at midnight." Though very unusual, such clouds are known to scientists and are called "noctilucent" from the Latin for "night-glowing." High in the atmosphere, around 50 miles, these clouds consist of ice crystals that reflect and scatter light. They can only occur if large amounts of water have been released into the air, as has been observed with some rocket launches. In 1908, people thousands of miles away from Tunguska reported seeing the mysterious night-time glow.

The radially flattened trees, the noctilucent clouds, the first-hand reports of the flash, and the lack of stony asteroid remains suggest that the Tunguska event was a massive air blast that released copious amounts of water. A large, icy comet could have exploded high in the atmosphere and left no crater or stony asteroid material while releasing its water, which could have either been carried away by the Tunguska River or floated westward by the upper air currents. Computer simulations carried out in 2007 supported this theory, showing that the mass of an exploding, water-rich comet could have been converted into an expanding fireball of high-temperature gases traveling at supersonic speeds. Therefore, many now believe that the Tunguska event was the explosion of a water-laden comet.

* * *

Despite the danger and fear associated with comets, they are among the most stunning of sights. A comet passing above us, that dramatic streak of light blazing across the sky, is breathtaking. The beautiful, fuzzy, glowing spheres with their iconic elongated tails are gorgeous, and they are so rare that a person can be counted lucky who sees even one bright comet in a lifetime. Once seen, the extraordinary event is forever embedded in one's mind.

Comets, however, are much more than a mystical mixture of beauty, danger, and dread. They are also a cosmic mystery. They conceal, deep within themselves, inscrutable secrets of the early Solar System. Scientists have been striving for centuries to unravel these secrets, but until very recently they were able to learn little because comets, those ancient, fleeting, often unobtrusive visitors from millions or billions of miles away, are notoriously difficult to study.

Since comets reflect almost no sunlight in deep space, they remain hidden, virtually invisible to our telescopes until they plunge toward the Sun, when they awaken and spew clouds of gas and dust as they glow like fireworks about to explode.

Most comets have orbits that last longer than 200 years. These are the long-period comets. Their orbits can take millions of years. Such comets traverse the far reaches of our Solar System outside the orbit of Neptune, aloof in an incomprehensibly vast sphere known as the Oort cloud, which stretches one-third of the way to the nearest star. Some comets, the short-period comets, are more punctual visitors, with orbital periods of less than 200 years. Their predictable timetables bring them on frequent jaunts close to the Sun. Comets with even quicker orbits are known to cruise close to the Earth every few years or hide in the asteroid belt, where they, too, can be jostled into a jaunt toward Earth.

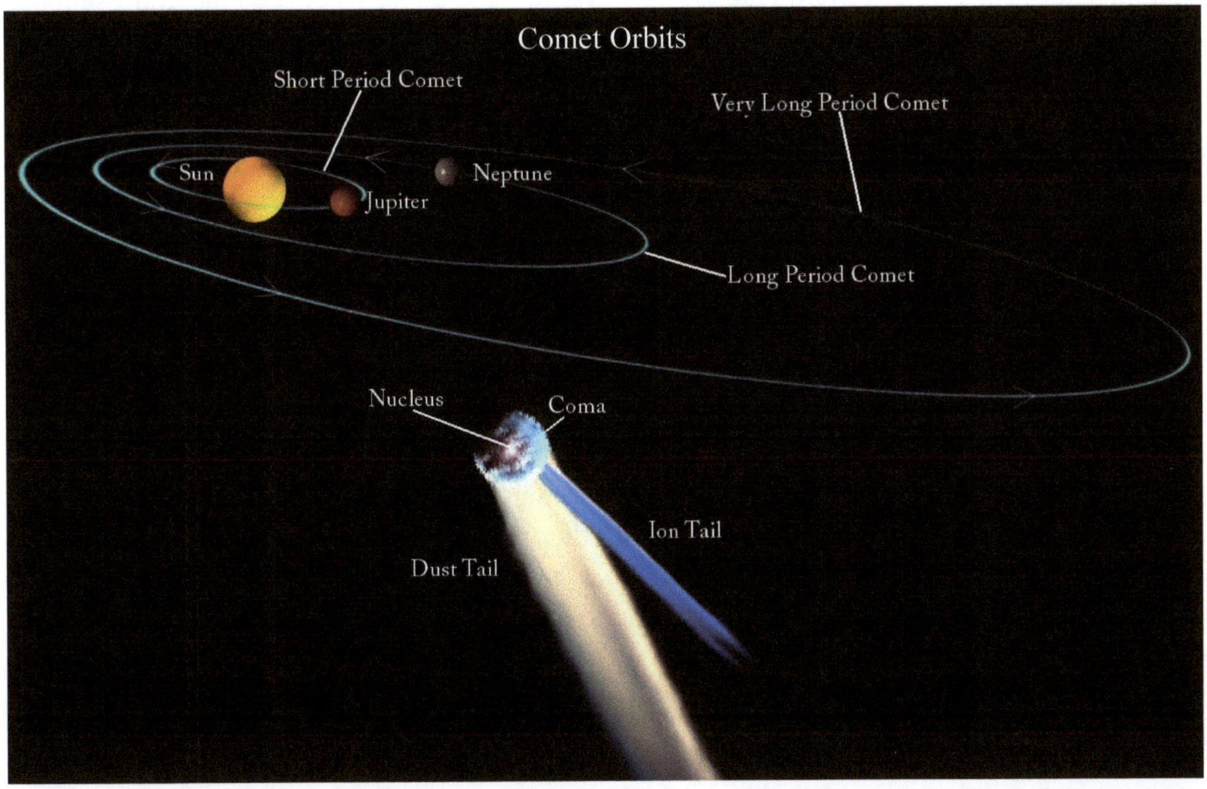

The comets populating the regions beyond the inner Solar System are cryogenic suspensions that are unpredictable, mostly anonymous, solitary wanderers floating through an immense ocean of space and peacefully waiting for an infinitesimal gravitational nudge, which may never come, to give them an impetus and redirect their path. When that nudge does come, produced by a minute gravitational ripple, the cloud ever so slightly trembles, and a sleepy wanderer can awaken in a new trajectory. That comet now initiates a steadily increasing process of transmogrification, transforming from a barely drifting, somnolent emissary of a kingdom of ice into a wildly somersaulting, agitated, and frothing beast that bursts into the inner Solar System with its tails ablaze. It will launch missiles of rock, fire jets of dust, and discharge clouds of gas as it streaks among the planets. The comets came into existence at the same time as the rest of the Solar System began to take form, some 4.5 billion years ago. In a diffuse, spinning cloud of molecules before the Sun and the planets were born, electrostatic forces gently pulled the grains of dust and crystals of ice together to form ever-increasing masses. Steadily these first frozen dust bunnies gained enough mass that gravity took over the building process to form larger clumps. The largest ones became planets; smaller ones became asteroids; and the smallest and most distant, out beyond the reach of most of the Sun's warmth and light, became the comets. The comets we see today are ice-covered remnants of the embryonic Solar System. Those which journey repeatedly close to the Sun eventually boil off their icy volatiles and become more asteroid than comet.

The comets of the newborn Solar System were flung about chaotically under the influence of the more massive planets' immense gravity. Many were undoubtedly ejected from the Solar System, while others were exiled to the outer darkness far beyond Neptune. Four billion miles from the Sun, the Kuiper belt (after Gerard Kuiper, who first suggested its existence in the mid-twentieth century) is home to many comets. However, most of them were hurled even farther to form an immense sphere around the Solar System, the Oort cloud (after Jan Oort, who first postulated its existence in 1950). If one were to stand in the Oort cloud and look towards the Sun, the Sun would appear as a tiny speck, just one of thousands of stars. Although the Oort cloud contains more than a trillion comets, its volume is so immense that each frozen comet, its temperature hovering near absolute zero (−459.67° F), probably has no neighbor closer than a hundred million miles. So little sunlight reaches this far that each comet flies virtually alone through a freezing realm of Stygian darkness. It is almost impossible to imagine any place colder, darker, and lonelier.

To this day, comets that have never experienced the Sun's warmth retain all the exotic compounds collected from the primordial solar nebula and remain almost unchanged. Comets, therefore, are ancient relics of a forgotten era, and if you could take the substance of one into your possession, you would be the owner of some of the very same atoms and molecules that were present at the birth of the Solar System and eventually formed the various bodies that populate it today. The innards of comets enclose and conceal the initial stages of our Solar System's formation and development, and astronomers are understandably eager to train their instruments upon these primeval pieces to ascertain their composition.

Some of those atoms and molecules, of course, eventually became us. Comets undoubtedly bestowed upon the young Earth gifts of all kinds of molecules, particularly water. The newborn Earth possessed little or no water when it first cooled down from the molten form of its fiery creation, but the impacts of thousands or millions of watery comets over eons may have supplied the water we and other living things on Earth rely on today. If this is the case—and scientists remain unsure—then we surely owe a debt of gratitude to comets. Scientists are also attempting to ascertain if there is a correlation between amino acids on comets and life on Earth. If the correlation is established, then humanity will have achieved a new understanding of one of the most profound and significant relationships in all of science.

A comet which journeys from its home in the Oort cloud to the neighborhood of the Sun undergoes a fantastic sequence of experiences. As it moves toward the Sun, its speed increases and it becomes warmer. Like a moth emerging from its cocoon, the comet transforms into a diaphanous wonder, apparently sprouting appendages similar to wings. These wings are called tails and are created by the emission of volatile substances which begin to effervesce, creating internal pressures that cause the speeding body to explosively eject fragments of itself and vent gases into space. The process is akin to bubbles escaping from champagne. When the champagne warms and the pressure is suddenly released, carbon dioxide gas comes out of the liquid and makes it bubbly.

The warming comet turns into a hazy, nebulous ball. Frozen materials in the comet can turn directly to vapor, a process called sublimation. The solid mass in the center of the comet, called the nucleus, becomes surrounded by a much larger cloud of material called the coma. "Coma" is Greek for "hair." Together the coma and nucleus of a comet are called the head of the comet. While the solid nucleus of a comet is generally less than 30 miles across, the coma may be larger than the Sun, a million miles or more across, and tails have been observed to extend three hundred million miles. This is greater than the distance all the way across the Earth's orbit.

A comet as viewed from Earth appears to hang motionless in the night sky, but this picture is deceptive. The comet only appears stationary because of its great distance from us. In fact, the comet is charging through space at approximately fifty thousand miles per hour and is about to enter the most violent phase of its journey.

As a comet passes the orbit of Earth on its approach to the Sun, it grows ever hotter, creating internal pressures which cause the surface to quake and volatiles to erupt. A typical comet passing the orbit of Earth is rocketing towards the Sun at 75,000 miles per hour. As it reaches a point close to the Sun, it can shed as much as three tons of ice, gas, and dust every second. Disgorging material at such a rate, the comet can become unstable, break apart, and shatter into pieces. Its existence hangs precariously in the balance.

A comet can transform into a series of miniature volcanoes hurling fountains of water, gases, particles, and boulders into space. This ejected matter forms the comet's tail. The venting fountains of matter can become powerful enough to cause the comet to gyrate wildly and wrench it onto a new path. This new trajectory could put it on a collision course with Earth, and although such an event is highly unlikely, it is in our species' best interest to keep an eye on exploding comets. Comets, like a deadly viper, can be exceptionally picturesque, and they can suddenly and unpredictably strike with overwhelming lethal force.

Comets that pass very near the Sun are called "sun-grazers." Solar observatories have taken numerous pictures of small comets swooping within a hair's breadth of fiery doom. Exposed to the Sun's immense gravity and torched by ferocious heat, only a fraction of them survive. The remainder are vaporized completely or swallowed whole, disappearing into the face of the Sun without a ripple, motes of dust sucked into an impassive, flaming tornado. They experience the Sun's atmosphere the same way a meteor experiences Earth's: with a flash of heat and light, followed by oblivion.

It is breathtaking to watch a comet as it dances through the twilit depths of space as if it were its last dance. A real showstopper, it rocks and rolls, spins and twirls, and jitterbugs towards the Sun. It is really moving now, hot, shedding its coat, growing a tail, and shaking with internal vibrations. However, this dance is being created by a pantomimist that is inanimate, artless and soulless, and the dance is performed in a void bereft of all music since the vacuum of space is unable to transmit any sound at all—not even the din of an explosion.

A comet heading away from the Sun is preceded by its tails because the solar wind pushes the dust and gas of the tails away from the comet faster than the comet is traveling away from the Sun. The surface of the comet begins to settle down. It becomes colder and less active. The jets turn off, and the coma blows away, leaving a quiescent ball of ice and dust typically no more than twenty miles across. Imperceptibly but continuously, a comet slows while traveling to the frigid outer reaches of the Oort cloud. Returning to deep space causes the comet to become dormant once more. A comet fades into obscurity as it recedes into darkness. It is like some exceedingly rare and iridescent gem gone into the heavenly vault for safekeeping. Only later will it be brought forth once again as a singular visitor to sparkle and glitter in the Sun's light.

With state-of-the-art telescopes, we can study comets in greater detail than ever before, but to truly understand them, we need to pierce the comet's glowing, gaseous shroud to get close to the glistening prize at its center, its icy heart: the nucleus. The holy grail of cometary science has been to discover and fully comprehend the configuration and composition of the nucleus of a comet. What is there on the surface? What are its chemical components, and what are the characteristics of the rocks and the volatiles that reside near the surface?

The nucleus of a comet is the fundamental building block of the Solar System. It is where the grandest and most captivating mysteries reside. In the past three decades, we have launched over a dozen missions to study comets, and we have learned about comets' chemistry, their interactions with the Sun, their physical surfaces, and how they are different from each other. Comets are worlds unto themselves.

Scientists used to think that comets were pure white, just like a snowball. By the 1950s, the prevailing view was that they were snowballs, but "dirty," that is, containing rocks, dust, and volatiles. All that changed in 1986 when the Giotto spacecraft beamed back images of Halley's Comet. The spacecraft (below) was named after the Early Italian Renaissance painter Giotto di

Artist's Conception of Giotto

Bondone. He had observed Halley's Comet in 1301 and was inspired to depict it as the star of Bethlehem in his fresco "Adoration of the Magi." Giotto flew past Halley's Comet's dusty coma, skirting its seething, sputtering nucleus at 43 miles per second. At that speed, a grain of sand can

Coma of Halley's Comet

pierce aluminum three inches thick. Thanks to the Giotto probe, for the first time in history we had a snapshot of the nucleus of a comet. This "world unto itself" turned out to be shaped like a peanut. Halley was no snowball. It had a thick layer of black dust covering its surface, and it also had pits, hills, and two lobes. Nine miles long, it was far bigger than anyone expected.

Scientists had assumed that all comets were similar, but they were mistaken. Just as every planet is different, so is every comet. Each one has experienced different processes and temperatures for different periods of time. Each one has a unique composition and a different chemical and geological history. What this means is that each comet is a one-of-a-kind worldlet waiting to be explored. Comet Halley with its coma is seen in the image above.

Though each comet is unique, some take uniqueness to the extreme and are truly strange. Comet Wild 2 was explored by Stardust, a 650-pound robotic space probe launched by NASA on February 7, 1999. Its main goal was to gather samples of dust from the coma of this strange comet, along with samples of interplanetary dust, and return them to Earth for analysis. It was the first sample-return mission of its kind. The mission was successfully completed on January 15, 2006, when the sample-return capsule returned to Earth.

Stardust Collector (tennis racket)

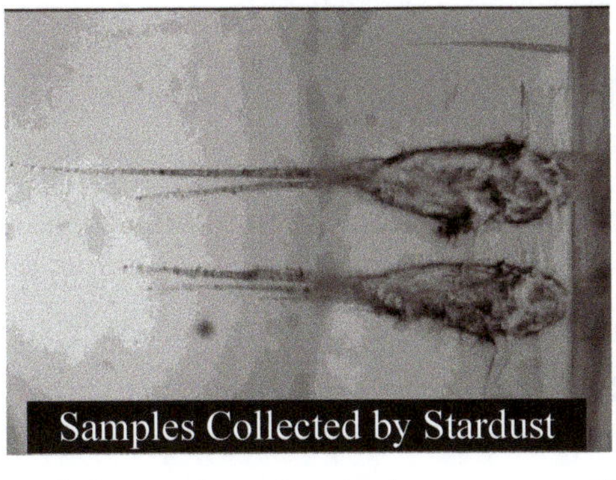

Samples Collected by Stardust

Extending from the Stardust probe as it flew into Wild 2's coma was a dust particle collector that bore a distinct resemblance to a tennis racket (above). The rectangular boxes of the "racket" contained aerogel, a silicon-based substance that is exceptionally spongy. The tennis racket with its aerogel captured hundreds of tiny particles (left). The collector then swung around into the sample-return capsule for eventual re-entry into the Earth's atmosphere. After the capsule made it back to Earth, it was retrieved, and the enclosed samples were studied by scientists who were amazed. The dust from Wild 2

Stardust Probe

Workers with Stardust Collector

contained compounds of titanium, aluminum, and calcium that will form only in the presence of heat (3000°F) and liquid water. This indicates that the comet very possibly was formed near the Sun before being flung outward to its present orbit. Wild 2 also held compounds of several inorganic materials such as the same calcium carbonate which forms the shells of snails, and organic hydrocarbons which are the building blocks of amino acids. These particular substances can readily combine when conditions are suitable to form the proteins which became life on our planet. At upper left is an illustration of the Stardust probe approaching Wild 2. The picture at upper right shows scientists preparing to examine the Stardust probe's dust collector.

On November 4, 2010, Hartley 2 (right) was the target of a flyby by the Deep Impact spacecraft. Hartley 2 was a comet that surprised even the most knowledgeable experts: it had snow and hail. Amazingly, golf-ball-sized chunks of dry ice were following the comet from up to a million miles away. With its enveloping cloud of hail particles, Hartley 2 presented the appearance of a frozen meatball floating through space enveloped in a swarm of buzzing flies. A hyperactive comet, it was observed to tumble faster and spit out more than most other comets its size.

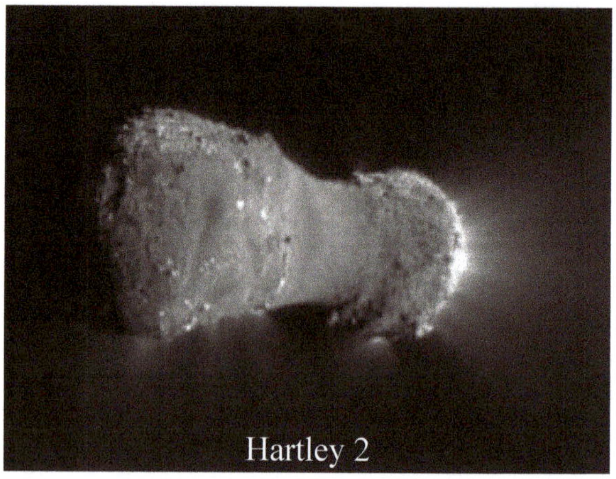
Hartley 2

Comet Tempel 1 was explored by the Deep Impact space probe. On Tempel 1, there are craters, plateaus, and high cliffs. Strata of rock lie atop one another, giving the appearance of accumulating layers of sediment. Scientists made their most significant breakthrough up to that time when, in 2005, NASA actually blew a hole in Tempel 1's surface. Deep Impact released a projectile, called an impactor, into the comet's nucleus. The impactor was an 815-pound copper-core projectile outfitted with a high-resolution camera to record the impact. The Deep Impact space probe successfully released its impactor on July 4, 2005, and it resulted in actual celestial fireworks. The projectile slammed into Tempel 1's rock-hard surface at a closing velocity of 22,800 miles per hour, creating an explosion equivalent to five tons of TNT. The explosion blasted a crater 150 yards across and ejected about 11 million tons of ice and 55 million tons of dust.

When scientists studied the material dislodged by the impact, they were, for the first time in human history, examining material from the interior of a comet. What they found was completely unexpected. There were precious and semiprecious stones, along with a variety of organic molecules. Such molecules are the ingredients from which life is assembled. Scientists, therefore, now suspect that comets performed a critical role in the evolution of the Solar System and theorize that comet collisions can be creative forces—they could even trigger the birth of life itself. Where do the ingredients of life originate, where were they all mixed together, and where did all this liquid water come from? Comets could hold the key to comprehending the origin and, therefore, the nature of life itself.

Comets have evidently shaped other worlds besides Earth. Many of the large moons of Saturn and Jupiter possess large quantities of water, and this water could have come from comets. The cratered surfaces of these bodies indicate with stark clarity the history of pummeling they have undergone over millions of years. The gases and water from successive comet strikes accumulated over eons to form the rich and varied atmospheres on these moons, including organic compounds, transforming these orbs into potentially life-bearing worlds. To find out more about this process, astrobiologists near the beginning of the twenty-first century arranged the landing of a probe on the surface of a comet.

Rosetta

The Rosetta mission was launched on March 3, 2004. The spacecraft (right) was named after the Rosetta Stone, an ancient carved relic that provided linguistic keys to ancient languages. Scientists hoped that the probe, like the stone, would prove invaluable for the solution of longstanding mysteries. The Rosetta spacecraft soaring through space with its solar panels fully extended resembled a giant Egyptian ibis with its wings fully outstretched in flight. After its launch in 2004, the spacecraft, which was about the size of an automobile, spent a year traveling to the Comet 67P/Churyumov–Gerasimenko, reaching speeds of approximately 20,000 miles per hour. Its target was the comet's nucleus (right), which was a mere three miles wide.

Comet 67/P

Rosetta brought to the comet a detachable robotic lander named Philae. Philae was to drop down to the surface and attempt the most detailed study of a comet ever. The Philae lander was named after the Philae obelisk, an ancient monument with multilingual carvings which also assisted in the decipherment of Egyptian hieroglyphics two decades after the Rosetta Stone was found.

The landing of Philae on the surface of the comet did not go as planned. Since the surface layout of 67P was unknown before Rosetta's arrival, the orbiter mapped the comet carefully before detaching its lander. After the best landing site had been chosen and named "Agilkia," Philae separated from Rosetta on November 12, 2014, and approached 67P at a relative speed of about three feet per second. After it first touched down, it bounced twice, coming to rest after about two hours. The bounces took so long because of the comet's weak gravity.

The probe was supposed to be secured to the surface by two harpoon-like devices that were to shoot out automatically into the body of the comet, but this did not work as planned. The surface at the initial touchdown site was relatively soft, covered with a layer of fine granular material. After its unplanned bounces, Philae landed at an angle, apparently under the shadow of an escarpment, which meant that it could not collect sufficient solar power. It lost contact with Rosetta when its batteries ran out after two days, well before many of the planned science objectives could be attempted. Contact was briefly and intermittently reestablished for four weeks several months later, and then contact was lost again. Nothing has been heard of Philae since. Although it did not perform optimally, Philae did manage to detect more than a dozen different kinds of organic molecules on the comet.

Rosetta remains in orbit around 67P. The probe will accompany the comet all the way around the Sun, during which time Rosetta's many instruments will be able to observe every phase of the cometary transformation as it approaches the Sun and then draws away from it. The probe will continually gather unprecedented amounts of up-close data. Before subsiding into somnolence, the probe will descend to the comet's surface, collecting data all the way, and at the end, it will join its substance to the dust and ice it was sent to study.

Comet Hale-Bopp suddenly emerged from the shadowy depths of space in 1997 to become the largest and brightest comet ever photographed. Analysis of the light revealed that this comet contained large amounts of the organic molecules that are found in life on Earth. This strongly suggested that the seeds of life may have been initially formed in the crucibles of deep space near a young star and then planted on Earth by a crashing comet. Such seeds may have not only carried the raw ingredients of life but also the water in which these valuable molecules could aggregate before morphing into simple life. As a falling seed buries itself into a hospitable piece of ground only to sprout later, so too life may have arisen when pluripotent molecules fortuitously found a benevolent sanctuary on the early Earth.

Conclusion

The discoveries made during investigations of Halley's Comet, Wild 2, Hale-Bopp, Tempel 1, 67P, Holmes, and Shoemaker-Levy 9 have transformed our understanding of comets. Scientists now believe that comets are more than just frozen time capsules containing samples of the ancient material from which the planets were assembled at the beginning of the Solar System. Comets played a central role in the history of each planet. From the results of the Stardust mission, we learned that comets are complex chemistry laboratories where the ingredients of life can form.

These materials did not arise naturally as a result of processes on the surface of the Earth but instead were delivered by these messengers from the outer Solar System.

When the story of life on the Earth began about four and a half billion years ago, our planet was a barren and hostile world. Comets that rained down on the Earth for two hundred million years released their gases and organic materials, creating an atmosphere and contributing their water to the Earth's oceans. Finally, life began on Earth. The dramatic story now being woven together by planetary scientists and those who study comets is not yet completely understood or accepted, but it likely contains at least a nucleus of truth.

There is much more to learn. If we keep studying these mysterious objects, whether or not we can really pin down the exact mechanics, we may learn how all life started. Researchers a hundred years hence will recall our century wistfully and refer to it as a golden age of cometary exploration. This is the time when the human race is taking its first tentative steps toward full understanding of these celestial messengers. We have sent probes to touch the comets themselves; we have held the icy comet-dust in our hands and studied their makeup down to the isotopes. Everything that future researchers do to advance knowledge in this field will be built on the foundation we are laying down today with our bold and clever probes.

Mark Twain (Samuel Clemens, below, left) was born as Halley's Comet appeared in the sky in 1835. The comet visits the inner Solar System once every 75 or 76 years, and as it neared Earth for the second time in his life, Twain said, "I came in with Halley's Comet…It is coming again… and

Mark Twain

Halley's Comet in 1910

I expect to go out with it…The Almighty has said, no doubt: 'Now here are these two unaccountable freaks; they came in together, they must go out together.'" Twain died on April 21, 1910, just as the comet (above, right) made its next pass within sight of Earth.

19
Jupiter: King of the Solar System

Earth compared to Jupiter

As soon as somebody demonstrates the art of flying, settlers from our species of man will not be lacking on the moon and Jupiter... Given ships or sails adapted to the breezes of heaven, there will be those who will not shrink from even that vast expanse.
Johannes Kepler

Jupiter is a colossus 318 times more massive than Earth and so large that over 1300 Earths would fit inside it. More than a planet, it is a mini-solar system swarming with its own moons, garbed in coal-black rings, and chased by asteroids. This primordial leviathan floating through the ocean of space is a planet of immense, raging storms, sheathed in perpetually turbulent clouds, streaked with fierce lightning bolts, and reverberating with planet-wide thunder. It is raked by a frenzy of furious winds, girdled in loops of lethal radiation, and shrouded in mystery. It sizzles with awe-inspiring auroras, swathed in an elongated, teardrop-shaped magnetosphere that is the largest single structure in the Solar System, and flaunts the bustling majesty of its immense bulk with splendid, polychromatic aplomb.

Ever since Galileo aimed a telescope at Jupiter, scholars have yearned to plumb the proliferating secrets of this enigmatic planet. A cosmic time capsule, Jupiter contains clues to the origins and secrets of the Solar System, and its moons could shed much light upon the possibility of extraterrestrial life. Scientists have longed to peer beneath Jupiter's opaque cloud layers and penetrate its moons' frozen exteriors, and in this they are beginning to succeed.

The Sun as viewed from Jupiter would resemble a tiny, glowing disk, significantly larger than the pinpoint stars encompassing it but much smaller than it appears to us from Earth. Likewise, from Earth, Jupiter resembles a simple point of light. The ancient people who named it could not have known its true size, but, whether by lucky guess or surprising intuition, they gave this colossal and majestic planet the name of the monarch of their gods. If our Solar System had a king, it would be this behemoth.

Jupiter is named for the king and ruler of the Roman pantheon of deities. The Greeks called him Zeus, and he has also been called Jove. Jupiter, the grandson of Uranus and son of Saturn, was the god of the sky, thunder, lightning, law and order, and fate. He was a capricious, jovial deity who delighted in frolicking with nymphs, demigoddesses, noblewomen, and commoners. A fierce, passionate, indiscreet god, he relentlessly pursued his heart's desires, and many lived to rue the day that they provoked his ire. His vengefulness was legendary, for he commanded the most awesome power known to ancient man, the lightning bolt. His scepter was topped by a golden eagle, representing his supreme authority. Ancient astronomers once used the symbol ♃ to represent Jupiter, and this symbol is still used today by those who practice astrology.

The ancients recognized, perhaps, that youth is speedy while age brings with it a slower pace. Mercury, then, observably the fastest-moving planet, was given the name of the youthful and celeritous messenger of the gods. Venus, endowed with radiance and creamy color, was the goddess of beauty. Mars, red as blood, was the fierce god of war. The slowest-moving planet the ancients could see was Saturn, father of Jupiter. Intermediate in speed between violent Mars and slow-moving Saturn would be the great king of the gods, Jupiter. Jupiter's light, though not the brightest among the planets, was the most constant, befitting a wise and steady ruler.

Juno

Jupiter's wife Juno, a breathtaking beauty, was a spiteful, rage-filled, jealous goddess who despised the infidelities of her constantly cheating husband. To salve her wounded ego, she would inflict her divine wrath upon any female who was tricked, seduced, or defiled by Jupiter. Jupiter, who surrounded himself in a fog of mystery, would often disguise himself to consummate his affairs. However, Juno had the power to disperse the gathering mists behind which Jupiter hid and thus reveal her husband's secrets and deceits. Today, humankind has learned to mimic the abilities of the ancient Roman goddess and has constructed a craft that will attempt to part the tumultuous clouds and let humanity virtually "see," by employing technology, to the heart of Jupiter and its moons. How fitting that we have named this spacefaring craft Juno.

The achievements of the Juno probe have been breathtaking. It traveled in a well-planned, hyper-exact orbit that first shot it outside the orbit of Mars and then summoned it back to Earth for a gravity slingshot assist. Suitably accelerated (the gravity assist Juno received from Earth added an additional 8800 miles per hour to the vehicle's velocity), it followed its planned path with pinpoint accuracy until it arrived at Jupiter after traveling 1.74 billion miles. Juno's position and velocity deviated from the planners' calculations so slightly that the scheduled 35-minute burn for deceleration had to be altered by only a single second.

The Juno spacecraft arrived at Jupiter on July 4, 2016, and the deceleration burn occurred early on July 5. The orbiter needed to slow down enough (by 1200 miles per hour) to allow Jupiter's gravity to capture it. At that time, Juno left its hyperbolic trajectory and entered the highly elliptical pole-to-pole orbit that would allow it to study the planet's exterior as well as much of its interior. After its five-year odyssey and the successful completion of its critical orbital insertion maneuver, the probe signaled NASA scientists on Earth with a musical tone. Scientists were ecstatic and relieved when the probe "sang its song of perfection," a simple sound by which the orbiter announced, "I am physically fit and ready to commence my mission of discovery." The song of perfection had traveled approximately 500 million miles through space, reaching Earth 45 minutes after being sent. Radio signals travel at the speed of light, but even at that tremendous speed, messages from Juno require anywhere from 36 to 54 minutes to reach Earth, depending on the positions of Earth and Jupiter in their respective orbits.

This mission would use an array of scanners, detectors, and cameras to amass data as the intrepid probe ran the gauntlet between a swirling maelstrom of raging storms on one side and a deadly torus of radiation on the other. That radiation, consisting of bullet-like, high-energy subatomic particles racing at almost the speed of light, is so intense that it would kill a human in less than a day. These speeding particles will eventually punch through the probe's circuitry, rearrange its molecules, and disable its electronic components. During the probe's projected flight schedule of 37 polar orbits over 20 months, it will be subjected to as much radiation as a million diagnostic x-rays. This amount of radiation will ultimately wipe the probe's software and burn out its hardware, disabling its instruments and leaving it a scorched, inert wreck. Mission planners expect this to take about two years, although traditionally their estimates have erred on the low side.

The persistent probe will peer through the planet's cloudy exterior, study its interior, measure its magnetic fields, and gauge its radiation levels. It will reveal the hidden truths behind the multihued rivers of billowing clouds that hitherto have shielded Jupiter from prying eyes. The scorned wife has appeared on the scene, prepared by hundreds of years of scientific advances, to capture, record, and report on her husband's hidden mysteries to eagerly waiting scientists. The information gleaned will furnish provocative glimpses of the planet's composition, structure, and attributes. This information will prove vital to planners of future missions to Jupiter and its moons. Light will be shed upon the origin and development of the Solar System and its first-formed planet, Jupiter.

According to the latest and most plausible theories of the origin of the Solar System, Jupiter formed first, at a distance from the Sun where liquid water could not exist, and then migrated inward. This migration caused the so-called "Late Heavy Bombardment," a rain of meteors the scars of which are still borne by our Moon, among other bodies. Jupiter then was drawn outward to its present location, partly by the gravitational force exerted by the planet Saturn, and Uranus and Neptune traded places. This complex model is referred to as the "Grand Tack Theory." Although it works reasonably well in computer simulations, there is much more to be learned. For example, the theory does not properly account for the present mass of Mars. Either Jupiter robbed Mars of some of its mass during the migration to its present location, or the theories need to be further refined. Juno will help us do that.

Juno Being Built

Juno Probe Artwork

The Juno probe, shown under construction at right, is equipped with a panoply of instruments capable of penetrating Jupiter's opaque outer layer of clouds. On board are a microwave radiometer, a flux gate magnetometer guided by an advanced stellar compass, a radio wave gravity science instrument, a device for measuring auroras, an energetic-particle detector, a radio and plasma wave sensor, an ultraviolet spectrograph, and, last but not least, the JunoCam, which is a visible-light telescope and camera which was included to facilitate education and public outreach. The JunoCam will be extremely vulnerable to the planet's radiation and may only function for a few orbits.

Previous missions to Jupiter alerted scientists to the fact that the planet is enveloped in the most powerful radiation belts and magnetic fields in the Solar System. Earth's radiation belts are of concern to astronauts who must pass through them, but Jupiter's belts are searing. These ferocious belts are capable of serving up a deadly mix of ionized particles a thousand times the lethal dose for humans. The Galileo probe was damaged by this radiation when it visited Jupiter, but thanks to experience gained during that mission, scientists were forewarned that Juno would need robust shielding to protect the satellite's delicate computers and instruments. Therefore, these sensitive systems are housed in a half-inch-thick titanium vault. In addition, the flight plan calls for Juno to "thread the needle." On each of its orbits, it will slip adroitly between the planet and the zone of intense radiation that surrounds its equator, passing a mere 2,580 miles above the cloud tops at a speed of 129,000 miles per hour.

Jupiter with Juno in Foreground

Juno (left and above, in artists' conceptions) is a spinning, hexagon-shaped, tri-winged probe over 11.5 feet in diameter. It spins for stability and is powered by the largest array of solar panels ever deployed into deep space. The panels employ 18,698 individual photovoltaic cells to generate the electricity needed to operate the vehicle's instruments, propulsion system, microprocessors, and communications equipment. The solar arrays, when unfolded for power generation, stretch the width of a basketball court. They are large because they have to be; in Jupiter's neighborhood, sunlight is one twenty-fifth as intense as the sunlight that reaches Earth. For protection against radiation, the panels are covered by a half-inch-thick sheet of glass. Another reason for the great size of the panels is that they were expected to lose 10 to 15 percent of their power output over the course of the mission due to radiation damage, so designers made the panels 15 percent larger to compensate.

Galileo Plaque on Juno

Lego Figures

Juno carries an aluminum plaque (right, top) bearing a portrait of Galileo and a facsimile of a page of Galileo's journal from the day he recorded his first observation of Jupiter's moons, which was January 10, 1610. In addition, three miniature aluminum figurines (right, bottom) were installed on the probe to inspire children to study science and math. The figurines, which were custom-manufactured by the Lego Corporation to endure the unforgiving conditions of space flight, depict the goddess Juno herself, with a magnifying glass to symbolize her inquisitive nature; Jupiter bearing his signature weapon, the lethal lightning bolt; and Galileo, exemplifying his scientific curiosity by bearing a telescope and globe.

To think that mere humans could fling a bus-sized object weighing almost four tons on a five-year trip almost two billion miles through space and have it arrive within one second of perfection; that we could sit on Earth and see the cloud bands swirl on Jupiter, view scintillating auroras larger than the planet of our birth, see thousands of miles below the layers of another planet's clouds, gaze in wonder upon the invisible, the beautiful, and the marvelous as we once again extend our hands, eyes, and minds into space on our unending quest for knowledge: these accomplishments stagger the imagination and will inspire future generations of astronomers, physicists, and planetary scientists to continue extending our knowledge, exploring the Solar System, and accelerating the pace of human achievement in space.

Physical Data

Even at Jupiter's distance from the Earth of between 400 million and 600 million miles, the planet still remains one of the brightest objects in our night sky. Only the Moon, Venus, and occasionally Mars are more vivid.

Jupiter is the largest and most massive planet in our Solar System and requires almost twelve Earth years to complete one orbit of the Sun. The sculptor Phidias, coincidentally, needed twelve years to complete his statue of Jupiter (or Zeus) in the temple of Olympia. That sculpture of gold and ivory was renowned as one of the seven wonders of the ancient world. Commentators have also chosen to recognize seven wonders of the Solar System, and the planet Jupiter boasts three of them: its auroras, its magnetosphere, and the Great Red Spot. (The other wonders of the Solar System are Saturn's rings, Earth's blue oceans of liquid water, Mars' Olympus Mons, and the Sun's nuclear furnace.)

Despite its high mass and long orbital period, Jupiter is the fastest spinning planet in the Solar System. It is 86,000 miles in diameter and rotates at a rate of 29,000 miles per hour. Moving this much mass around its axis so quickly creates a pronounced bulge at the equator, making Jupiter an oblate spheroid. Jupiter takes almost ten hours to complete one full rotation on its axis. However, the planet is not a solid ball, so different parts of it rotate at different speeds, and this stirs up its turbulent cloud formations.

Jupiter's immense mass enables it to act as a gravitational garbage scow, scooping up stray debris such as comets and asteroids that drift too close. This was incontrovertibly proven when Jupiter shredded the comet Shoemaker-Levy 9, converting it into a string of glowing fragments that then pummeled Jupiter. The scars in Jupiter's atmosphere were visible for weeks. One of those scars was larger in diameter than the planet Earth, which, momentarily at least, convinced some residents of Earth to pause, rethink, and develop a deeper appreciation of the raw, unbridled power unleashed in such catastrophic comet collisions. Many people did not seriously consider or comprehend the full fury of the fearsome force of such a simple, routine collision until it was demonstrated before their startled eyes.

We on Earth should be deeply appreciative of the attractive powers of Jupiter. Its gravitational pull has dragged millions of comets and asteroids to their demise or hurled them out of the Solar System. This is far greater than the number of fragments of flotsam it has let float through to careen among the inner planets. Without Jupiter, the inner Solar System would be much less safe for us. Humankind's ancestors would most likely have met the same destiny as the dinosaurs, obliterated by a hemisphere-smashing meteor that would have scattered our prehuman ancestors' bones and then buried them in a layer of soot, dust, and iridium.

Jupiter orbits the Sun at a distance of approximately 5.2 astronomical units, or almost half a billion miles. That far from the Sun, Jupiter's exosphere of clouds is quite cold, with an average temperature of about –166°F. Yet the planet radiates twice as much heat as it receives from the Sun. One reason for this is that Jupiter still retains heat from its formation. Additional heat comes from the decay of radioactive elements in Jupiter's interior. Also, the planet is slowly shrinking, which raises internal pressure and creates still more heat. Jupiter's heat, no matter the source, is transported outward through the gaseous atmosphere to the outermost cloud layers where it dissipates into space.

Researchers believe Jupiter formed in the same way as stars: by condensation. Thus, Jupiter and the Sun were born from a single solar nebula, gathering to themselves more and more matter (mostly gases) with their ever-increasing gravity. Current models of Jupiter suggest that an Earth-sized, solid, rocky inner core of about ten Earth masses formed first. This core then gathered an ever-increasing atmosphere. As the planet's massive heart grew, it attracted more and more gas from the surrounding space until it reached the gargantuan size we see today.

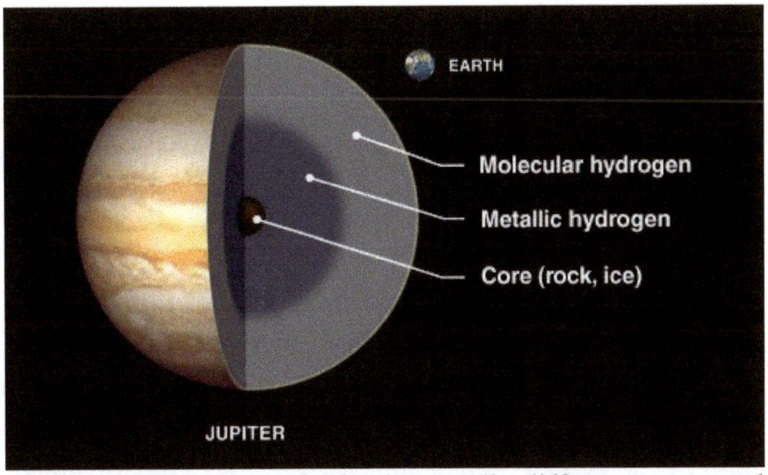

Scientists have been able to draw some conclusions about Jupiter's interior from the presence of its powerful magnetic field. Planetary magnetic fields are thought to be created by the motion of electrically conductive material in the core. Jupiter is mostly hydrogen, which does not conduct electricity under normal conditions. However, because of the intense pressures near Jupiter's center, matter behaves in fundamentally different ways, and the hydrogen there transforms into an electrically conductive soup resembling liquid mercury. Scientists refer to hydrogen in this state as "metallic hydrogen." While metallic hydrogen is an elusive rarity on Earth, it must be common inside Jupiter, and this would explain why Jupiter has the strongest magnetic field of any planet in our Solar System.

Probes

Most of our knowledge about Jupiter has been transmitted to us from spacecraft dispatched to explore, study, and survey the great planet. These probes have proven themselves to be reliable, resilient, and long-lasting. The earliest Jupiter probes were small and flew fast, which meant they could not linger near Jupiter. However, during their time in the vicinity of Jupiter, these intrepid over-achievers captured amazing photographs of the Jovian system and measured the physical characteristics of the planet and its moons.

These probes display all the attributes and behaviors of human heroes: they are our robotic avatars, launched on a hero's journey from which they will never return, sacrificing themselves on a selfless quest for knowledge which must surely culminate in either a death-plunge into a distant planet's super-dense atmosphere, or eternal exile in the cold depths of uncharted interstellar space. They dedicate themselves to gathering as much information as they can in the short time they are allotted and often exceed the expectations placed upon them by mission planners.

Pioneer 10 was the first spacecraft to fly past Jupiter, passing within 124,000 miles of the tops of Jupiter's clouds in December 1973. This probe detected a great deal of radiation, but it also demonstrated that our probes could survive the radiation, at least for a time. Pioneer 10 studied Jupiter's magnetic field and atmosphere and took the first close-up pictures of Jupiter. Pioneer 11 did even better, flying a mere 22,000 miles above the Jovian cloud-tops. In December 1974, it took more pictures, including pictures of some of the moons.

Following the success of the Pioneer series of spacecraft, the outer planets were approaching a favorable alignment that only occurs once every 175 years, and NASA wanted to take advantage of this rare and fortunate circumstance. NASA had learned with the Mariner 10 mission that went first to Venus and then to Mercury in 1973-74 that the "gravity assist" method was viable and that it could reduce flight times by years. Gravity assist is a process that allows a spacecraft to swing easily from planet to planet. It is like having an extra, virtual engine that requires almost no fuel. As the spacecraft approaches a planet, the planet's gravity grabs it and in effect hurls it onward toward its next destination at increased velocity. NASA's mathematicians and mission planners were able to map out trajectories that would enable the next batch of probes to visit all the outer planets in succession.

Unfortunately, funding for such missions was tight, and the expense of a trip of this magnitude could not be borne. To save money, NASA chose to visit only Jupiter and Saturn. From among 10,000 possible trajectories for the two Voyager spacecraft, two were selected that would enable the probes to scoot past Jupiter and its large moon Io, subsequently visiting Saturn and its largest moon Titan. NASA, anticipating the success of this first phase of the mission, preserved the option of sending Voyager 2 onward to Uranus and Neptune.

Voyagers 1 and 2 arrived at Jupiter in March 1979 and July 1979, respectively. They took more than 18,000 pictures of the planet and its moons, in the process finding three new moons as well as a thin dark disk around the planet: a ring. Both craft continued on to Saturn, where they were also spectacularly successful. Voyager 1's path carried it out of the Solar System, but Voyager 2 appeared to have the capability to fly to the outer planets. To NASA scientists this was irresistible. Voyager 2 was reprogrammed, and the two-planet mission became four; the expected five-year lifetimes of the Voyagers have now stretched to 40 years. In total, the two spacecraft took pictures of the four planets, 48 of their moons, and their systems of rings.

Galileo Orbiter

The Galileo spacecraft (right) was the next probe to visit Jupiter. Its mission lasted three times longer than scientists originally planned, delivering significantly more information than anticipated. Aboard the main spacecraft was an atmospheric unit named the Galileo probe. This titanium probe weighed 750 pounds and was protected by a robust heat shield, while a rugged parachute was attached to slow its descent. On December 7, 1995, the gallant probe dove into Jupiter's upper cloud layer at 106,000 miles per hour, plummeting more than 100 miles into the roiling gases. It continued to transmit data for 58 minutes before succumbing to the intense heat and pressure. Those 100 miles represented only a tiny fraction of Jupiter's 43,000-mile radius, but having penetrated only to the depth of "the skin of a peach," the probe was already measuring atmospheric pressures 24 times greater than the Earth's at sea level, wind speeds around 400 miles per hour, and temperatures over 300°F.

After the descent of the atmospheric probe, the Galileo orbiter remained in orbit for almost eight years. Powered by a long-lived nuclear battery, it continued surveying Jupiter and its moons and broadcasting its findings back to Earth. After eight years, radiation from Jupiter had damaged the spacecraft, and its fuel supply was rapidly diminishing. After Galileo completed its 35th orbit of Jupiter, it was directed into the planet's atmosphere to avoid accidentally contaminating the Jovian moons and endangering any life there. This death plunge occurred on September 21, 2003.

As a result of Pioneers 10 and 11, Voyagers 1 and 2, and the Galileo orbiter and probe, our knowledge of the Jovian system and the other outer planets grew dramatically. The Pioneer missions served as scouts for the Voyagers. The Voyager probes mapped out the main characteristics of the planet and its environment. The increased sophistication of the instruments on Voyager 1 and 2 provided enough information that researchers were still analyzing it when the Galileo mission began. The Galileo mission was followed by the Cassini spacecraft in 2000, which sent us more data as it passed Jupiter on its way to Saturn.

Following Cassini, the spacecraft Ulysses, which was designed to study the Sun from a polar orbit, and New Horizon, which was created to explore Pluto and beyond, utilized Jupiter for gravity assist maneuvers. While flying past Jupiter, these two missions communicated additional data that added to our understanding of this remarkable planet. Among the pictures taken by New Horizon were the clearest images ever captured of Io's Tvashtar volcano, whose sulfurous fallout was seen covering an area larger than France.

All of the probes, from Pioneer to New Horizon, were precursors to the astonishing achievements of the Juno probe, outlined above. And Juno will not be the last. Ambitious plans are afoot that will send even more capable and sophisticated probes to the fifth planet and its moons. For example, the European Space Agency is planning the Jupiter Icy Moons Explorer (JUICE) for launch in 2022.

Jupiter's Atmosphere

The most unusual, striking, and beautiful feature of this lord of the Solar System is its conspicuous gaseous envelope. Jupiter's vibrant covering of parallel stripes is more intricate, gaudy, and vivid than any other in the Solar System. Although the psychedelic, ever-evolving geometric patterns might strike a casual viewer as elegantly simple, closer examination reveals that the seemingly clear-cut striations are composed of turbulent whorls, eddies, waves, and vortices that flow in different directions, merge, change colors, and otherwise behave as if governed by undecipherable rules of chaos.

The colors (yellow, saffron, beige, lemon, citron, mustard, orange, auburn, ginger, apricot, peach, tangerine, titian, red, pink, and more) resemble the artwork of an impressionist; the mosaic of powerful, swirling storms resembles a planetary interpretation of Van Gogh's "Starry Night" and seems to demand an emotional reaction. The chaotic, shifting colors can vary from latitude to latitude, and from one day or hour to the next. Through gaps in the upper layers, bluish clouds can be glimpsed deep below. Jupiter's poles are bluish, while its equator is reddish-orange.

The colors come from a witch's brew of chemical compounds drawn up from the mysterious, unexplored depths to burst forth in colorful sprays on the visible outer fringes of the sphere. What chemicals they are and how they produce the varied hues they do in the light of the Sun remain unclear. Such quantities of elements, in such frigid conditions and fleeting light, interspersed with jets and plumes of other substances as they refract, reflect, and radiate electromagnetic energy, create a mix of materials and conditions that has never been studied on Earth; in fact, it could not be studied on Earth. The question of the colors of Jupiter's belts and bands will take a long time to answer, not least because they vary with the temperature, altitude, and composition of the clouds.

Jupiter resembling "Starry Night"

On Earth, clouds are always made of water—there are no other options. Jupiter's colored clouds of ammonia, sulfur dioxide, and other chemicals are interspersed with generous quantities of hydrogen, helium, methane, and hydrogen sulfide, along with more exotic substances like germanium hydride and even arsenic.

The clouds form and flow in bands that can be wider than the Earth. By convention, "belts" are dark in color while the stripes that are light in color are referred to as "zones." The cloud tops are higher in a belt and lower in a zone. Measurements show that winds within a belt or a zone blow at hurricane speeds of about 225 miles per hour.

The dark and light bands alternate and are separated by powerful jet streams. Although the bands are relatively stable, with time they shift or fade temporarily and are nowhere near as straight as they appear in the illustration at right. Roughly speaking, zones tend to flow eastward while belts tend to flow westward. Why this happens remains a mystery. On Earth, winds can, of course, blow in different directions in adjacent regions, but they do not keep blowing in the same orderly pattern for centuries. Earth's winds and jet streams meander across the surface from day to day in ways that are impermanent and chaotic. On Jupiter, some force of nature overrides chaos and randomness, creating the appearance of order. We do not yet know what that force or mechanism is, but scientists are searching it out, trying

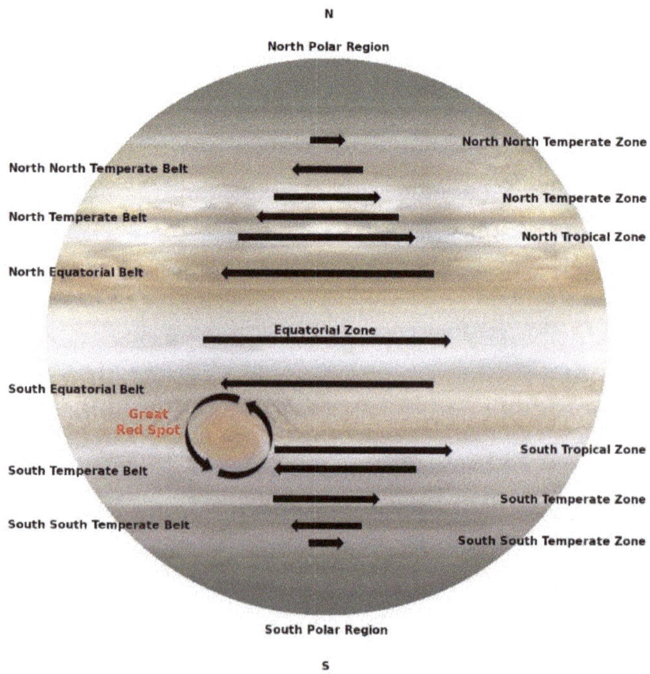

various algorithms and hypotheses in supercomputer simulations to fathom the intricacies of Jupiter's weather patterns. However, some internal power-pump must be driving the winds, and that source of internal power must vary in such a way as to continuously energize those alternating, multidirectional flows. Possibly the bands are keeping pace with rotating cylinders of fluid concealed deep within the planet. Complex convection currents could transfer heat from the center to the upper layers in closed loops running north and south or east and west.

For many years, scientists had no idea how deep Jupiter's colored stripes went. Now, thanks to the Juno probe, it is known that the atmosphere of Jupiter is striped to a depth of thousands of miles. Since Jupiter's uppermost clouds contain crystals of ammonia, there must be a rain of ammonia snowflakes on a lower stratum of the atmosphere. Deep within, there are water clouds like the ones on Earth, but here they represent just one stratum, a single sliver from the many and varied species of clouds that envelop this gaseous sphere.

Ammonia would not be the only rain. Neon dissolved in helium droplets might be showering down onto the inner hydrogen sea, lit up by electric currents in the gaseous medium like a trillion descending neon bulbs and then, perhaps, rising and falling again, solidifying into dense neon hail as the pressure skyrockets.

Jupiter's atmosphere consists mostly of hydrogen gas, with some helium and a smattering of other gases, one of which is methane. The planet's swirling gases produce powerful lightning bolts (possibly the largest in the Solar System) that constantly blast their way through the cloud layers. These ferocious electrical discharges forcibly separate carbon atoms from methane molecules. The free carbon atoms can crystallize under the intense pressure of Jupiter's interior, forming diamonds. (This process has been reproduced by scientists who created slices of pure diamond in a special chamber filled with methane and hydrogen gas. The gases are blasted with electricity and pressurized just as on Jupiter.)

As the diamond crystals grow in size and weight, they fall towards the interior of the planet. Alas, on Jupiter a diamond is not forever. Plummeting toward the planet's core, the diamonds experience such extreme heat that eventually, at 14,400°F, the solid can liquefy, returning the once-precious gem to a state of liquid carbon, perhaps accumulating to form a molten diamond sea.

Storms

The great heat emanating from Jupiter's core, in concert with the planet's furious rotation, spawns monster cyclones. On the Earth, spiral cloud systems can form due to the interaction of high- and low-pressure regions in our atmosphere as they slide past each other. When the winds in such a formation reach 74 miles per hour, it is classified as a hurricane (in the Atlantic Ocean) or a typhoon (in the Pacific Ocean or the Indian Ocean). These storms lose energy and break up when they move over land; since Jupiter has no solid surface to break up these storms (or if it does, it is too far below to have much effect), Jupiter's storms do not die as quickly.

Although they can display a variety of colors, Jupiter's storms usually appear as milky white ovals that speckle the surface. They travel either with or alongside the atmospheric bands in which they are embedded. They are in constant motion, appearing and disappearing on time scales that vary with their sizes and locations. The storms are accompanied by lightning bolts that would dwarf any seen on Earth.

During Juno's passes over Jupiter's poles, it observed huge storms. It found twelve cyclones, each the size of Texas, spinning at the South Pole, while a single cyclone at the North Pole was observed to be as large as the United States. These cyclones are most likely transient phenomena, and future probes will almost certainly detect different storm formations.

The Great Red Spot

The most famous storm in our Solar System is Jupiter's Great Red Spot. It is the most impressive storm we have witnessed on any planet. Measuring about 7,500 miles along its short diameter and 12,000 miles along its long diameter, it is large enough to fit Earth and Mars side by side. Here, heat rises up from the center of the planet, churning a cocktail of volatile chemicals and stoking the "fires" of this beast.

This enduring storm has persisted since at least 1830 and possibly since 1664. Robert Hooke reported observing a great red spot when he turned one of the first telescopes toward Jupiter in 1664. Giovanni Cassini likewise saw a red spot the following year. However, no one reported seeing anything like it again until 1830, and Hooke's spot was evidently in the wrong hemisphere. At any rate, we know the storm has lasted at least 187 years and most likely much longer. No such titanic tempest has been observed anywhere else in the Solar System to have raged so long.

This Great Red Spot appears to us as a very large anticyclone. (A southern-hemisphere cyclone rotates clockwise, while the Great Red Spot rotates counterclockwise, making it an anticyclone.) This great storm reaches five miles above and well below the adjacent cloud layers. Intriguingly, it travels at a variable rate of speed, indicating that it is probably not linked to a single source of convective heat lower down. The color of this storm has been attributed to the presence of complex organic molecules (possibly including red phosphorus), but the color has not remained constant. Over time, it has varied from deep red to pale brown, pink, and even white.

Astronomers do not know what brought this storm into existence or even how long it has existed. Their best guess is that the storm began in a way similar to how storms form on Earth. Pictures taken between 1995 and 2014 indicate that the storm is slowly shrinking, but even so, it will more than likely last for centuries. UCLA spin specialists believe that Jupiter's rapid rate of rotation contributes to the Great Red Spot's longevity.

The Great Magnetosphere

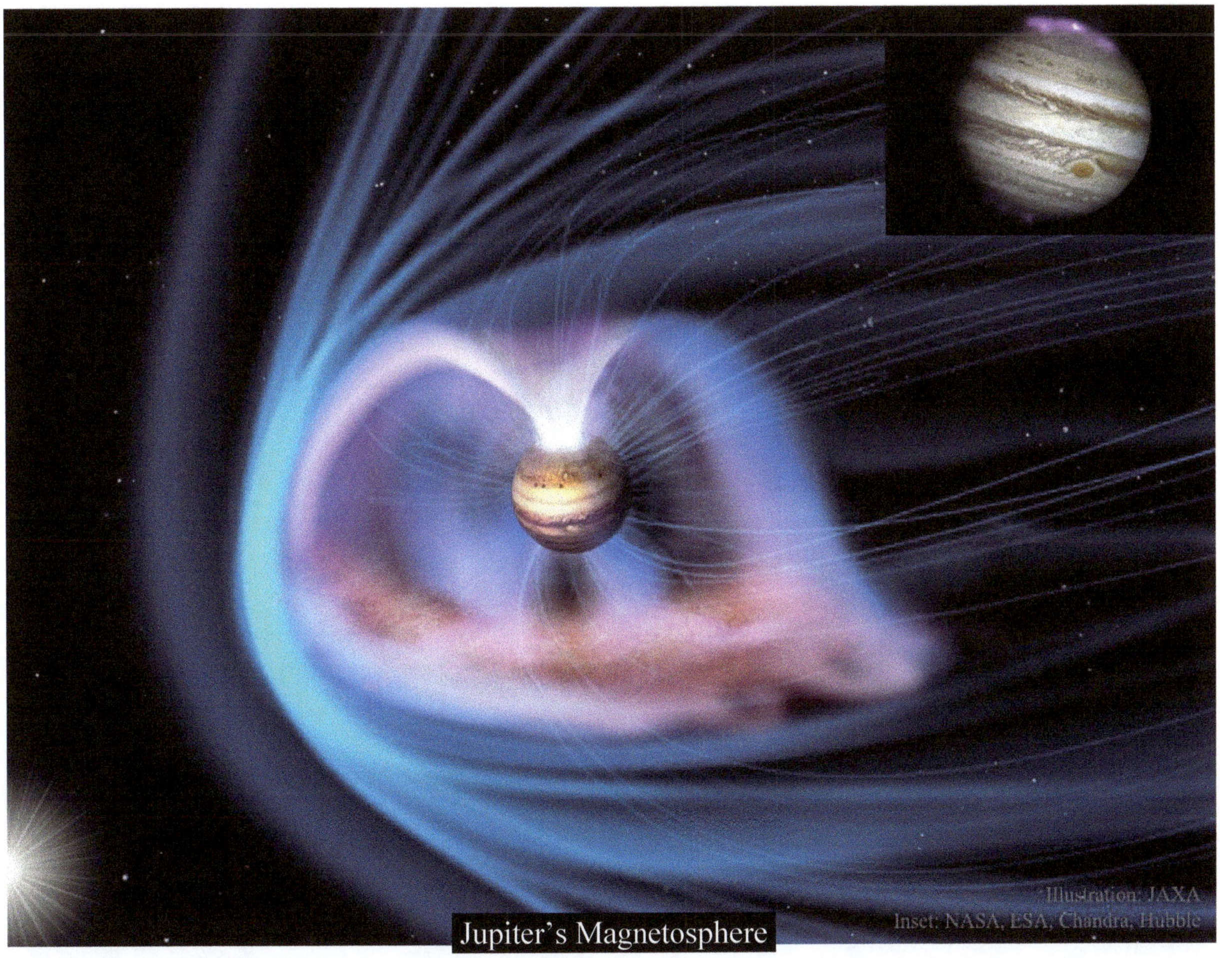

Jupiter's Magnetosphere

Jupiter is in effect an enormous, spinning dynamo, creating a magnetic field that is about 19,000 times greater than Earth's, so strong that it is detectable by Earth-based instruments, although it has also been measured and verified by spacecraft. Small charged particles interacting with this magnetic field may be what makes Jupiter the strongest radio source in the Solar System, although scientists are not yet certain what causes the radio wave emissions.

Jupiter's magnetosphere (above) is the gigantic cavity in the solar wind created by this mighty magnetic field. It has the shape of an extended teardrop enveloping the planet. The round side of the teardrop, called a bow shock, faces the Sun. Here the solar wind is redirected away from and around the planet. On the far side, the magnetosphere tapers off and becomes a thin tail, extending further than the orbit of Saturn. These dimensions make Jupiter's magnetosphere the largest permanent structure in our Solar System. If it were visible, then from Earth it would appear three times larger than the full Moon.

Jupiter's Awesome Auroras

Charged particles trapped in Jupiter's magnetic field create auroras on Jupiter similar to those created here on Earth within our own Van Allen radiation belts; however, Jupiter's auroras are larger than the planet Earth and appear ring-shaped. Jupiter's powerful magnetic field accelerates charged particles to near the speed of light as they spiral into the poles. There, the particles interact with Jupiter's atmosphere and release energy. Cameras on both Voyager and Galileo succeeded in imaging ultraviolet aurora arcs on the night side of Jupiter. The Hubble Space Telescope has taken pictures of ultraviolet auroras on Jupiter's dayside. Earth-based observations have recorded infrared emissions at both poles and imaged the associated polar auroras. The auroras show up best in the ultraviolet portion of the electromagnetic spectrum (see image on previous page) because that is the range in which the solar-wind particles are most likely to fluoresce. However, the auroras are striking at visible wavelengths as well, and are powerful enough to be observed from Earth.

Jupiter's Rings

In 1974, the Pioneer 10 spacecraft was making its closest approach toward Jupiter, about 77,500 miles from the planet, when the probe detected a sudden decrease in the density of charged particles just inside the innermost known moon. The explanation was unexpected: Jupiter has rings.

Jupiter's ring system consists of very dark material and cannot be seen from the Earth. In 1979 the rings' existence was verified by the first Voyager spacecraft. The second Voyager recorded additional pictures of the rings.

Jupiter's ring system is a complex structure spanning 270,000 miles in four distinct sections. There is an outer gossamer ring bounded by the orbit of the moon Thebe and an inner gossamer ring bounded by the orbit of the moon Amalthea. The main ring is about 100 feet thick and extends inward from the orbits of the moons Adrastea and Metis to an inner edge at about 80,000 miles. The final ring is a toroidal halo of particles that is about 13,000 miles thick and extends from the main ring inward to a little over 60,000 miles from the planet's center.

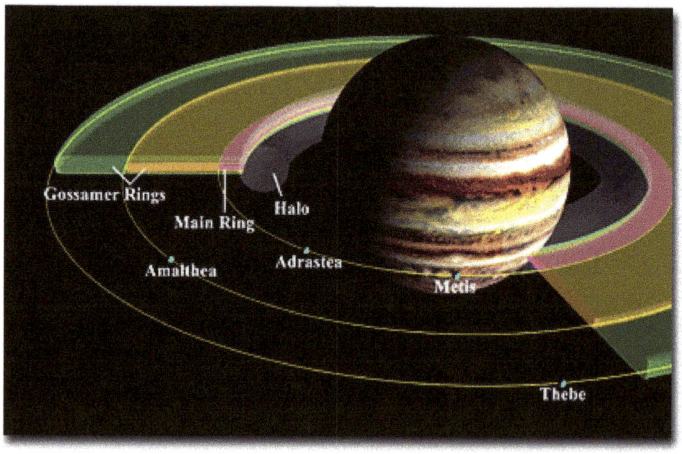

Rings require a constant source of new material. Unless they are regularly replenished, the already tenuous rings would eventually attenuate into nothingness. The four moons named in the preceding paragraph are the apparent source of the rings. The ring particles are generated by impacts on these moons by micrometeoroids and other cometary debris. It is possible that volcanic activity on Io contributes to the rings as well. Because their gravity guides and controls the flow of ring particles around the planet, Amalthea, Adrastea, Metis, and Thebe are referred to as "shepherd moons." Yet undiscovered additional moons may also be contributing to the process. The shepherd moons are not to be confused with the much larger Galilean moons, which orbit farther out from Jupiter.

Jupiter's Moons

The moons of Jupiter occupy an important place in the history of astronomy and thus in the history of humanity and its quest for knowledge. When Galileo first turned his revolutionary telescope on Jupiter, he saw something that startled and astonished him: four tiny dots of light spinning around the mighty orb of Jupiter. Unbeknownst to Galileo, these four mini-planets, until then unseen and unsuspected, were the elfin heralds of a new age, the dawn of modern astronomy. He wrote:

> But what exceeds all wonders, I have discovered four new planets and observed their proper and particular motions, different among themselves and from the motions of all the other stars; and these new planets move about another very large star [Jupiter] like Venus and Mercury, and perchance the other known planets, move about the Sun. As soon as this tract, which I shall send to all the philosophers and mathematicians as an announcement, is finished, I shall send a copy to the Most Serene Grand Duke, together with an excellent spyglass, so that he can verify all these truths.

According to the most widely accepted scientific theories of the early 1600s, all heavenly bodies were supposed to move in circular orbits around the Earth. This accorded nicely with prevailing theological ideas, which acknowledged the Earth and human beings as the centerpiece of God's creation. Galileo's discovery of moons clearly and undeniably orbiting another planet yanked the Earth from its erstwhile pride of position, relegating it to the mediocre status of just another planet. Those observations in 1610 forever altered humankind's perception of the cosmos, putting to rest once and for all the idea that the universe revolved around the Earth. Undeniably, the Earth is special and unique, but for a very different reason: it is our home. Galileo's telescope and diagrams convincingly demonstrated that the universe does not revolve around us. Unfortunately, it took a century or two for all scientists to be convinced.

Jupiter's moons played a crucial role in another discovery: the first reasonably accurate measurement of the speed of light. In 1676, the Danish astronomer Olaus Römer observed a delay in the eclipse of a moon of Jupiter depending on whether the Earth was moving toward Jupiter in its orbit or away from it. He reasoned that the delay was due to the additional distance that light had to travel to reach Earth when it was receding from Jupiter. Since he had a rough idea of the size of Earth's orbit from others' observations, he divided distance by time to estimate the speed. Römer arrived at a value of 140,000 miles per second, a remarkably accurate finding, considering that the accepted value today is exactly 299,792,458 meters per second, which works out to about 186,000 miles per second. Before Römer's time, few believed that light had a speed; instead, the most widely accepted view was that light traveled instantaneously from point to point. This was the view of Aristotle. Even those who, like Galileo, suspected that the speed of light was finite had no idea how to measure that speed. Galileo evidently tried to measure the speed of light with lanterns on widely separated towers, but of course, this experimental methodology was too crude to yield meaningful results.

What are these moons that taught us so much even when they were barely distinguishable specks of light on the lens of a four-power telescope of the 17th century? In fact, they are a little bit of everything—big, small, dense, light, dry, wet, sterile, or, just maybe, teeming with life. Today these bodies are frigid agglomerations, remnants of the embryonic Solar System, shards of icy mountains, and shattered fragments of wandering rocks. They have been transformed over time by heat, chemical interactions, and solar radiation into the wondrous worldlets we observe, study, and probe. They represent a fascinating range of conditions, characteristics, and environments, and it is this incredible variety that draws us to them. Even so, they are not static. In time they will evolve, just as the Earth and the planets did, into something more. In the far future, these moons will become increasingly hospitable to life. Four billion years from now, as the Sun begins to expand in its final epoch, they may achieve their fullest life-harboring potential as they transform into warm water-worlds in the aging Sun's radiant embrace.

The Four Galilean Moons

The Galilean moons, the ones discovered by Galileo in 1610, are huge and impressive—truly worlds in their own right. From top to bottom in the image at right, they are Io, Europa, Ganymede, and Callisto. Ganymede is larger than the planet Mercury, and three out of four of these moons are thought to possess internal oceans of liquid water, more water even than Earth. Their underground oceans could be heated by a process called tidal friction, providing a suitable and stable environment for life.

The question of the existence of life in the liquid environments of Jupiter's Galilean moons is of great interest to scientists and laypersons alike. For many, the urge to search for life there is irresistible. As mathematician David Hilbert so cogently and succinctly expressed, "We must know. We will know." The human impulse to explore and discover will carry us inexorably on toward those moons, and we will persist in our investigations until we attain the knowledge we seek. Once we have had the chance to explore those distant, alien oceans, then whether we find life there or the absence of life, it will represent a watershed moment and have a profound effect upon our understanding of the universe and humanity's place in it.

The Galilean moons have magnetic fields similar to Earth's. Each of them orbits Jupiter under the powerful influence of Jupiter's magnetic field, which stirs up electrical currents within a conductive layer inside the moons. The magnetic fields thus created are large enough to have been detected by the probes that we have sent.

All four Galilean moons are tidally locked. In other words, they orbit with the same face always directed towards Jupiter. The three innermost orbit in a resonance pattern of 4:2:1. That is, for each orbit of Ganymede, Europa completes two orbits and Io completes four. These four moons were named by the German astronomer Simon Marius, Galileo's contemporary and rival, who is believed to have discovered them independently of Galileo. The Galilean moons, in order from innermost to outermost, are Io, Europa, Ganymede, and Callisto. By convention all of the Jovian moons are named after lovers or favorites of Jupiter; for example, Amalthea was his (non-human) nurse.

In addition to the "big four," Jupiter is orbited by a diverse assortment of smaller moons. In fact, any rock, no matter how small, that orbits a planet qualifies as a moon, since the astronomical authorities have neglected, as of yet, to establish strict guidelines. So far, 69 bodies, 53 with names and 16 without, have been designated as moons of Jupiter, and more are surely waiting to be discovered.

Io

Io

Io, about the size of Earth's Moon, has a startling and eye-catching appearance. Its bizarre surface resembles a spattering of melted cheeses among daubs of tomato sauce on top of a hot pizza. The splotchy, cheesy hues (white, orange, red, yellow, and brown) are the work of over 400 volcanoes whose depositions have tinted the surface with the colors of a variety of sulfurous chemical compounds, especially sulfur dioxide. Sparse fumes of gaseous sulfur dioxide, vaporized by the hot lava, also make up Io's tenuous and fleeting atmosphere.

In March of 1979, Voyager I captured images of an umbrella-shaped crescent rising above Io. A similar crescent is visible in the image below (next page). At first, scientists were baffled, but eventually, they determined that the crescent was a dazzling volcanic plume rising more than 2,000 feet above the tormented surface. This salient event gave invaluable insight into the true nature of this moon: Io is the most volcanic body in the Solar System.

For the past quarter century, our probes have continually captured images of Io erupting. Probes have even sailed through the volcanic plumes and sampled the space-borne ejecta, which consists mostly of ionized sulfur and oxygen. The particles either fall toward Jupiter or career outwards toward the orbits of the outer Galilean moons. Neighboring Europa has been a beneficiary of Io's generosity; large quantities of sulfur and oxygen, presumably from Io, have been detected upon its surface. Some of the particles may also find their way into Jupiter's tenuous ring system.

The volcano called Surt (after the king of the Norse fire giants) can throw dust and ash 300 miles into space from Io. Surt has been observed to erupt with as much as 80 terawatts of power, more powerful than all the rest of Io's volcanoes combined. Its eruptions are 5,000 times more powerful than the eruption of Mount Etna in 1992.

Loki Erupting on Io

Loki, the largest volcano on Io, shown erupting in the image at left, pours out more than 10,000 square yards of lava per second during its most energetic outbursts. Loki, the mischief-maker, was a shape-shifting trickster god of Norse mythology who was sometimes helpful, at other times malicious, but always unpredictable.

The volcano Prometheus can throw its plume 62 miles into space. It sits in the middle of a lake of lava 160 miles long, colored red with bright sulfur and sulfur dioxide. This mighty extraterrestrial volcano is named after a Titan of Greek mythology who was harshly punished for tricking Zeus. Zeus had Prometheus chained to a mountain, where an eagle daily devoured his liver, which would regenerate itself in time for the torture to be resumed on the following day. It was Prometheus who, in opposition to the wishes of the gods, gave humanity the gift of fire.

The silicate lava spewed forth by Io's volcanic vents is heated to over 2500°F. This lava is of a type known as "ultramafic." Its viscosity is so low that it flows like water. Such lava occurred on Earth four billion years ago. Thus, Io is believed to be a present-day model of the early Earth. Once ejected into space by Io's numerous volcanoes, some of the lava returns and slowly rains down upon the surrounding terrain. So many eruptions occur that the outer face of the moon may be completely resurfaced as often as every few thousand years.

The energy for this volcanic activity begs explanation since radioactive heating is inadequate for a moon as small as Io. The favored explanation is heating due to tidal friction. Io is pulled on by the gravity of Jupiter and the other three Galilean moons, and the rhythmic gravitational kneading of Io's interior by this system of orbiting bodies repeatedly deforms Io's crust. This constant flexing and deformation produces an enormous amount of heat which must be released, and the result is volcanism. Another consequence of this tug-of-war on Io is that Io has a more eccentric orbit than it would follow if Jupiter alone were influencing its orbit.

Io has the highest density of all the moons in our Solar System. The density of Io implies that this tortured moon has a rocky composition similar to that of our own Moon, with very little water. Salt deposits suggest that Io may once have had an ocean, but any ocean it had would have evaporated long ago. Scientists suspect that Io is the driest body in the Solar System.

Io's thin atmosphere undergoes striking daily transformations. For two hours of Io's day (which is 1.7 Earth days long), Io is eclipsed by Jupiter. The blocking of the Sun's light causes temperatures on Io to fall precipitously, and the sulfur dioxide gas in Io's atmosphere freezes and collects on the surface in the form of frost. This causes the atmosphere to "deflate." When the sunlight returns, the atmosphere redevelops, replenished from the frozen sulfur dioxide as well as gases from Io's volcanoes.

Io contributes significantly to Jupiter's magnetic field. Charged particles spewed out by Io are vacuumed up by Jupiter's overpowering gravity, making the field stronger than it would otherwise be. Dust and gas consisting of sulfur, oxygen, and other substances originate from Io's volcanic activity. They bubble up out of the hot lava into the atmosphere of Io, and the portion that does not rain back down onto the surface of Io is eventually carried across the gulf to Jupiter in moving helices.

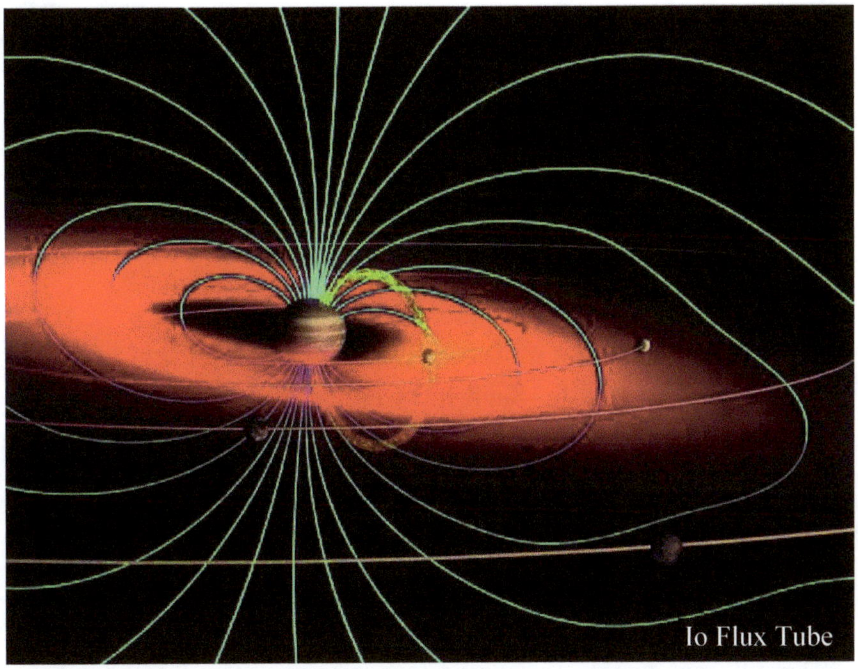

Io is joined to Jupiter by a surprising and powerful feature that makes it unique in the Solar System—and uniquely dangerous. It is an immense flux tube carrying a magnetically-powered stream of energized particles all the way across the gap between Io and Jupiter, connecting them near their poles. This flux tube creates the most intense auroral activity of any moon in the Solar System, more than 100 times as powerful as Earth's auroras. A glowing streak appears on Jupiter's atmosphere where the flux tube touches it. The glow is the result of electrons striking atmospheric molecules. The electric current through the flux tube is a mind-boggling three million amperes at 400,000 volts. Stated another way, this is over a trillion watts of power, which is about one-sixth of the total electrical power produced by generating stations on Earth. The illustration (above) shows Jupiter's magnetic field lines, in green, converging at the poles. The flux tube of Io, the innermost moon in the picture, is highlighted in yellow. The flow of current occurs because Io's proximity to Jupiter and the incredible strength of Jupiter's magnetic field. Also shown in the equatorial plane of Jupiter are the orbits of the Galilean moons. In order, starting with the innermost, they are Io, Europa, Ganymede, and Callisto. The doughnut-shaped red cloud, as seen at right, is a field of exceptionally intense radiation that would be entirely capable of swiftly killing a person and simultaneously frying Juno's circuits, delicate optics, and sensitive instruments. It is this region of lethal radiation that the intrepid probe is avoiding by deliberately "threading the needle" as it passes between the doughnut of death and the cloud-tops of Jupiter's atmosphere.

Europa

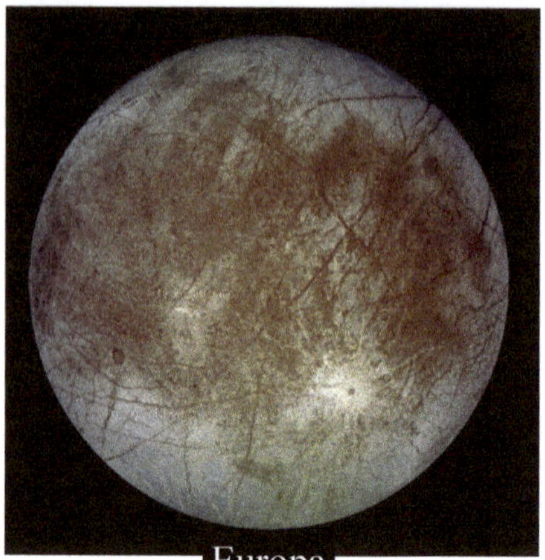

Europa

Europa, the smallest of the Galilean moons (right), is frigid and covered with a frozen shell of ice. The surface of this icy princess is crisscrossed with deep fissures that can be more than a thousand miles long, as well as curved ridges, grooves, and stripes. Beneath all that ice, a salty ocean more than sixty miles deep may slosh and churn under the influence of tidal forces and convection currents. Earth's oceans, by comparison, reach a depth of a mere seven miles at their deepest.

The surface of Europa is marred by just a few impact craters. There are so few because the surface is continually being reformed by fluid extruded from below. A weak magnetic field is present, partly induced by Jupiter's powerful nearby field, with the rest possibly created by flows of conductive saltwater in Europa's subsurface ocean. Deep below the ice and putative ocean lies a dense, rocky core.

Organic compounds and some oxygen have been detected on this moon, which are tantalizing hints of the possibility of life. Such hints continue to motivate planetary scientists to intensify their search for life by probing and investigating the features and attributes of this miniature world.

In November 2013, the Hubble Space Telescope spotted two giant geysers shooting water 24 miles into space near the South Pole of Europa. Scientists believe that since water is coming out of Europa, it stands to reason that material like sulfur and oxygen can get in. These compounds could be an energy source ("food") for primitive life resembling organisms found on Earth near deep ocean thermal vents. An exploratory mission slated for between 2022 and 2025, the Europa Clipper, will fly by Europa dozens of times and investigate its life-bearing possibilities. The Europa Clipper will also scout out possible sites for a Europa lander.

Ganymede

Ganymede

Ganymede (right) is the greatest of all the moons in the Solar System. Its diameter of over 3,200 miles makes it as large as a planet, and it is more than 1,000 miles larger in diameter than our own Moon. Although Ganymede is larger than Mercury, it is less dense, which suggests the presence of water. In fact, there is strong evidence that Ganymede hides a salty, 60-mile-deep global ocean 95 miles beneath its crust of rock and ice.

Ganymede's surface exhibits distinct patches of dark and light terrain. The dark regions are believed to be older areas. It is in these areas that we find the largest concentration of craters. There are also complex patterns of parallel and intersecting ridges and grooves on the surface.

Ganymede is mottled with craters that occur on both the light and dark terrain. Some larger craters appear unusually bright and are sometimes surrounded by systems of rays. The white coloring suggests that meteors slammed into Ganymede and broke through a thin layer of dust, uncovering a highly reflective layer of water ice below.

Substances found in small amounts in Ganymede's icy surface include solid carbon dioxide, hydrogen peroxide, and sulfur compounds. Evidence for the existence of molecular oxygen and ozone trapped in the ice has also been found. Ganymede, like Earth, has an iron-rich core and a permanent magnetic field that is strong enough to create its own magnetosphere as well as auroras analogous to our northern and southern lights.

Callisto

The great moon Callisto (right), second largest in the Jupiter system, has a unique appeal for human explorers. Of the four Galilean moons, it is the farthest from Jupiter, and thus it lies outside Jupiter's fearsome radiation belts. Humans could never expect to survive on Europa or Io—the hapless explorers' DNA would be scrambled in the first week unless extreme measures were taken to protect against the radiation. But on fair Callisto, we may find a comfortable and relatively serene haven for human pioneers and colonists. Here spacecraft could be serviced and provisioned before heading farther out into the Solar System, likely with a gravity assist from nearby Jupiter.

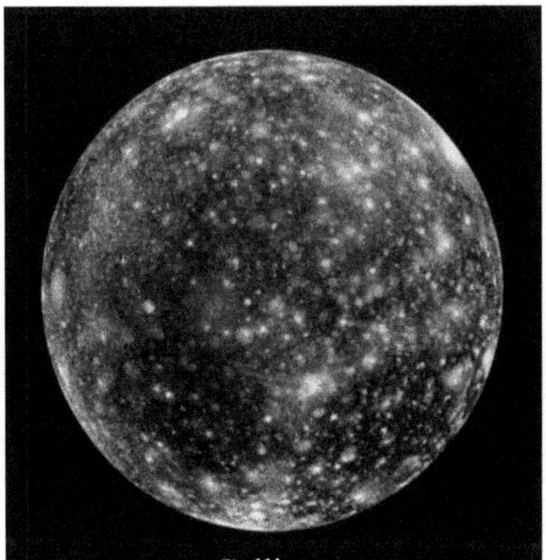

Callisto

Density data indicate that Callisto may consist of as much as 50% water ice. Water ice and solid carbon dioxide are both present on the surface of this moon. There is also evidence that Callisto has a tenuous carbon dioxide atmosphere that is slowly escaping into space. Deep below the icy, rocky surface, Callisto probably harbors a layer of liquid water. This ocean is believed to lie at a depth of approximately 120 miles. The presence of such an ocean would help explain Callisto's magnetic field.

With no smooth plains like those found on our own Moon, Callisto is scarred all over by numerous impact craters. The Galileo spacecraft discovered craters smaller than seven miles in diameter that are hidden by drifts of fine, dark material resembling a mixture of minerals consistent with clay. It is likely that these impact craters were created during the first 500 million years after the formation of the Solar System. This coincides with the period of time when scientists believe our own Moon's craters were formed. The fact that these craters still exist is evidence that very little global heating and melting has occurred on Callisto. There is thus very little in the way of tectonic activity, so human settlements here would be less likely than on other moons to be swept away in a moonquake or volcanic eruption.

Other Moons

Jupiter's outer moon system is made up of a swarm of small, lumpy satellites a few miles wide or less. Some of these bodies orbit in groups, some have retrograde orbits, and many may be the remnants of larger objects. The eight inner moons almost certainly originated with the planet, while the outer moons may have once been asteroids or possibly comets that were snatched out of the cosmos by the grasping gravitational hand of this great planet. The highly eccentric orbits and inclinations of these moons support this view; objects that formed together as a system would lie in roughly the same planetary plane. Evidently, many of these moons once roamed free but have become unintentional vassals of their overlord Jupiter, bound and restrained by unyielding gravitational fetters.

Such small objects become part of Jupiter's community when they are going just the right speed at just the right distance, reminiscent of how NASA recently inserted Juno into orbit around Jupiter. This is, of course, not always what happens. We know that Jupiter often slingshots objects away or simply devours them as it did Comet Shoemaker-Levy 9.

Conclusion

The salient characteristic of Jupiter and its system is variety. Each moon, each atmospheric band, and each spherical shell within the planet's atmosphere is unique, with its own properties of temperature, pressure, and chemistry. With so many environments, so much potentiality, the idea that somehow life has arisen and found a niche in this vast and varied set of environments is not in the least implausible.

Life, as far as we know, requires water as well as energy and the atomic building blocks of life. All of these are richly abundant in the Jupiter system. Europa, Ganymede, and Callisto almost certainly possess deep oceans. For energy, there is ample heat present, as evidenced by Io's surfeit of violent volcanos. And the elements needed for life, whether carbon, hydrogen, oxygen, sulfur, nitrogen or almost anything else, can be found here. Whatever ingredient the recipe book for cooking up life may call for, it is abundantly available on Jupiter or one of its moons. Isaac Asimov expressed the opinion that the clouds of Jupiter may well be the most favorable environment for the existence of life in the entire Solar System.

For this reason, humanity has sought to fathom the arcane, deeply hidden mysteries of chemistry, physics, meteorology, and (just possibly) biology that are locked away within this astonishing system, although its beauty alone is sufficient to lure us on to endeavor to sate our curiosity. A relic from the early Solar System, guardian of the building blocks of life if not life itself, this ancient world and its moons will continue to attract humankind's attention, ambition, and probes for a long time to come.

Mythology of Jupiter's Moons

Io	A river goddess beloved of Jupiter; jealous Juno tormented her with fierce gadflies
Europa	A human princess whom Jupiter abducted and raped
Ganymede	A young prince who was snatched up and carried to Olympus to be cupbearer to the gods
Callisto	A nymph beloved of Jupiter; Juno turned her into a bear
Amalthea	A goat who supplied milk for the infant Jupiter
Thebe	A naiad who bore Jupiter a son
Metis	Cousin and former spouse of Jupiter; mother of Minerva
Adrastea	A nymph who cared for the infant Jupiter

Descending Umbrella of Lava on Io

Io (left) and Europa passing in front of Jupiter

Huge storms, including the famous Great Red Spot, are seen scattered across the face of Jupiter.

20

Saturn: Crown Jewel of the Solar System

I saw Eternity the other night
Like a great ring of pure and endless light.
Henry Vaughan

Enceladus, they say...
as often as he turns his weary sides,
he shakes the solid isle,
and smoke the heavens hides.
Virgil

Ten times farther from the Sun than the Earth, yet only halfway to Uranus, roves the most distant planetary wanderer known to the ancients, the mighty titan Saturn. To the unaugmented human eye, this brilliant point of light is magnificent, but through a telescope the planet's true splendor starts to emerge as it morphs into an awe-inspiring, golden-banded, ringed colossus. In recent times, the powerful cameras of Cassini have snapped images of this creamy opal spheroid at a level of intimate detail and clarity never before captured. No other planet in our Solar System can rival Saturn's polychromatic opulence. It floats serenely through the bleak, obsidian ocean of space, resembling a stunning oval sapphire containing sectors of amber and aureolin, accented by bands of saffron and citrine, and framed by a wafer-thin disk of trillions of highly reflective ice particles that sparkle like precisely cut diamonds. This mesmerizing planet is unquestionably one of the most sublime, stimulating, and singular sights in the cosmos, a haloed pearl characterized by symmetry, mystery—and incongruities. The juxtaposition of contradictory elements that distinguishes this majestic and unparalleled world creates a profound sense of surrealism. Deceptively calm in its enameled appearance, this planet is a swirling maelstrom of frigid gases capable of spawning globe-girdling storms. It holds under its gravitational sway a contingent of utterly normal yet intriguingly unique satellites, along with a seemingly stable whorl of glittering particles that unexpectedly and inexplicably array themselves into grooves, gaps, and spikes.

Saturn is one of the brightest lights in the Earth's night sky, at times the third brightest after the Moon and Venus. Like a star, Saturn is composed mainly of hydrogen, making it the type of planet referred to as a gas giant. However, unlike a star, Saturn does not fuse hydrogen.

At a distance of about nine hundred million miles, Saturn is the sixth planet from the Sun and takes 29.4 Earth years to complete one orbit. It is the second largest planet in our Solar System in both size and mass: 764 times the volume of Earth yet only 95 times as massive, making it the least dense of all the planets. Saturn is tilted substantially, by almost 27 degrees to its orbital plane. (This is slightly different from the tilt of the rings.) This tilt gives Saturn seasons like Earth's, but on Saturn, each season lasts about seven years.

Although planets are usually thought of as spheres—indeed, spherical shape is part of the International Astronomical Union's definition of the word "planet"—Saturn is less spherical than any of the other planetary bodies of the Solar System. This oblateness is caused by its rapid rate of rotation—one rotation every ten hours—which flings Saturn's diffuse, gaseous bulk outward, creating a sizable equatorial bulge. Though only second in mass, bulk, and rotational speed in our Solar System, Saturn is arguably first in the hearts and minds of astronomers and anyone who gazes upon its unique loveliness.

Mythology

This mighty orb was named after one of the most powerful figures of ancient Roman mythology. The original Saturn was a titan, one of a race of beings even more powerful than the gods—until, that is, Saturn and his fellow titans were toppled from their position of power as the Olympian gods violently seized their place atop the divine hierarchy.

Saturn, son of Uranus (Heaven) and Gaia (Earth), was informed by prophecy that he would suffer defeat and humiliation at the hands of his offspring. To circumvent this insidious threat to his reign, he conceived the straightforward scheme of swallowing each child as it was born. Thus, the first five of his offspring were consigned to his belly. Jupiter avoided the fate of his older siblings thanks to a clever stratagem concocted by his mother, Cybele. Unbeknownst to her husband, she substituted a swaddled rock in Jupiter's place, which Saturn duly swallowed.

Saturn's attempt to nullify the threat posed by his children proved futile. His offspring eventually emerged unscathed from his digestive tract, and under Jupiter's leadership commenced a great war to establish their supremacy. Saturn was defeated in this war (known as the "Titanomachy"), and Jupiter seized dominion over the cosmos. The rest of Saturn's regurgitated children arose to constitute a new divine nobility, and together they ruled the universe. These included Juno, goddess of marriage; Neptune, god of the sea; Pluto, god of the dead; Ceres, goddess of grain; and Vesta, goddess of the hearth.

Although in the aftermath of this cosmic civil war the titans were relegated to secondary status, eventually a reconciliation occurred, and Saturn was permitted to rule over a so-called Golden Age, a legendary period of perfect peace and happiness in Italy. In light of this myth, Saturn was revered by some as a god of agriculture and wealth.

One of the more prominent festivals of ancient Rome was held in honor of Saturn. It was the Saturnalia, a winter festival beginning on December 17 and lasting seven days, which included the winter solstice. During this cheerful festival, sumptuous feasts were consumed, all business came to a halt, and executions and military operations were postponed. At this season, masters traditionally demonstrated good will toward their servants, who were given first position at the head of the table and served by their masters. The Christian Church later wove these rituals and traditions, along with others from diverse cultures, into the celebration we call Christmas.

The ancient symbol that astronomers and astrologers use for the planet Saturn is ♄, which represents a sickle, and recalls Saturn's status as an agrarian deity. Similarly, more than 2700 years ago, long before the time of the ancient Romans and Greeks, the ancient Assyrians inhabited the fertile land between two rivers in the country we call Iraq today. These early Mesopotamians assiduously studied the heavens and created intricately inscribed clay records from which modern-day investigators can deduce that they identified the bright planet Saturn with Ninib, their god of spring.

Rings

Today, the rings are the single feature for which Saturn is best known, but until Galileo turned his improved version of the newly invented telescope on Saturn, in the year 1610, human beings had no inkling of the existence of Saturn's wondrous rings. Until that fateful day, Saturn appeared to even the sharpest-eyed Earthbound observer as nothing more than a bright point of light. Upon first viewing the rings through his primitive telescope, Galileo could not discern the rings' exact shape. All he could make out was a pair of stray, round smudges flanking the planet. He thought they might be two large moons, like the moons he had previously discovered orbiting Jupiter. However, two years later, the two mysterious moons had disappeared, and then, amazingly, in another two years they returned. By this time the puzzled astronomer was referring to the outer structures as "handles."

Perhaps fearful of ridicule, condemnation, or plagiarism, Galileo shared his astonishing finding in a secret letter to his friend, the scientist Johannes Kepler. Galileo wrote that he had discovered that Saturn was "triple-bodied," concealing his message in an anagram. Soon after this, however, he made his discovery public in "The Starry Messenger," a book in which he also for the first time described the mountains of the Moon and the four largest satellites of Jupiter. For the next 40 years, theories of handles, ovals, and triple planets whirled through the stargazing intelligentsia of Europe until the Dutch astronomer Christiaan Huygens revealed that the structure surrounding Saturn was a ring.

By 1655, Huygens had developed a better way to grind and polish lenses, and with these improved lenses inserted into a four-inch diameter, 15-foot long tube, he was obtaining better views than Galileo ever did. It was this superior optical instrument that empowered Huygens to be the first to spot a moon orbiting Saturn. Huygens kept meticulous records of his observations, and in 1656, on a hand-drawn map of recent discoveries, he labeled a satellite orbiting Saturn as "meus," which is Latin for "mine." The moon he so identified was Titan, the largest of Saturn's natural satellites.

In addition to accurately observing the shape of the rings, Huygens also discovered that the rings were inclined to Saturn's orbit. However, Huygens assumed that the rings were a single solid disk with a substantial thickness and wrote that he saw Saturn "girdled by a thin flat ring, nowhere touching, inclined to the ecliptic."

Approximately two centuries later, in the nineteenth century, James Clerk Maxwell proved mathematically that in order for the rings to exist in a stable configuration, they must consist of a multitude of small particles. Some hundred years after that, the Voyager 2 spacecraft captured images conclusively proving that the rings were indeed composed of billions of minuscule particles. The rings as revealed by Voyager were more complex and amazing than previously supposed, consisting of more than 100,000 individual rings.

Saturn's Rings

Saturn's rings show themselves to observers on Earth at angles of inclination varying from zero to 27 degrees. Thus, an observer on the Earth sees the northern face of the rings for about fifteen years, followed by the southern face for the next fifteen years. In between, the rings are facing us edge-on, and are virtually invisible due to their incredible thinness. It is this cycle that explains Galileo's conundrum about the initial disappearance of his "handles" in 1612. In 2017, the rings were at their maximum tilt, offering Earthbound observers the best view possible.

Saturn's visible rings stretch out to a distance of about 90,000 miles from the center of Saturn, while their breadth from the inner edge to the outer edge is approximately 38,500 miles. They are as thin as ten feet in some places and rarely more than about 300 feet thick. Even at their thickest, the rings are a million times thinner than they are wide. But the strangeness of the rings does not stop there. Close up, the rings, though superficially resembling a grooved phonograph record, show clumps and waves, ripples and twists, spokes and swirls: myriads of shapes constantly evolving as if alive.

Some scientists suspect that the rings are newborn babes, less than 100 million years old. According to this view, the rings exist, for the moment, because Saturn's moons have developed ice-water volcanism, in which the magma is molten ice that is subsequently blasted by powerful eruptions into space. This liquid water then flash-freezes into sparkling flakes of ice that either fall back to the moon's surface or continue to orbit giant Saturn as newly minted members of the ring family. Data from the Cassini probe supports this hypothesis. Cassini found that the sparkling ice crystals of the rings were much too clean to have been in orbit since the birth of the Solar System. If the rings were that old, they would have accumulated much more dirt and dust by now or else fallen into the ravenous Saturnian gravity well.

Given that the rings are relatively young, how much longer will their stately, elegant, and breathtaking dance continue? A few million years? Scientists estimate that material from the rings is raining down upon Saturn at a rate of ten tons per second, and this cannot continue forever. Some experts say the rings may last as few as 10 million more years; fortunately for humanity we have developed technology that allows us to explore our Solar System and photograph its wonders, and with impeccable timing, we have done so at the moment in time when these magnificent, iconic, and ephemeral structures exist for us to behold.

Even with the best telescopes, it is not possible to see the sizes of the granules that form the rings. However, conclusions can be drawn from observations of how the rings scatter light. Thus, it is known that the particles around Saturn most likely range from less than an inch to several yards in diameter, with larger ones being much fewer in number than smaller ones. Some larger bodies, called ring moons, may be as large as several miles in diameter. These small moons exist embedded in the major rings, but only a few have thus far been detected.

The rings are named alphabetically in order of their discovery, and from the planet outward they are known as the D, C, B, A, F, G, and E rings. The B ring is the brightest, thickest, and broadest, extending from about one and a half to two Saturn radii. This ring is noticeably separated from the major outer ring by the Cassini division, which is easily spotted through a telescope. This gap was named in honor of the Italian-born French astronomer Gian Domenico Cassini, who discovered it in 1675.

Other gaps in the rings were named after individuals who made significant contributions to knowledge of Saturn. Thus, there is the Huygens gap, the Encke gap, the Keeler gap, and the Maxwell gap. Johann Encke was a German researcher who discovered the gap named for him. James Keeler was an American astronomer who photographically confirmed Maxwell's deduction about the particulate nature of the rings 40 years later. At least one moon exists in the Keeler gap, and it is speculated that most if not all of these gaps contain moons. The gaps may have been created by these moons because of orbital resonance. Orbital resonance occurs when the orbital periods of two or more bodies form whole-number ratios. Resonance creates regularly repeating gravitational tugs that can cause significant changes in the distribution and motion of orbiting materials, ultimately producing the gaps and bands.

The two small moons Pandora and Prometheus have been dubbed "shepherd moons" because of their influence on the F ring. During Voyager I's flyby, these two moons were found to be orbiting on either side of the F ring, which itself had been discovered only a year earlier by Pioneer 11. Prometheus, which is shaped like a potato, pushes the ring outward while Pandora, the oblong moon outside the ring, pushes the ring inward. Cassini obtained a spectacular video record of these two moons in action prodding and nudging their ring into shape.

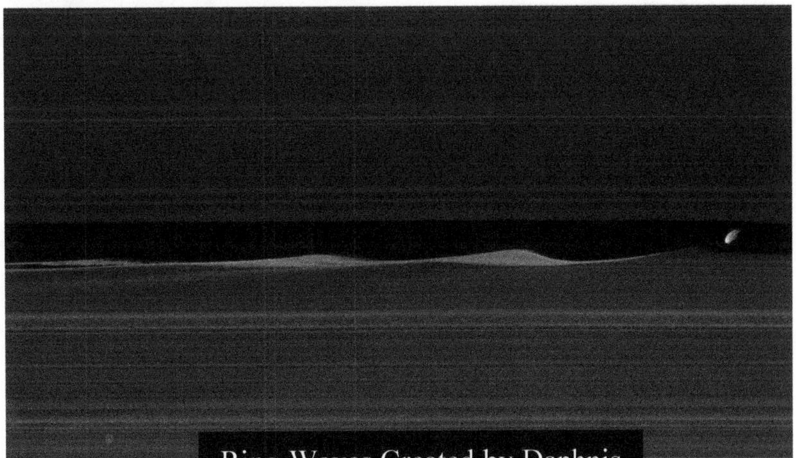

Ring Waves Created by Daphnis

The picture at right, digitally photographed by the Cassini probe, is a remarkable record of the unusual and mysterious shapes and patterns that can arise amidst the extraordinary complexity of the ring system. In the image we can see the tiny moon Daphnis as it weaves and bobs through its gap, climbing and descending like a roller coaster. Sometimes above the plane of the rings and sometimes below, it unceasingly exerts its gravitational influence on the nearby ring particles, and the effect is dramatic. Waves and ripples are created, clearly visible in the image above, where the rings undulate like the surface of a tranquil ocean that gently rises and falls.

The ring particles near Daphnis are traveling at almost but not quite the same speed as the moon. Outside the moon's orbit (farther from Saturn) they move a bit slower, and inside the moon's orbit (closer to Saturn) they move a bit faster. When Daphnis passes the slower-moving, outer edge particles, it pulls on them, and as the speedier particles on the inside edge pass Daphnis, they pull on it. The combined motions are convoluted and add together to produce the effect seen in the image. The visible ripples are up to a mile high—quite small compared to the immensity of the rings, but very large indeed if they were ocean waves on Earth.

Football-field-sized "propellers" (see page 234) are formed in the rings as ice particles flow over tiny moonlets. It is thought that this is a demonstration of the process that formed moons and planets from smaller particles in the early Solar System. However, here this process cannot proceed to fruition due to the shearing forces and tidal stresses caused by Saturn's gravity. The closest that a moon can exist near a large planet without being pulverized by the effects of tidal forces is called the Roche limit, and Saturn's rings spin within this limit.

A still stranger phenomenon was observed by the Cassini probe—a bizarre aberration never anticipated in Maxwell's calculations. Unbelievably, there are two-mile-high piles of ring rubble that form a 20,000-mile-long chain of spikes that project perpendicularly from the outer edge of the B ring. The spikes were detected and their size deduced from the shadows they cast on nearby rings (right). There is no explanation for this anomaly, although it must be due to the gravitational effects of so-called "shepherd moons," small bodies whose gravitational pushing and pulling affect the movements of the B ring's particles.

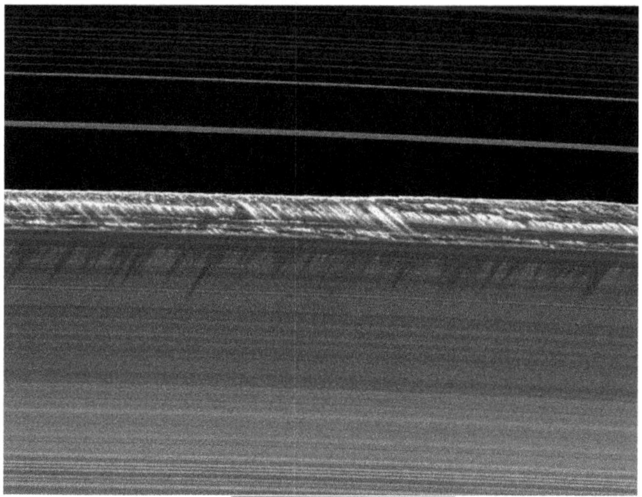

Ring Spikes

Perhaps the most amazing thing about the rings of Saturn is that they are a dynamic system. Even now, they are undergoing perceptible, continuous, and unpredictable change. The rings form a frenzied speedway where particles jostle for position at 100,000 miles per hour, and, at times, although consisting of tiny particles, the ring material appears to flow around and over obstacles like a fluid. The mass of this chaotic assemblage has been estimated at 15 quadrillion tons—an impressive number, though only one hundred-billionth the mass of Saturn, or one ten-millionth the mass of Saturn's largest moon. Even the tiny moon Pandora contains more material than all the rings put together. If Pandora were to wander inside the Roche limit and be torn to bits, it would form rings half again as broad as the picturesque rings we see today. The rings of Saturn are a fertile field for the study of gravitational effects in large systems with many particles. A better understanding of the ring system could provide valuable insights into the development of our own Solar System and the formation of solar systems around other stars.

The Moons of Saturn

Sensitive Earth-based telescopes have detected many moons orbiting Saturn. As of June 2017, 53 of these moons have been given names. The names are drawn from the titans of Greek and Roman mythology, and also from giants of Inuit, French, or Northern European traditions. An exception is Pandora, named for a woman who was married to a titan. A dozen or so other moons bear only numerical designations.

Saturn's moons continually supply new material for the rings. Many scientists believe that without this ongoing replenishment, the rings would eventually disappear because their particles are constantly being dragged down into the clouds of Saturn by the planet's inexorable gravity or escaping to the barren depths of space.

Nine of Saturn's moons are more than 60 miles in radius and were discovered with telescopes before the twentieth century. Most of the more recently discovered moons were identified when astronomers began analyzing the data that Voyager sent back in the early 1980s. Other moons have been found in a variety of ways after 2000, including detection by the Cassini spacecraft beginning in 2004.

All of the inner moons have low-inclination and low-eccentricity orbits with respect to the planet. The eight largest moons probably formed around Saturn in the same way that the planets in our Solar System formed around the Sun, while the others are most likely captured asteroids.

Titan

Not quite half a century after Galileo's world-shaking discovery that Jupiter possessed moons, Huygens was the first to find a moon circling Saturn. The moon he found was Titan (right), the largest and certainly the most famous of Saturn's moons. At 3,200 miles in diameter, it is the second largest natural satellite in our Solar System, larger even than the planet Mercury. Orbiting at a distance of 450,000 miles from Saturn, Titan takes about sixteen Earth days to orbit Saturn once. Unique among all the moons of the Solar System, Titan is a frigid sphere that possesses auburn clouds, a thick atmosphere, and flowing liquids.

Titan

The surface of Titan as seen from the Earth is permanently veiled in a bright brownish-red haze. The picture above was taken by the Cassini spacecraft on November 13, 2015. Bright methane clouds are visible at the bottom of the photograph, near Titan's South Pole. The brownish-red haze is Titan's thick atmosphere, which consists mostly of nitrogen, with traces of methane, ethane, acetylene, ethylene, hydrogen cyanide, and carbon. Ultraviolet radiation from the Sun drives chemical reactions that produce the orange-brown haze. The temperature on the surface of this exceptional moon is about -300°F.

As part of the Cassini mission, a detachable probe named Huygens was released and entered Titan's atmosphere at 12,000 miles per hour before descending by parachute to the moon's surface. It landed near Titan's equator on a rock-strewn floodplain that resembled the surface of the red planet Mars. During the 150-minute descent, the lander captured the roar of Titan's 275-mile-per-hour winds, detected hundreds of organic compounds, and took astounding photographs. The red-gold landscape of Titan resembles the burnished copper of a new coin. Like Earth and no other moon, Titan has river channels, wind erosion, and dunes. The atmosphere is very dense, 1.6 times as dense as Earth's, and is about 200 miles deep as compared to Earth's at 60 miles deep. Mountains of water ice form rigid, rock-like features over which liquid hydrocarbons can flow. The image at right was returned on January 14, 2005, by the Huygens probe during its descent to the surface of Titan.

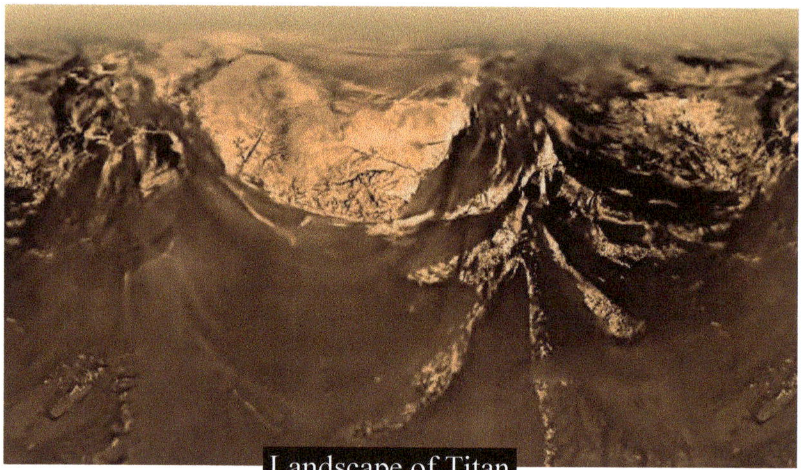

Landscape of Titan

The dunes of Titan are larger than any on Earth, over 300 feet high (see page 242). Their ridges run parallel, often miles apart, and the dune fields stretch across hundreds of miles of the desert-like terrain. Unlike dunes on Earth, these dunes are cohesive and do not flow or shift easily. The material of which they are made is sticky, possibly due to electrostatic charges like those that make polystyrene beads stick together on Earth. This would mean that Titan is the home of electrified dunes.

The dunes can be sculpted almost instantaneously by noisy and unpredictable winds—noisy because Titan's thick atmosphere conducts sound better than Earth's atmosphere does, and unpredictable because the winds can change direction suddenly. The dunes are not made of sand, at least not the familiar inorganic sand of Earth; the most likely candidate for the dune material is a mixture of hydrocarbons, that is, substances like methane, ethane, butane, and similar compounds. On Earth, these substances are usually encountered as gases, but in the frigid temperatures of the outer Solar System, where frozen water is harder than stone, these materials can form sand-like solid grains. It has been calculated that the quantity of methane on Titan is equal to 300 times the Earth's total natural gas reserves.

Five years after Huygens landed near the equator of Titan, the Cassini orbiter caught sight of the first liquid seen on another body in our Solar System besides Earth. Cassini had discovered Kraken Mare, the largest sea on Titan. Instead of water, the seas on Titan are filled with liquid hydrocarbons. Later, Cassini discovered more hydrocarbon lakes near Titan's North Pole.

A similar lake of simple hydrocarbons, dubbed Lake Ontario, is located near Titan's South Pole. It is the largest lake on Titan, although there are many others, particularly in the northern hemisphere.

Among the many features found on Titan are islands, rivers, dry lakebeds, and fjords. Some islands, nicknamed "magic islands," pop into existence in the middle of a lake and then disappear. Scientists speculate that these mysterious islands may consist of masses of nitrogen bubbles effervescing out of solution.

On the beaches of Titan, solid, liquid, and gas meet in an environment with the chemical complexity and energy necessary for the evolution of life, although there is as yet no evidence of life there. The whole system evidently fits together in a hydrocarbon cycle not unlike Earth's water cycle, but with methane clouds, acetylene rain, and lakes filled with liquid natural gas. However, the drops of acetylene rain would be bigger than raindrops on Earth. They would also fall more slowly and make a bigger splash when they plop onto a surface. These differences are due to Titan's lower gravity and higher atmospheric pressure. On Earth, such large reservoirs of flammable methane or acetylene would present a constant danger of explosion, but on Titan, no such danger exists. For the hydrocarbons to burn, free oxygen would be required, and there is virtually no free oxygen on Titan.

Because methane breaks down in the presence of sunlight on Earth, it is believed that Titan's methane should have decomposed and dissipated long ago. Therefore, there must be some process that replenishes it, perhaps an underground, aquifer-like ocean of methane from which methane slowly escapes onto the surface.

Titan Rocks with Haze

Another oddity discovered by the Huygens lander was that although hydrogen was relatively plentiful at higher altitudes in Titan's atmosphere, it is missing at lower altitudes. This is a tantalizing hint of the possibility of life because one explanation for the disparity is that something might be consuming the hydrogen.

Titan's thick organic haze is visible in the image at right of its Mars-like, rocky surface. To investigate the question of whether the haze might indicate the presence of life, scientists have replicated Titan's atmosphere in laboratories on Earth. They discovered that the hazy orange color of Titan's skies could be reproduced by exposing methane to ultraviolet light. Carbon dioxide was then added, and an organic haze was observed to form. Such a haze could exist in varying light conditions and over a wide range of concentrations of methane and carbon dioxide.

The surprising conclusion: Bio-haze could have been present on Earth during its early history, and could have remained for millions or even billions of years. Thus, the atmospheric chemistry of Titan may enlighten us about the chemistry of the young Earth. Titan's atmosphere is thick with organic aerosol particles. A similar haze over the early Earth would have meant 100 million tons of ground-settling organic material each year. This would have provided an abundant food supply for living organisms. A bio-haze could also have shielded prehistoric life from the deleterious effects of ultraviolet radiation and even regulated the climate. Such findings are typical of the broadening understanding science can gain by studying the planets and moons of our Solar System.

Enceladus

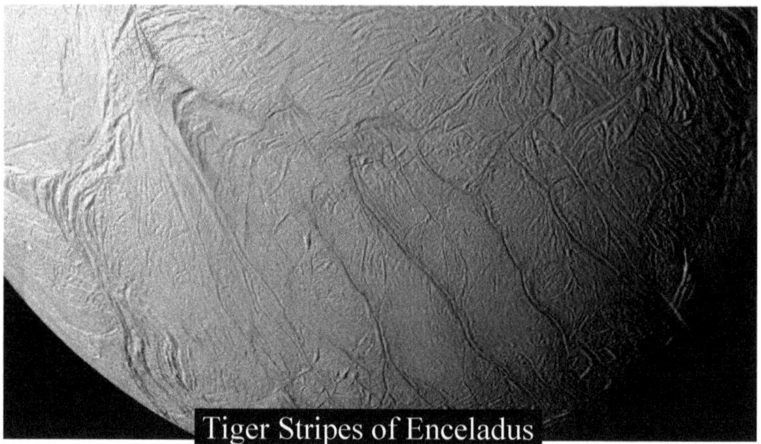

Tiger Stripes of Enceladus

Enceladus was discovered in 1789 by the English astronomer William Herschel and named after one of the giants in Greek mythology who rose up and fought against the Olympian gods. Enceladus was the son of Gaea and Uranus, making him the full brother of Saturn.

Enceladus is an icy, watery world that boasts hundreds of powerful geysers. It is one of the most visible moons of Saturn because the frozen water in which it is covered reflects light well, so well, in fact, that Enceladus is the most reflective body in the Solar System. The surface of Enceladus reflects 90% of the sunlight that hits it. For comparison, Earth's Moon reflects only about 7%.

Pictures taken by Voyager during its rapid flyby showed few large craters. The Cassini orbiter, on the other hand, passed near Enceladus numerous times and was able to capture images clearly showing that the moon's North Polar Region is pockmarked by many craters while the South Polar Region is remarkably free of craters. This inconsistency in Enceladus's surface topography is a strong indicator of recent tectonic activity capable of resurfacing the moon's southern hemisphere.

Dramatic eruptions of water vapor and ice from geysers on Enceladus (below) were observed by the Cassini space probe. This ice volcanism was especially apparent in the South Polar Region, where elongated fractures called "tiger stripes" stretch for more than 80 miles and appear on photographs as bluish-green fractures on a snowy white landscape (above). Cassini imaged about 20 plumes shooting miles into space from the fissures. These jets of liquid, vapor, and ice crystals derive their heat from tidal friction caused by the immense gravity of the planet Saturn. The resonant harmonic orbit of the more-distant moon Dione contributes to tidal heating within Enceladus.

Geysers on Enceladus

In October 2015, Cassini flew through the geysers 30 miles above the moon's surface and detected ammonia, frozen salt water, hydrocarbons, and nano-silicates. In effect, Cassini used its plasma spectrometer to "taste" the watery geyser. Since nano-silicate grains can only form in the presence of hot water, there must be hidden hydrothermal vents deep within Enceladus. In addition, Cassini captured images showing that Cassini's South Pole was radiating more heat than anywhere else on the moon's surface. Earth's South Pole is the coldest region on our planet, yet the south pole of Enceladus is, intriguingly, that moon's hottest region.

Enceladus discharges thousands of tons of ice from its geysers into space each year, and while much falls back as snow, a considerable amount permanently leaves the moon and enters the surrounding space, creating Saturn's broad, diffuse E ring. Scientists therefore hypothesize that the E ring will survive only as long as Enceladus remains geologically active.

The Cassini probe also discovered a slight rocking in Enceladus' rotation. Such wobbling could only occur, scientists say, if there were an outer ice shell surrounding a liquid interior and a solid core. Thus, a shallow global ocean, perhaps only a mile deep, probably lies between Enceladus' rocky center and its 20-mile-thick icy crust. The crust is likely thinner at the South Pole where the geysers erupt. In addition to water, carbon and oxygen also flow in those geysers, and the possibility of extraterrestrial life there cannot be ruled out. Astrobiologists urge persuasively that Enceladus ought to be one of the highest priorities for future exploration and study. It has energy (from tidal heating), organic material, and liquid water. That is a textbook-ready list of the ingredients necessary for Earth-type life. Moreover, nature has provided astrobiologists with a surprising and convenient source: jets that spurt Enceladus's organic material into space, making it relatively easy to sample. An ideal project (artist's conception below) would include a lander to drill into the ice and even release a mini-sub with the capability to search *in situ* for living cells, taste the water, and listen for whale chatter, all to determine if Enceladus has crossed the threshold of life.

Futuristic Enceladus Mission

Other Moons

The moons of Saturn vary widely in size, from greater than 3,000 miles in diameter to as small as 10 miles. They consist mostly of the lighter icy substances that are ubiquitous in the outer Solar System, such as frozen water, ammonia, or nitrogen. The following is a miscellany of Saturn's marvelous moons.

Mimas is heavily cratered and possesses one of the largest impact craters in relation to body size that has ever been discovered. This crater is named Herschel in honor of Mimas's discoverer William Herschel. The crater is 80 miles across although Mimas itself is only 250 miles in diameter. The large crater creates an uncanny resemblance to the Death Star in *Star Wars* (see page 240).

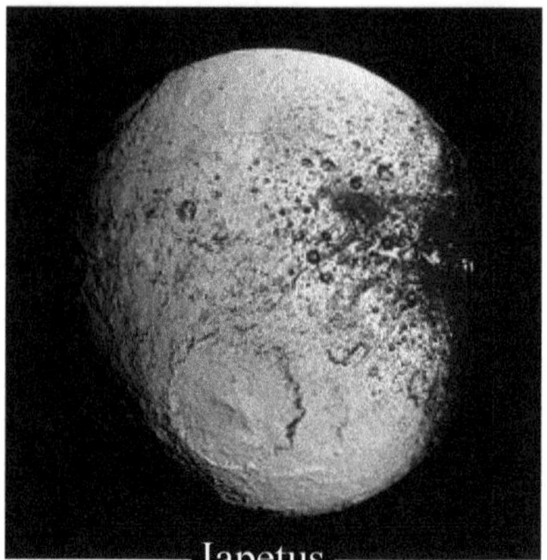
Iapetus

Iapetus (right) has a white side and a black side. As it orbits Saturn, therefore, from the point of view of observers on Earth, Iapetus seems to blink. A scientist would say that it shows differences in reflectivity between its leading and trailing hemispheres. The trailing face is ten times as reflective as the leading face and is heavily cratered, consisting mostly of water and ice. Evidently, any water on the front of Iapetus heats up and becomes a vapor that wafts around the moon and falls onto the back, where it freezes into fresh, shining ice.

Phoebe is Saturn's most distant moon, and it is the only moon of Saturn that moves in a retrograde orbit. This retrograde orbit leads many to believe that Phoebe is a captured cometary body. Phoebe has an orbital period of about three Earth years, which is the longest for any of Saturn's moons. (The shortest period, for one of the close-in moons of the rings, is 12 hours.)

The moon Hyperion resembles a sponge and has a highly eccentric orbit, both of which are unusual for such a large moon. Its behavior is described using chaos theory. This moon, which has a period of about 21.25 Earth days, effectively tumbles in a chaotic pattern with no set motion or stable orbit. Hyperion is the only orbiting moon known to behave in this manner.

Janus and Epimetheus are co-orbital moons, meaning they share the same average orbital distance. Their close proximity causes a dance-like gravitational interaction. As they transmit angular momentum from one to the other and back again, the gravitational force is enough to drive one of the moons into a higher orbit and the other into a lower. At the next close approach, the process is reversed. This process will probably continue until the two moons collide, either creating a single, larger moon or billions of new ring particles. The moons Tethys and Dione are also co-orbital, as are Telesto and Calypso. Janus has a texture that bears a resemblance to a meatball, while the moon called Pan looks like a ravioli.

The Exploration of Saturn with Spacecraft

Carl Sagan wrote, "How lucky we are to live in this time, the first moment in human history when we are in fact visiting other worlds." Indeed, today we can observe Saturn with space telescopes and speedy spacecraft equipped with high-resolution electronic cameras, filters, and instruments capable of "seeing" beyond the visual spectrum to capture the subtle and transcendent characteristics of this remote world. Four probes in particular have added to our knowledge of Saturn, its rings, and its moons. They were Pioneer 11, the two Voyagers, and Cassini.

Pioneer 11 was launched in April of 1973 to explore the outer planets. This audacious spacecraft flew close to Jupiter to obtain a gravity assist, and when it reached Saturn, it made a daring flyby in which it passed through the rings at a distance of only 13,000 miles from Saturn's atmosphere. The craft captured images of two new rings, identified several new moons, and detected the presence of Saturn's radiation belts.

Voyager 2 was launched on August 20, 1977, 16 days before its twin, Voyager 1. However, Voyager 1 flew a faster, more direct route and was therefore the first of the two to reach Jupiter and then Saturn. It cruised by Saturn on November 12, 1980, a feat duplicated by Voyager 2 nine months later. Unfortunately, Voyager 1's faster route precluded it from flying on to Uranus and Neptune, and it was doomed to leave the Solar System altogether after passing Saturn. NASA still hoped that the other twin, Voyager 2, could continue on, eventually flying by the last two known giants of our Solar System, and complete the "Grand Tour." A Grand Tour of the outer planets is only possible when the giant planets are on the same side of the Solar System, which occurs only once every 175 years. Voyager 2, thanks to its longer, slower trajectory, was able to complete the Grand Tour. (The data that it gathered on Uranus and Neptune today constititute most of what we know about these ice giants.)

Together, the two Voyagers captured and transmitted staggering amounts of information about Saturn's atmosphere, rings, and moons. What these probes revealed about the rings was fascinating. In some of Saturn's rings, unexpected structures such as kinks and radial spokes were discovered. Since first being observed, the spokes have apparently disappeared, demonstrating the transient nature of the ring system.

Both Voyager spacecraft used sensitive instruments to gather information on electromagnetic radiation, including the intensities and polarizations of visible, ultraviolet, infrared, and radio waves, as well as magnetic fields and charged particles. However, they could not see through the haze of Titan's atmosphere, and Titan's surface remained a mystery. The task of lifting the veil and penetrating that dense, opaque haze, was later assigned to Cassini's radio wavelength sensors and the Huygens lander.

Cassini-Huygens Spacecraft

Cassini-Huygens (right) was a 2.5-ton spacecraft, as big as a van, launched in 1997 as a joint project between NASA and the European Space Agency. The orbiter portion of the spacecraft was named after Giovanni Cassini, an Italian-born French astronomer who discovered four of Saturn's moons and the largest division in the rings. The Huygens lander was named after Christiaan Huygens, the Dutch astronomer who discovered Titan.

The Cassini-Huygens spacecraft packed a 15-foot antenna for transmitting data and receiving instructions. Its exquisitely sensitive magnetometer was mounted on a

35-foot boom to avoid having its data tainted by stray magnetic fields from the normal functioning of the probe's instruments and power supply. After following a complicated, seven-year, gravity-assisted trajectory, flying by Venus (twice), the Earth, and Jupiter, Cassini-Huygens finally reached Saturn in mid-2004, guided in its path and orientation by a 2400-star map programmed into its memory. By the time it reached Saturn, Cassini's computer hardware was antiquated by Earth standards. Six times a week, Cassini uploaded its precious accumulated data to NASA servers, requiring nine hours to do so each time. A modern smartphone vastly outstrips Cassini when it comes to data transfer speeds.

One of the largest and most expensive probes ever built at that time, it was designed to survey, investigate, explore, and photograph the planet, rings, and moons for several years. This it did with remarkable success. The Huygens probe successfully descended through Titan's atmosphere on a parachute and landed on the surface of Titan in 2005, as described above. The lander transmitted measurements and images to Cassini, which Cassini then beamed back to Earth at the speed of light across a billion miles of space, requiring an average one-way transmission time of 80 minutes. Huygens was the first spacecraft to land on a moon of another planet.

After spending a total of 20 years in the frigid depths of space, including 13 years of yeoman's service near Saturn, the Cassini probe ended its mission in September 2017. It had flown by Titan 120 times, orbited Saturn 294 times, and collected over 450,000 images. As its penultimate maneuver, Cassini spent five months executing 22 breathtaking dives between Saturn and its rings. Skirting the rings so closely was risky, since at such high speeds even the tiniest collision could be disastrous, but planners decided that the risk was worth it as the craft was almost out of fuel and was nearing the end of its mission regardless. The final orbits lasted seven days apiece, each one culminating in the nerve-racking transit which was anticipated to be "seven seconds of terror." Researchers soon came to realize that the region between Saturn and its rings contained fewer dust grains and other particles because they could hear the particles striking Cassini's dish antenna and, in essence, count them. The region Cassini was passing through in its final perilous passages was in actuality relatively free of particles, and this led researchers to describe the region between the rings and the cloud tops as "The Big Empty." The fruit of these hair-raising maneuvers was a set of unique photos of Saturn's rings from the inside out, along with close-up pictures of Saturn's turbulent atmosphere and a treasure trove of data on the planet's interior.

The final maneuver, the "Grand Finale," was a fiery death-plunge into the atmosphere of Saturn, assisted by a gravitational nudge from Titan dubbed "the kiss goodbye." The reason for the self-immolation was to eliminate the possibility of Cassini's colliding with and possibly contaminating one of the moons that might harbor life. It was too likely that, despite all precautions, some earthly microbes may have accidentally attached themselves to Cassini, endured the ride through space, and could also survive a crash-landing on a moon. Any hardy Earth bacteria gaining a foothold on a moon of Saturn would present a deadly threat to any form of life already there, and the unintentional destruction of that life would be an irreparable tragedy. Therefore, Cassini was directed towards the cloud tops of Saturn to meet its final destiny.

Cassini entered Saturn's atmosphere traveling at 76,000 mph, and atmospheric friction heated it from -300 degrees to 2,000 degrees in a matter of seconds. Even as Cassini took its final dive, its antenna was pointed toward Earth and it was still transmitting data, revealing secrets about the composition of Saturn's atmosphere until its signal was finally lost at 4:55:46 PST on Friday, September 15, 2017.

In its last moments, the probe presumably melted and its remaining nuclear fuel combusted. Its final moments must have been spectacular, resembling the descent of a fiery meteor, as its metallic essence was united forever with the gases of the magnificent planet it was sent to study.

Planetary scientists eagerly await the opportunity to surpass the phenomenal achievements of the Cassini mission. Future missions to Saturn may include the Titan Saturn System Mission (TSSM), which will explore Saturn, Titan, and Enceladus. This project is expected to cost approximately $2.5 billion, has an estimated launch date in 2020, and would include the circumnavigation of Titan by a hot-air balloon.

Saturn's Atmosphere

Saturn appears hazy when viewed from the Earth, but the bands, streaks, and haze are really layers of clouds. These cloud formations appear as subtly colored patterns of light and dark bands, which, like the bands on Jupiter, flow both east and west. Winds here have at times been observed to have velocities over 1,000 miles per hour, which is astonishing in itself, but especially in view of the fact that this is almost two-thirds of the planet's rotational speed. The tremendous speed and violence of these winds make them some of the most ferocious in the Solar System, outstripped only by the mega-storms of Neptune.

Saturn's atmosphere is mostly hydrogen (90%), but methane, ammonia, hydrogen sulfide, water, and a host of other substances have been detected. It is rich in helium-3, an isotope which one day might be very valuable as a fuel for fusion reactors. Atmospheric temperatures on Saturn have been recorded as low as -310°F. The visible layers of Saturn's clouds are formed from molecules of minor compounds that condense in the hydrogen-dominated atmosphere. Photochemical reactions give rise to some of the colors seen high in the atmosphere. The highest clouds of Saturn are thought to be made of ammonia crystals and are colored in various shades of yellow, brown, and red.

Scientists have speculated that on Jupiter, Saturn, and the outer ice giants, it rains diamonds. Lightning turns methane into soot, which hardens into flakes of graphite as it falls. As these pieces descend, the temperature and pressure increase until the carbon transforms into a diamond. This process has been modeled at the University of Cardiff in Wales.

Saturn has great white spots, huge storms similar to Jupiter's Great Red Spot. The white spots, which can be as large as the Earth, seem to appear and reappear over thirty-year cycles which closely match Saturn's orbital period around the Sun. Storms on Saturn, which have been observed since 1876, can last for years. These weather patterns appear to be driven by internal heat.

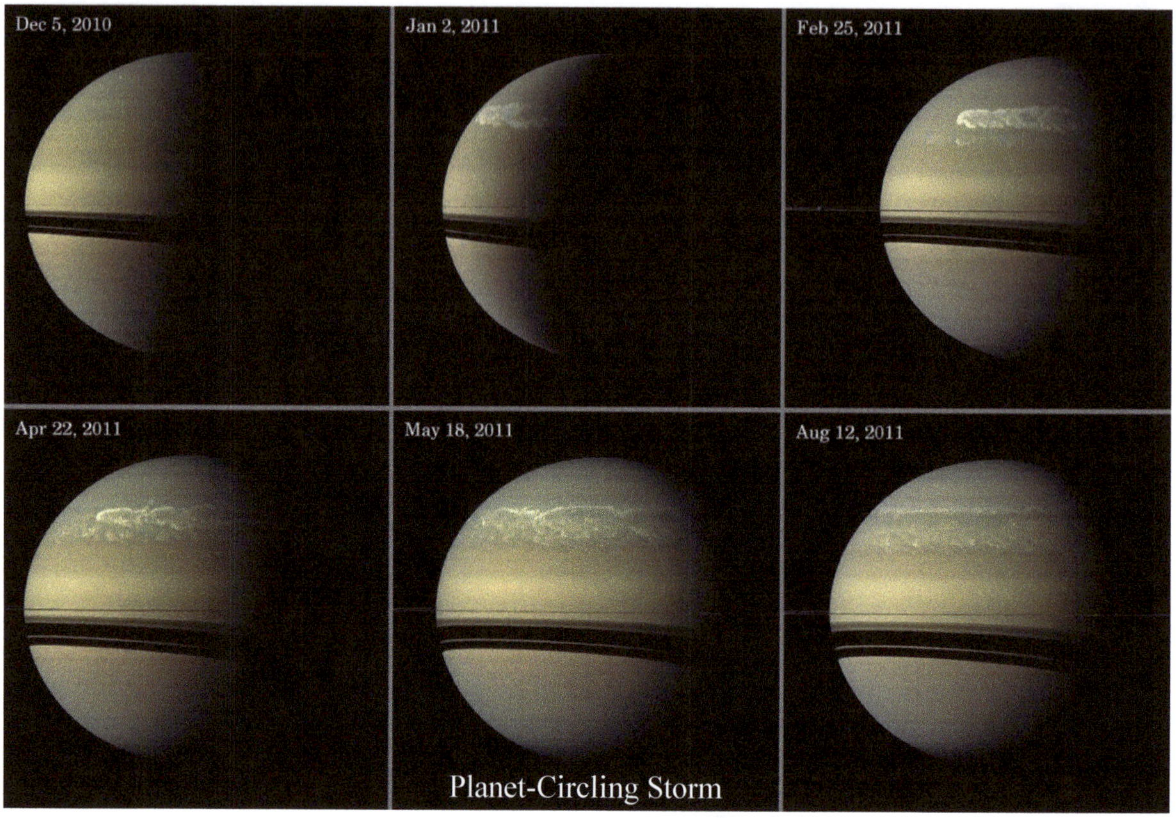

A gargantuan storm in 2010 grew so large that in eight months it completely encircled the planet (above). Finally, its head caught its tail and then the storm quickly dissipated. Scientists are still seeking explanations for the storm's rapid growth, colossal size, and eventual disappearance. Mythologies throughout history have included tales of giant serpents that curl around until they bite their own tail; examples are Ouroboros from ancient Egypt and the Midgard Serpent, a nemesis of Thor from Norse tradition. To behold a real-world phenomenon that mirrors these ancient tales on a distant planet is astounding.

The Interior of Saturn

Since Saturn consists mostly of the very lightweight gas hydrogen, it has a very low density. Saturn experts are fond of noting that if there were a large enough ocean to dunk it into, then Saturn would float. Deep inside the planet, however, 500 miles below the cloud tops, the gaseous hydrogen becomes liquescent and, farther down, turns liquid. The temperature at this depth is about 2,000°F. Even as a liquid, molecular hydrogen is compressible, and to achieve Saturn's mean density of roughly 0.69 grams per cubic centimeter requires pressures over a million times the atmospheric pressure humans are accustomed to on Earth. Pressures this intense occur beginning at about 12,000 miles below Saturn's clouds, approximately one-third the distance toward the center of the planet.

Information about the internal makeup of Saturn is obtained from the study of Saturn's gravitational field, which is not spherically symmetrical but occurs in the shape of a teardrop with the rounded part facing the Sun. This strange gravitational field, not to be confused with the magnetic field, results from Saturn's oblate shape and rapid rotation. The gravitational field was first measured by observing how spacecraft travel in the vicinity of Saturn as well as the shapes of some of the constituents in the rings. Intriguingly, the data show that Saturn is substantially more centrally condensed than Jupiter. Thus, there must be much denser material than hydrogen located near the core of Saturn.

The enormous pressure of Saturn's atmosphere causes hydrogen to be compressed into so-called "metallic hydrogen," which is an electrical conductor. Electric currents in this metallic hydrogen are responsible for the planet's magnetic field. Deep inside the planet, the temperature of this fluid reaches 20,000°F, hotter than the surface of the Sun. Heavier elements in a molten state have probably settled into a small rocky core with a temperature of about 27,000°F. The mass of this core may be equal to about 10-30 Earth masses, but evidence is scant.

The Magnetic Field of Saturn

Saturn has a magnetic field that is almost exactly the same as would be produced by a gigantic bar magnet. The poles of this magnet are aligned almost perfectly with Saturn's axis of rotation (off by less than one degree). Saturn's magnetic moment (in effect, the strength of its magnetic field) is 600 times the magnitude of Earth's. This may seem huge, but it pales in comparison with Jupiter's magnetic field, which is some 20,000 times stronger than Earth's.

Saturn is surrounded by a magnetosphere. A planet's magnetosphere is the region of space in which its magnetic field repels the charged particles of the solar wind. Saturn's magnetosphere, like that of Jupiter, is in the shape of an elongated teardrop. The rounded side of the teardrop faces toward the Sun. On this side, the magnetosphere measures about 20 Saturn radii from Saturn's center, while on the opposite side it stretches much farther in the direction of the orbit of Uranus. Some of the charged particles trapped in Saturn's magnetosphere form belts around Saturn similar to the Van Allen radiation belts which envelop the Earth.

The moons Titan and Hyperion orbit at distances close to the magnetosphere's minimum dimensions. Thus, they occasionally travel outside it and can be pelted by the solar wind. In Titan's upper atmosphere at these times, energetic charged particles from the outer rim of Saturn's magnetosphere collide with neutral atoms, energizing the particles and causing erosion of the atmosphere. A halo of such energetic atoms was observed by the Cassini orbiter.

Saturn possesses ultraviolet auroras produced by the impact of energetic particles from the magnetosphere onto atomic and molecular hydrogen in Saturn's polar atmosphere. The Hubble Space Telescope has taken ultraviolet images of the auroral rings around the poles of Saturn. These images show the symmetry of Saturn's magnetic field and reveal details of the way the auroras respond to the solar wind and the Sun's magnetic field.

Surrounding Titan and extending to the orbit of Saturn's moon Rhea is an enormous doughnut-shaped cloud of neutral hydrogen atoms. A disk of plasma, composed of hydrogen ions and possibly oxygen ions, extends outside the orbit of the moon Tethys and almost to Titan. The plasma rotates in nearly perfect synchrony with Saturn's magnetic field.

The Hexagon Storm

Saturn's Hexagon Storm

At Saturn's North Pole is found one of the strangest phenomena in the Solar System. It is a huge, rotating storm in the shape of a hexagon (right). Truly gargantuan in size, one side of the hexagon is longer than the diameter of the planet Earth. Like Titan, Saturn's gases react to sunlight, hence during Saturn's winter the hexagon appears bluish, while in summer it warms and develops a golden tinge. But what is most baffling about this immense storm is its geometric regularity. It is almost impossible to believe that there are natural processes that could generate such a long-term, well-formed atmospheric disturbance in the shape of a hexagon, but the photographic evidence is conclusive. Researchers have been hard-pressed to explain the phenomenon. One hypothesis is that six regularly spaced cyclones, whirling away beneath Saturn's cloud cover, exert equal pressure on all six sides of the polar storm, resulting in the astonishing geometric configuration. It seems apparent that the hexagon is formed by powerful jet streams; furthermore, the surprising shape may turn out to be as transitory as Venus' twin polar cyclones or the rings' radial spokes. However, the perplexing hexagon has been visible for several decades, which is as long as we have been observing Saturn's poles.

Conclusion

For beauty and wonder, Saturn is genuinely unique in our Solar System and possibly in the galaxy or even the universe. To gaze upon these magnificent panoramas, these cosmic vistas of pure loveliness, delights the mind. Like Ptolemy of old, we can say we are consuming divine ambrosia at the feet of the gods as we relish the rich visual sensations of this most intriguing and captivating world.

The swirling colors are reminiscent of the masterworks of impressionist painters. The originality, uniqueness, and transcendence of this powerful vision reverberate throughout our minds, exuding and exemplifying aesthetic elegance. Every stroke of artistic élan and panache, every bold nuance of tint and shade, chiaroscuro, perspective, shape, texture, and motion pleasurably excites the observer's imagination. In the Saturn system, the majestic splendor and profound wonder of the universe acting in its roles as consummate artist, designer, and builder are on full display.

Saturn's peculiarities fill us with the joy of discovery. To go, to see, to feel this astonishing planet first-hand—the same tug at the heart must have compelled every human explorer from the dawn of time. But Saturn's wonders are unique. Nowhere else do we see such spectacular, intricate rings, a storm in the shape of a hexagon, or a moon with a rich gaseous atmosphere.

The Cassini-Huygens probe was hurled along a five-billion mile trajectory to Saturn and tasked with the mission of seeking answers to a specific set of scientific inquiries. During this intrepid probe's two decades of existence, it performed impeccably and answered many of those questions. However, it did not learn all there is to know about Saturn and its environs. While exploring the Saturn system this vessel exposed further mysteries that pose profound questions, stimulate our curiosity, and exhort humanity to set sail on a return mission sooner rather than later.

Among all the accomplishments of the Cassini mission, perhaps the most significant was the finding that virtually all the Solar System's moons and planets are sown with the raw ingredients of life. Those seeds may be just waiting for the proper conditions before they burst forth and blossom into life.

Could there be a nascent form of life simmering under Saturn's swirling bands of frigid clouds, shadowed by the delicate rings? Or could roiling Titan harbor the complexities of living organisms under its orange haze? Could frosty Enceladus contain a second genesis of life thriving in nutrient-rich salt water at the mouth of a hydrothermal vent at the bottom of a subsurface global ocean? When or if we discover this to be so, our lives and our world will be forever changed. But even then, it would be just one more wonder to add to the list of marvels that characterize Saturn and its system of moons and rings.

PICTURE GALLERY OF SATURN

Saturn photographed in infrared. Reddish regions are warmer, blue are colder.

The delicate shadows on Saturn in this photograph are being cast by the rings.

Saturn from Above

Inside-out view of the rings taken by Cassini just before plunging into Saturn's atmosphere

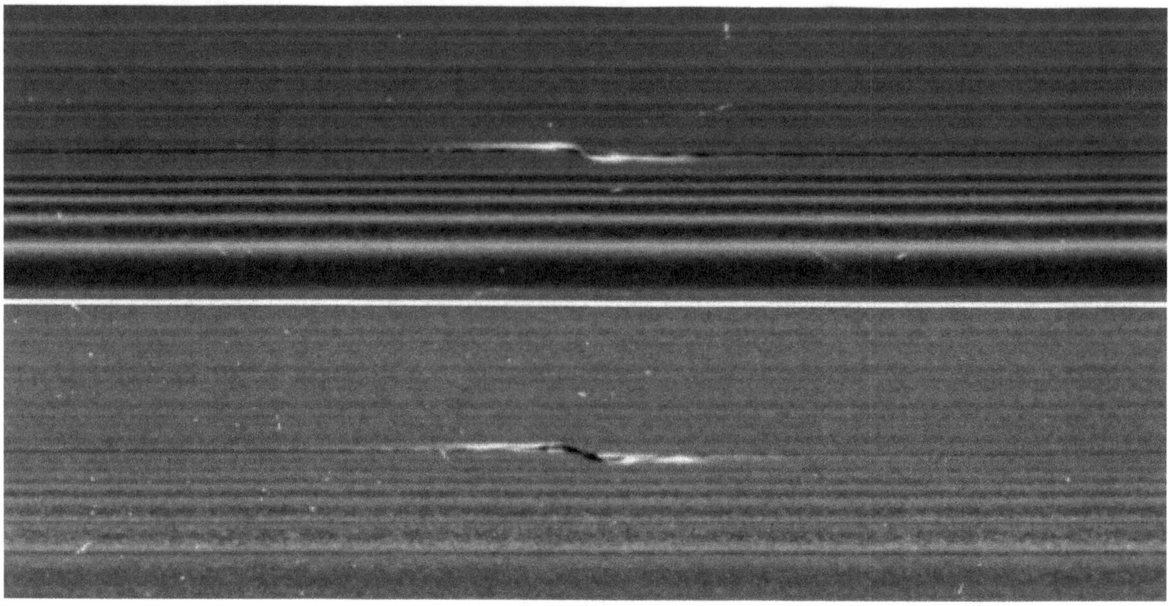

“Propellers,” anomalous features of Saturn’s rings

The bright point of light is the star Aldebaran. As it passes behind the rings, information from how its light flickers can be used to make deductions about the density and other characteristics of the rings.

Prometheus, Pandora, and Epimetheus among the rings

PICTURE GALLERY OF SATURN'S MOONS

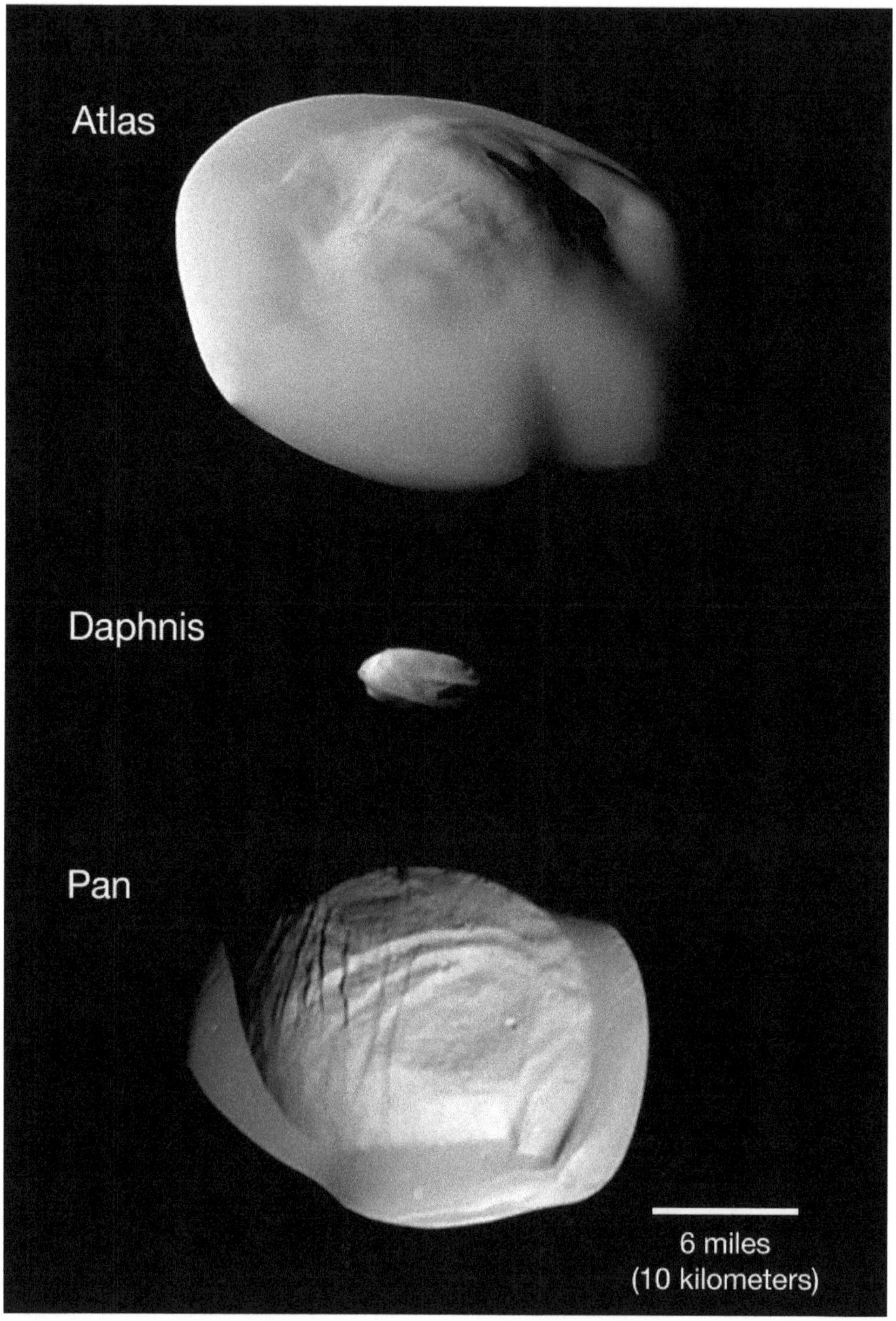

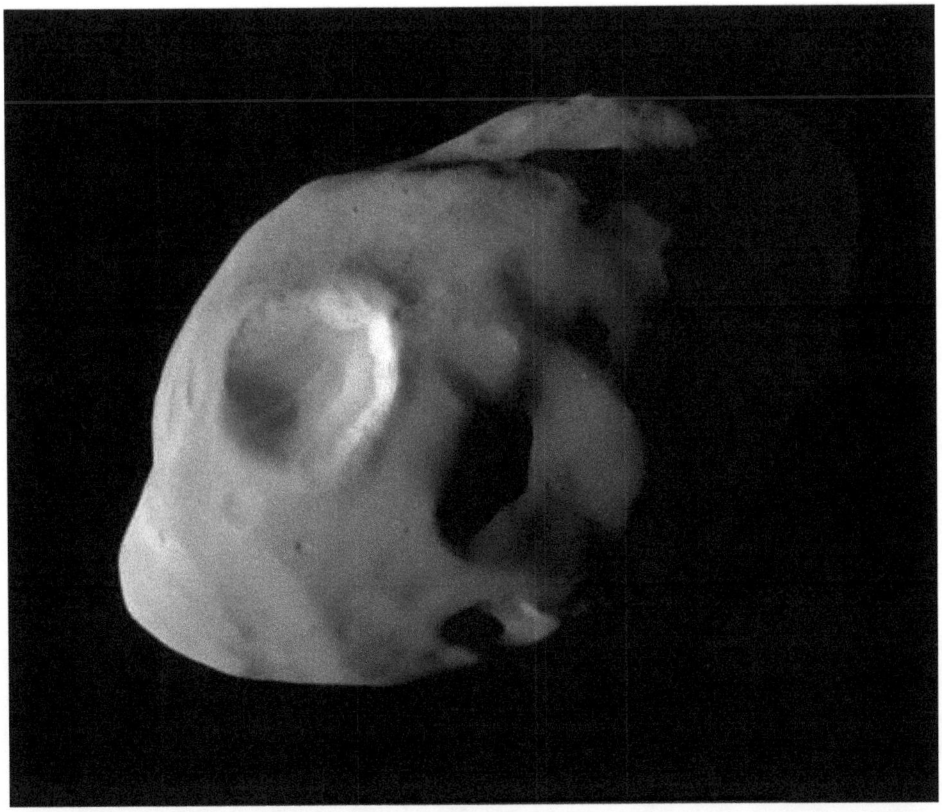

Pandora

Left to right: Prometheus, Rhea, and Janus among the rings

Odysseus Crater on Saturn's moon Tethys

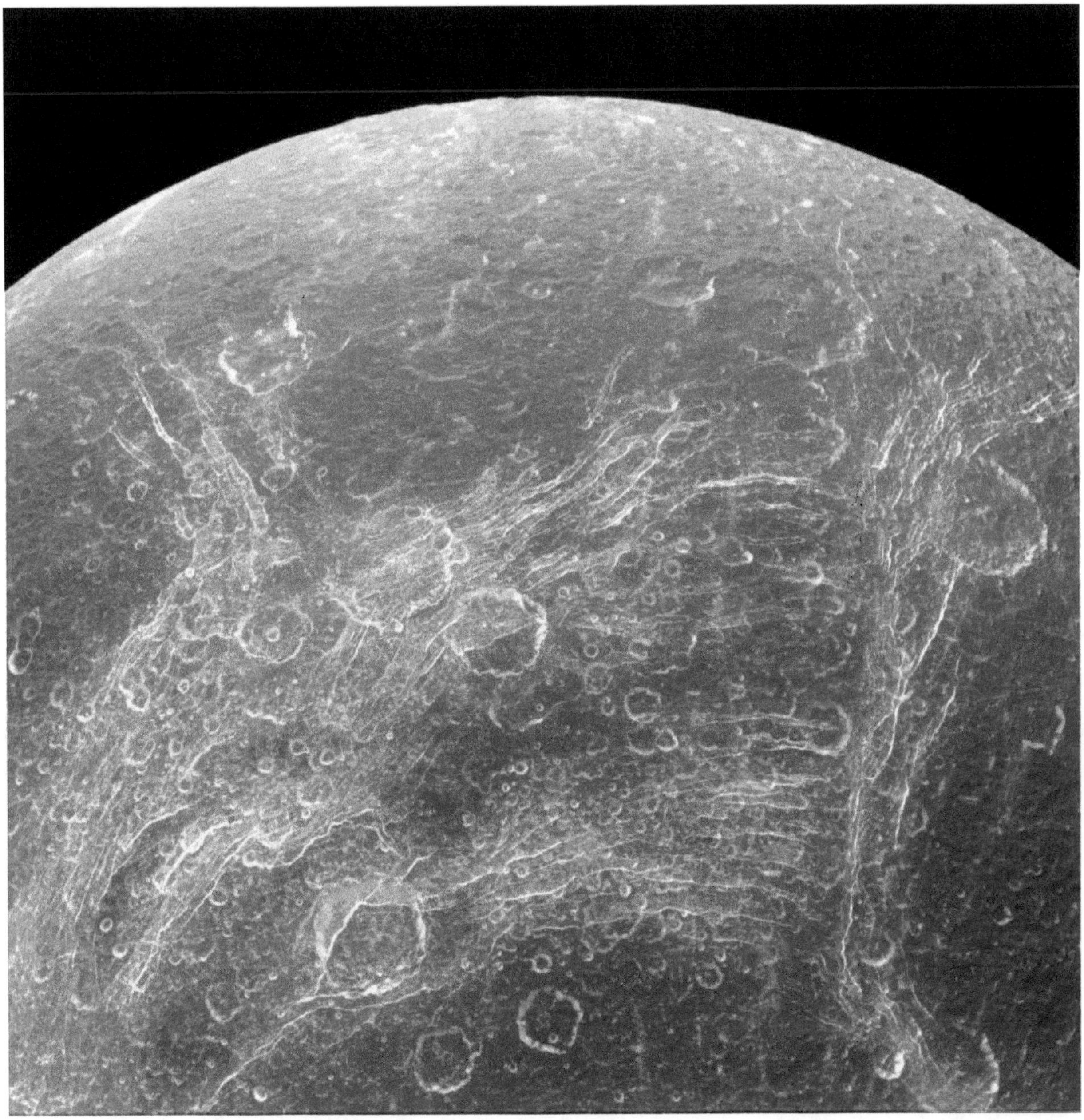

The moon Dione

Mimas as the "Death Star"

Titan in front of Saturn and its rings

Dunes on Titan (above) and Earth (below).

21
The Ice Giants

Neptune

We see it [Neptune] as Columbus saw America from the coast of Spain. Its movements have been felt, trembling along the far-reaching line of our analysis with a certainty hardly inferior to that of ocular demonstration.
William Herschel

...by the shore of Triton's Lake...
I will clear my breast of secrets.
Euripides

For thousands of years, people believed there were five planets: Mercury, Venus, Mars, Jupiter, and Saturn. Those five were easily differentiated from stars because the stars followed a constant, fixed, and predictable course, while the planets looped and wandered, speeding up and slowing down as if they had a will of their own. It was believed for millennia that these exceptional, translocating illuminations were the homes of the gods—or even the gods themselves.

When it eventually became apparent that the Earth was not the center of the universe, careful thinkers came to the inescapable conclusion that the Earth was just another planet that orbited the Sun. Belief in a six-planet Solar System then became dogma for over a century, backed by observation and conviction, and bolstered by arcane theories.

Uranus

The ascendancy of the six-planet model came to an abrupt end in the second half of the eighteenth century. Although no one at the time had any inkling that there might be more planets, an amateur astronomer cataloging dim stars saw something he could not fathom. It was a faint, diminutive disk in a place where it did not belong. At first, he mistook it for a comet, but the object's disk-like appearance made him suspicious. After further study of its trajectory, he came to the realization that he had found a new planet. William Herschel, a teacher, musician, and composer who in his spare time hand-polished lenses for his homemade 20-foot telescope, had just discovered Uranus (left).

That newly discovered planet, along with another discovered 65 years later (Neptune, at top of previous page), are different from the first six. Circling the Sun at greater distances and slower speeds than any of the inner planets, these nearly identical orbs are virtual twins in size, shape, and composition. True colossi of ice, rock, and gas, they are the third and fourth largest planets in the Solar System and consist mostly of hydrogen and helium. They both possess systems of tenuous rings, a small flotilla of moons, and oddly shaped yet substantial magnetic fields. The atmospheres of both planets most likely envelop oceans of superheated water surrounding a solid core the size of Earth. Like their larger brethren Jupiter and Saturn, both Uranus and Neptune spin at a furious rate.

For millennia, these frigid behemoths were either invisible to the naked eye or so dim as to escape notice. Uranus existed at the limit of what the naked eye could see while the possibility of Neptune's existence was not entertained in even the most whimsical musings of heterodox astronomers or the stray thoughts of astrologers.

These blue-green marbles are the least visited and the least studied of all the planets. In our modern epoch, dozens of human-made machines have orbited Mars or rolled across its deserts, and robotic probes have boldly darted among the rings of Saturn and skirted Jupiter's cloud bands. Yet Uranus and Neptune have been visited only once each, and those visits were just momentary flybys, quick glimpses from a daring probe destined to depart our Solar System forever. Those glimpses, however, caught stunning images of unforeseen rings, new moons, and peculiar, unearthly atmospheres.

Detection and Observation

Many years before Herschel identified Uranus, our forebears surely saw it in the night sky among the myriad points of light, but to them it would have looked just like any other dim star. From two billion miles away, this blue-gray wanderer's motion was too slow, its reflected light too feeble for the ancients to differentiate it from the fixed stars. Like a tiny diamond lost amid countless cubic zirconia, Uranus remained hidden in plain sight. Only after centuries of steady advances in telescope design were optical tools available that were of sufficient quality and power to enable an observer to determine that Uranus was a planet.

Neptune, meanwhile, is utterly invisible from Earth without a telescope. With an apparent magnitude of about 7.8, it is approximately one-fifth as bright as the faintest stars visible to the unaided eye. The famous Italian astronomer Galileo, one of the first to telescopically investigate the heavens, actually discovered Neptune, although he did not identify it as a planet. He saw it through his telescope on December 28, 1612, and again on January 28, 1613, recording it in a sketchbook. Neptune was relatively close to Jupiter at that time. Since he rarely mentioned or

Milky Way as seen through the Delicate Arch

drew other stars, it is possible that Galileo suspected there was something special about that particular "star."

The first observer to officially chart Uranus, as opposed to just jotting it down in a notebook, was John Flamsteed in 1690. (He, too, thought it was a star.) When William Herschel found Uranus, in 1781, it became clear after meticulous observations and after input from other astronomers that he had in fact discovered a new planet. Herschel suggested naming his discovery Georgium Sidus (Latin for "star of George") after King George III of England, his benefactor, while the French favored the name Herschel. In due course, the name Uranus was settled upon, which was a logical continuation of the naming conventions for planets up to that time, since Uranus was the father of Saturn, who was in turn the father of Jupiter. Hershel, catapulted to fame and financial security by his discovery, continued to observe Uranus with increasingly sophisticated telescopes, eventually building one that was forty feet long. In addition to Uranus, he also discovered Uranus' two largest moons, Titania and Oberon.

Astronomers observing the orbit of Uranus noticed a peculiarity in its orbit. The planet seemed to be violating the law of universal gravitation which had been formulated by Isaac Newton over a century before. Something was jostling, or perturbing, its orbit. The discrepancy between the predicted orbit and the actual orbit was too great to be ignored, causing many to suspect that a hitherto unidentified planet existed beyond Uranus. In 1845, the young and unknown English astronomer John Couch Adams completed his calculations predicting the position of this unseen planet. He sent word of his findings to the Royal Society, which at the time was the foremost English scientific organization, but no one who examined his report recognized its value. However, less than a year later the French astronomer Urbain Jean Joseph Le Verrier, who was older and better known than Adams, announced his own, strikingly similar, calculations. Under Le Verrier's guidance, on September 23, 1846, the German astronomer Johann Galle observed the new planet, which came to be called Neptune, very close to where Adams' calculations had predicted it should be, and less than a degree off from Le Verrier's calculations. Seventeen days after the finding of Neptune, the English amateur astronomer William Lassell discovered Neptune's largest moon Triton, and in 1851 he discovered the Uranian moons Ariel and Umbriel.

Even with large telescopes, almost nothing of Uranus or Neptune can be seen from Earth beyond their shape, color, and orbit. Therefore very little was known about either of these planets until they were visited by Voyager 2 during the latter part of the twentieth century. Voyager 2 revealed more about Uranus and Neptune in the few hours it spent near each of these planets than had been learned in all of previously recorded history.

After doing yeoman's work at Jupiter and Saturn, Voyager 2 sped through the Uranian system on January 24, 1986. Passing within 50,000 miles of the planet, it took dramatic pictures and gathered data that allowed for accurate determinations of the masses and radii of the planet and its major moons. During its five-hour flyby, Voyager 2 detected Uranus's magnetic field; discovered the moons Cordelia, Ophelia, Bianca, Cressida, Desdemona, Juliet, Portia, Rosalind, Belinda, Perdita, and Puck; studied the planet's unique atmosphere; and found two new rings.

Voyager 2 then ventured onward, making use of a gravity assist from Uranus to speed up and alter its course to intercept distant Neptune, a billion miles further away. Over three years later, on August 25, 1989, Voyager 2 reached Neptune and sailed within 3,000 miles of the planet's cloud-tops. The probe was then directed to fly by Triton, Neptune's largest moon, even though this meant a radical alteration in its trajectory. This change in course, however, was deemed irrelevant, since

Voyager was destined to leave the Solar System regardless. Voyager 1 had taken a similar tack on a close flyby of Saturn's largest moon, Titan, which flung the spacecraft out of the plane of the Solar System. Today, Voyager 1 is traveling upward away from the plane of the Solar System, while Voyager 2 is traveling downward. They are now both more than one hundred times farther from the Sun than the Earth is, vectoring through the interstellar medium of sparse gas and dust. They are more than ten billion miles along on their journey to eternity, dutifully serving humankind as diminutive heralds from the 20th century.

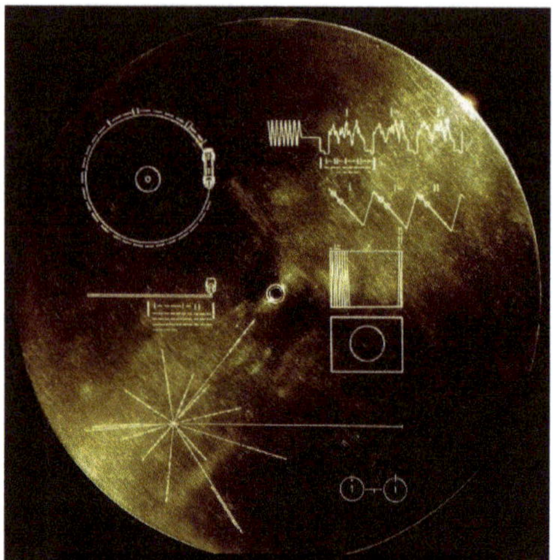
Voyager II's Golden Record

Planners of the Voyager missions were keenly aware that the Voyagers are humankind's interstellar proxies. As the two spacecraft hurtle outward into the space between the stars, they may encounter an alien, space-faring civilization, or alternatively, they could be retrieved by advanced humans of a future age. Therefore, both Voyager spacecraft are equipped with a specially prepared time capsule: a twelve-inch gold-plated copper phonograph record (left) containing a rich selection of sounds, images, and data. The disc holds music and greetings in many human languages as well as other information intended to communicate to any beings who may find it the abundance, beauty, and diversity of the human species, human cultures, and the planet Earth. Included among the musical selections are pieces, by Beethoven, Bach, and Chuck Berry, as well as an Indian raga and folk music from places like the Solomon Islands, Bulgaria, and Azerbaijan. Voyager 1 will drift within 1.6 light years of Gliese 445 in 40,000 years, and Voyage 2 will reach Sirius in about 296,000 years.

Planetary Data

Although Uranus and Neptune resemble each other in a number of ways, they also have many differences. Uranus gives off a blue-green or teal glow, while Neptune is slightly more bluish. These colors are present because the two planets' atmospheres contain trace amounts of methane, which absorbs red light. Uranus does not radiate much heat, while Neptune, like Jupiter and Saturn, radiates more energy than it receives from the Sun.

Uranus spins on its side, while Neptune rotates upright on its axis like all of the other known planets of our Solar System, including Earth. The spin of Uranus is not perfectly parallel to the ecliptic; one of its poles is directed slightly above the ecliptic and the other below. According to international convention, the north pole of a planet is defined as the pole that is above the ecliptic regardless of the direction in which the planet spins, so Uranus, in fact, spins clockwise, meaning that it rotates in a way which seems backward (retrograde) to us from our vantage point on Earth.

As a result of Uranus' near-90° tilt, its axis of rotation points almost directly toward the Sun twice during its year. This means that in Uranus' northern or southern hemisphere, a single "day" lasts a full Uranian year. This is equivalent to 42 Earth years of light followed by 42 Earth years of darkness. At the equator during Uranus's spring and fall, a day lasts about 17 hours as measured by Voyager 2. Twenty-one years later, these same points near the equator approach a solstice and experience virtually no sunrise or sunset. Of all the planets in our Solar System, only Uranus experiences days of such drastically varied durations.

The question of why Uranus has such an extreme tilt is intriguing. Perhaps the presently observed tilt is the result of normal, though large, fluctuations in the planet's inclination. (Evidence suggests that Mars too has experienced such wild swings.) On the other hand, it is also possible that something pushed Uranus over. The best theory to date is that early in its history, Uranus experienced a collision with an Earth-sized object. The enormous impact knocked Uranus over onto its side, and it has been rolling along like a barrel ever since. Uranus, like the other planets, was born from material revolving counterclockwise in the plane of the Solar System, and the fact that it now rotates in a fashion so spectacularly different from all its siblings adds to the allure and mystery of this blue-green, hulking giant.

The currently sunlit pole of Uranus bears a thick layer of haze. This area also radiates ultraviolet light, a phenomenon known as "dayglow." The average atmospheric temperature is about -350°F. Surprisingly, when Voyager 2 flew by, the temperature of Uranus's cloud tops measured about the same everywhere, whether at the illuminated pole, the dark pole, or the equator.

Uranus is located 1.8 billion miles from the Sun, and its diameter is approximately 32,000 miles. The winds blow along latitudinal bands, reaching speeds as high as 450 miles per hour, though few cyclones disturb the serenity of its turquoise envelope. The magnetic field is 50 times stronger than Earth's.

Despite Uranus' great mass, its gravity is only nine-tenths as strong as Earth's. On rocky planets, the force of gravity is traditionally measured at the surface of the planet. Since the giant planets are worlds of fluids, probably with a comparatively small solid center, they do not have a meaningful surface. Therefore, the scientific community has chosen to measure the gravity of a planet at the point where the atmospheric pressure equals the atmospheric pressure on Earth at sea level. This can be misleading, since on giants such as Uranus and Neptune, the pressure of one atmosphere is achieved relatively near the cloud tops. At any rate, the gravitational force measured at this point on Uranus is remarkably close to the gravity at Earth's surface. The additional mass of Uranus, which increases the gravitational force, is counterbalanced by its greater diameter, which reduces the gravitational force.

Neptune's diameter of about 30,000 miles makes it slightly smaller than Uranus. Neptune is thus the fourth largest planet in the Solar System, but since it is denser than Uranus, it ranks third among the planets in mass. Its rapid rotation, once every 16 hours, and the fluidity of its gaseous envelope mean that its shape is flattened at the poles.

Neptune, the farthest planet from the Sun, surprisingly generates the most violent storms anywhere in the Solar System, with wind velocities of over 1100 miles per hour. Enormous storms appear and vanish frequently. The rotating storms seem to be vortices that open wells into the interior, sucking material upward. Compared to Neptune, Uranus is a tranquil world. This may be changing, however, as Uranus is now leaving its winter and heading into a possibly more turbulent spring.

Uranus' and Neptune's possess numerous volatile gases (e.g. ammonia, methane, and carbon monoxide) that often freeze to form ices in these planet's outer atmospheres, which hover at temperatures near – 370°F. It is for this reason that these planets are referred to as "ice giants."

Both planets have deep, salty oceans and surely contain more water than rock. The oceans probably contain dissolved gases and salts, and these flowing fluids render them electrically conductive, generating magnetic fields. These, since they are caused by such shallow movements, are oriented strangely. Our magnetic field on Earth is oriented almost exactly north and south, tilted at only 11 degrees from the axis of rotation, while Uranus' is at an angle of 58.6 degrees and Neptune's is inclined at 47 degrees. Furthermore, on neither Uranus nor Neptune do the north and south magnetic poles align with the center of the planet.

Uranus Showing Rings

Neptune receives less than half the sunlight Uranus does, but Neptune oddly emits more heat, and scientists are not sure why. This anomalous heat, possibly due to the decay of radioactive nuclei in Neptune's core, may be responsible for Neptune's mammoth storms.

Moons and Rings of Uranus

Uranus has 27 known moons and is banded by no fewer than 10 rings. The rings of Uranus are not as famous or showy as those of Saturn, but they are visible from Earth with a telescope. Because of the vertical orientation of the rings, when viewed from Earth they can at times resemble a giant bull's eye (right).

Uranus was the second planet discovered to have rings. The rings, which are narrow and rather opaque, were detected from Earth in 1977, some nine years before Voyager 2 would pass the planet. Voyager 2 photographed Uranus' rings, and it also sent radio signals through them so scientists could study the effects. The particles of the main rings were revealed to be mostly larger than five feet across. Saturn's rings, by contrast, consist of much smaller particles. Some of the particles composing Uranus's rings may be the remnants of a moon that drifted inside the planet's Roche limit and was pulverized by its tidal forces. Stretching 61,000 miles in radius, Uranus' rings circle the equator and are slightly tilted. The outermost ring, dubbed R1, is distinctly blue and appears to consist of ice particles from the moon Mab. The color may be the result of extra-fine ice particles that reflect blue light and absorb other wavelengths.

The five largest moons of Uranus measure from about 300 to 1000 miles in diameter and were discovered with the use of telescopes between 1851 and 1948. The other 22 moons were discovered much later and are very small. Cordelia, for example, has a diameter of only 32 miles. Many of these small moons orbit very near the planet.

The four largest moons of Uranus, in decreasing order by size, are Titania (which is half the size of our Moon), Oberon, Umbriel, and Ariel. These moons are not very dense, only slightly denser than water. They do, however, have impact craters similar to those on Earth's moon. Water ice appears in the spectra of these moons. Thus they may, someday, make an excellent last stop for humans planning to depart for deep space.

Miranda

Miranda (left) is the oddest of Uranus's moons. Its surface is gouged, cratered, and jumbled. Cryovolcanoes spew slushy ice, canyons drop twelve miles, and the rough terrain includes ridges, valleys, faults, and crevasses. This 290-mile-diameter moon's tortured surface defies explanation, but scientists hypothesize that Miranda has been smashed to pieces and re-formed, maybe more than once. One result of these violent events is evidently the creation of the tallest known precipice in the Solar System, Verona Rupes. It is over 33,000 feet tall, which is eight times as high as Earth's highest cliff, located on Mt. Thor in Nunavut, Canada. However, Miranda's gravitational tug is so slight that a pebble falling from the pinnacle of Verona Rupes would take approximately 12 minutes to fall those six miles. The pebbles speed at impact would be about 60 mph, more than enough to be fatal to a human.

Uranus's four outer moons have retrograde orbits with high inclinations and are probably stray asteroids that Uranus has captured.

Moons and Rings of Neptune

Neptune has 14 moons, the largest of which is Triton. The iconic photograph below, taken by Voyager 2 three days after its flyby, shows the paired crescents of diminutive Triton and huge Neptune.

Triton is a super-cold body, the coldest known in the Solar System, and has temperatures that hover around -390° F. It has pink snow, a frosty methane cap, and vast plains resembling the maria of Earth's Moon, while other areas resemble the skin of a cantaloupe. Triton is about the same size as Pluto and orbits in the opposite direction from most of Neptune's other moons. Triton's strange orbit and similar composition to Pluto strongly suggest a common origin.

Neptune and Triton

Triton possesses a wispy atmosphere of nitrogen and methane, and its surface bears a coating of frozen water, nitrogen, and methane. The south polar cap is salmon-colored due to nitrogen ice and snow. The long Tritonian seasons mean the South Pole has been in constant sunlight for over a century.

As Voyager 2 passed it, Triton sprayed jets of nitrogen six miles into space, probably vented through fissures. These plumes, mixed with carbon particles, evidently blow across the surface and leave dark streaks. Triton is spiraling inward, and in perhaps 100 million years it will reach the Roche limit and be ripped to pieces. Because of the amount of material, the result may surpass Saturn's rings.

The six inner moons of Neptune are prograde, but are lumpy and irregular; they lack the necessary mass to pull themselves into a spherical shape. Of the outer six moons, three have retrograde rotations. Nereid, the outermost moon, has the longest orbital period—360 Earth days.

Neptune has five rings. As revealed by Voyager in 1989, they are wafer-thin and diffuse, yet in some places, they can appear lumpy or twisted. The rings are named in honor of astronomers (from innermost to outermost): Galle, Le Verrier, Lassell, Arago, and Adams. The Adams ring bulges distinctly in four locations around its circumference. These swollen arcs are named Liberté, Egalité, Fraternité, and Courage. The gravitational influence of the moon Galatea confines this ring, and there are probably other shepherd moons. When Voyager 2 spotted them, the arcs seemed to be decaying. Scientists speculate that unless the rings are supplied with new material from some source, they may disappear within a century.

Mythology

Uranus is named after a titan of Greek mythology, Ouranos. This mighty being was associated with the sky and air. The planet was not named after the Muse of astronomy, Urania (Greek Ourania), although that would have been fitting. In turn, element 92 of the periodic table, uranium, was named after the planet. Uranus is the only planet with a Greek name rather than a Roman one.

Neptune with Trident

When moons were discovered around Uranus, they were named after magical beings from English literature, all of them spirits of the air. The idea was that Uranus was a deity of the air and ought to be accompanied by characters from the same domain. The original names were bestowed by John Herschel, who followed in the footsteps of his illustrious father William Herschel and likewise became a well-known astronomer. The spirits after which the moons were named were characters in Shakespearean plays or in Alexander Pope's mock-epic poetic masterpiece *The Rape of the Lock.* As more moons were discovered, additional names were drawn from these sources, and the original idea that they should be the names of supernatural beings was discarded. Thus, we have the moons Miranda, Ophelia, Juliet, etc., named after decidedly mortal Shakespearean characters.

In Roman mythology, Neptune was originally the god of fresh water but soon became identified with the Greek god Poseidon and was transformed into the trident-wielding god of the sea. Neptune was one of the principal Olympian gods, a son of the mighty titan, Saturn; brother to the king of the gods, Jupiter; and brother of the ruler of Hades, Pluto.

Since Neptune was the supreme ruler of the sea, it was decided that the moons of the planet Neptune would be named after other sea deities. There is no danger of running out of these; the

Nereids alone numbered 50. Daughters of the minor sea-god Nereus, they included Thalassa, Galatea, Despina, and their 47 sisters.

The names of other moons were also drawn from the mythology of the sea. Triton was named after a merman who was a son of the sea god Poseidon. The moon Naiad was named after the naiads, long-lived, beautiful goddesses who inhabited springs, rivers, fountains, and lakes. Proteus was a sea-based prophet and shape-shifting sea-shepherd who was subject to the sea god Poseidon. Proteus knew all things—past, present, and future—but would only share his knowledge under duress.

Conclusion

Mythical ice giants were dangerous, mysterious, legendary creatures, the protagonists of frightening stories told around campfires on cold nights by our ancestors. Today's ice giants remain mysterious, but less so. These modern monsters of the outer reaches occupy our thoughts, fill our telescopes, and may one day provide a way station to the distant stars. These far-off worlds themselves may never host a human being, but on the landscapes of their frigid moons, our descendants may one day tread dreaming of explorations still to come and pass an evening or an hour around a Cybernetically AMPlified low Frequency InfraRed Emitter, admiring the heavens and regaling companions with more tall tales.

The farther we travel away from our Sun, the less warmth and light it provides, according to an inverse square law. Close to the Sun, water is warm and fluid, while farther out it is frozen and harder than stone. At distances of billions of miles from the Sun, molecules that behave as gases nearer the Sun transform into fluids and, if sufficiently far away, become solids. Uranus and Neptune are water worlds blanketed in layers of hydrogen and helium. If these worlds had been closer to the Sun, the Sun's point-blank energy might have blasted away these lighter gases, and then these two planets could have given rise to living organisms similar to those found on Earth. Both possess sufficient mass to maintain an atmosphere and both possess enough water to have global oceans. Both have strong magnetic fields capable of protecting incipient life. However, both just happen to be in the wrong place in our Solar System. Someday in the far future, this will change. The Sun will expand and create substantially more heat and light than it does today. The ice giants' lighter gases could be blown away, and their water could become warm and liquid.

Will the sky god Uranus and trident-armed Neptune ascend? Will the Sun breathe life into these frosty, lifeless worlds and transform them into orbs of life? Will they boast oceans of liquid water, continents of dry land, and breathable atmospheres? With time, heat, chemistry, and human ingenuity, these worlds could certainly be inhabited by future generations of our descendants.

If our destiny is to sail to the stars, then we will need to make use of every resource along the way. The moons of the ice giants possess wonders and oddities: steep cliffs, pink snow, and slushy nitrogen ice, for example. They also hold much that could be exceptionally useful, including exotic chemical compounds, plentiful water, and an abundance of minerals. As we explore, learn, and discover, and our horizons broaden, today's worlds of ice and gas will become living spheres hosting ports, ships, and space arks. Upon such arks, our progeny will sail to the stars.

In the past, we have succeeded in hurling probes outward to the outer Solar System at over 100,000 miles per hour. At that speed, it would take 18,000 years to reach the nearest stars. Therefore, the next step in our journey is to explore the outer reaches of our Solar System. From bases in the systems of Uranus and Neptune, we will be able to visit the dwarf planets of the Kuiper Belt and the Oort Cloud. From there, we will sally forth to the stars, galaxies, and filaments beyond. Space is vast, and our voyage of discovery is just beginning. Uranus and Neptune may very well be our last local stop on the way to the rest of the universe.

Woodcut of Observatory in Alexandria at the time of Hipparchus

22
Trans-Neptunian Objects

Or bid the soul of Orpheus sing
Such notes as, warbled to the string,
Drew iron tears down Pluto's cheek.
John Milton

Charon or the grim King's dog could
not prevent me then from carrying you
up into the fields of light.
Euripides

Beyond the ice giants Uranus and Neptune exists a cold, dark, stygian void of near-perfect vacuum, an ageless corridor of emptiness so barren and vast that even though trillions of icy rocks orbit there, only rarely will a rock come within a million miles of its nearest neighbor. Some of the occupants of this sparsely populated region are as small as a grain of sand. Others, slightly larger, are found in a variety of shapes: ovoids, shards, or lumps. Only the largest have sufficient mass to crush themselves into spherical shapes that resemble small planets. The ices with which these rocks are covered consist of frozen nitrogen, carbon dioxide, ammonia, and other ordinarily volatile substances never found in solid form where humans thrive. The frozen material also includes large quantities of rock-solid water ice harder than steel. Speeding through their territory faster than any bullet, these shadowy bodies are so distant from Earth that when we happen to spy one, it appears as a virtually stationary pinpoint against the panoramic spectacle of stars behind it. These dirty ice balls barely reflect light and thus were only recently discovered; in fact, their existence was not even suspected until the mid-20th century. They were discovered only because these bodies move slightly compared to the amphitheater of relatively fixed stars behind them. Astronomers examining photographs separated by a period of weeks can detect and measure the minuscule movements and then make the determination that what they are seeing is a newly discovered island of rock sailing in the cold black sea of space near the periphery of our Solar System.

These distant forsaken relicts are leftover bits and pieces that were cast to the borderlands of the early Solar System. At times these errant, gravitationally guided missiles can appear unexpectedly, and then they may precipitously and disastrously rain down on the inner Solar System, threatening the Earth and other planets with havoc, death, and destruction. Thankfully, such events are quite rare, although knowledgeable observers understand that it is only a matter of time before a life-menacing object plummets toward us once again. These potentially devastating, frigid fragments have remained unchanged for five billion years, ever since they first coalesced from the disk-like primeval nebula of dust and gas from which our Solar System was formed. Hence, these castoffs are of immense scientific value, not only because they are natal specimens that hold the clues to the formation of the Solar System, but also because understanding them may save humankind from an extinction-level event. It is imperative that we study them.

Since our most powerful telescopes can barely resolve these far-flung objects, in order to learn about them we must dispatch robotic emissaries equipped with specialized instruments to investigate them firsthand. Such probes, like the famed American explorers Lewis and Clark, journey far to explore the unknown, then report their discoveries to eagerly waiting citizens and luminaries. One modern-day explorer, the New Horizons interplanetary spacecraft, after nearly a decade of unpowered and silent travel, has entered this remote wilderness. As it approached Pluto, it was summoned from its deep slumber, awoke, and began recording and broadcasting its findings. Unlike Lewis and Clark, this intrepid traveler will not return. It will dart off further, studying ever more distant rocks and ice until it runs out of fuel and becomes a castaway itself.

In the past, cognoscente and commoner alike believed that the outer edge of the Solar System coincided with the orbit of the planet farthest from the Sun, at approximately 30 or 40 astronomical units (AUs). Today's experts prefer to place the outer edge of the Solar System some three light-years from the Sun, or about 200,000 astronomical units. In the more distant expanses of this region, an invisible frothy foam of magnetic bubbles forms, with each bubble a hundred million miles across. In this quasi-fluid region between the nearest stars and our Sun, a large portion of the solar wind comes to a standstill. Charged particles and magnetic fields from the opposing stars wage cosmic war, pushing against each other like colossal sumo wrestlers. Here the effervescence of magnetic energy creates bow waves that ebb and flow at the front lines of the conflict and form a shifting boundary known as the heliopause. The importance of this layer of protection has only recently been discovered. This Sun-created border is our first line of defense against the bombardment of rays from the rest of the universe. Cosmic rays are particles traveling at nearly the speed of light, and without some protection against them, fragile DNA molecules could be damaged, resulting in mutations that could be detrimental to life or make life as we know it impossible.

Beyond this glacial fringe lies a vast ribbon of dust and gas trapped between the various heliopauses. Adjacent stars, each with their own solar wind, compress and distort our Sun's sphere of influence, sometimes creating unexpected crescent-shaped features delineated by streams of charged particles.

Oort Cloud and Kuiper Belt

Meanwhile, between the orbit of the outermost planet and the heliopause are two distinct sets of space rocks, the Kuiper Belt and the Oort Cloud. The Oort Cloud is a spherical shell that surrounds the Kuiper Belt, which in turn surrounds the orbits of the planets. Its spherical shape, made possible by the sparseness of its contents, contrasts with the disk-like configuration of the Kuiper Belt and the orbits of the eight planets. The Oort Cloud is a collection of remnants from the early Solar System and is the home of many long-period comets, which can bide their time for millennia between visits to the inner Solar System. Although the Estonian astronomer Ernest J. Öpik had previously proposed the existence of a distant, vast reservoir of comets, the Oort Cloud was named after the Dutch astronomer Jan Oort, who in 1950 presented an estimate of the cloud's size and an explanation of how it was affected by the gravitational tugs of nearby stars.

The Kuiper Belt is a donut-shaped region containing objects that orbit beyond Neptune but closer to Earth than the Oort Cloud. It lies in the same plane as the orbits of the planets and is named after

Gerard P. Kuiper, a Dutch-American astronomer who hypothesized its existence when the only known object at that distance was the planet Pluto. (Ten years earlier, the Irish astronomer Kenneth Edgeworth had made a similar prediction, and the Kuiper Belt is sometimes called the Edgeworth-Kuiper Belt to recognize both astronomers.) It was not until the 1990s, more than twenty years after Kuiper's death, that the orbiting bodies he predicted were found. By 2012, 1,500 had been discovered, and the pace of discovery is accelerating. Just in the second half of 2014, over 1,745 Kuiper Belt objects (KBOs) were identified.

Because of their location outside Neptune's orbit, more than 30 AUs from the Sun, these bodies are also referred to as trans-Neptunian objects (TNOs). The outer edge of the Kuiper Belt has not been rigorously defined, but most authorities exclude objects that never get closer to the Sun than 47 AUs, which is the location of the 2:1 Neptune resonance where an object makes one orbit around the Sun for every two of Neptune's. Short-period comets, those with orbital periods less than 200 years, also dwell in the Kuiper Belt between their sunward jaunts.

Certain Kuiper Belt objects have been designated as dwarf planets, including Pluto, Makemake, Haumea, and Eris. Ceres, in the Asteroid Belt, has also been accorded this status, but Ceres orbits well inside the orbit of Neptune and is not even close to being trans-Neptunian. To qualify as a dwarf planet, an object must orbit the Sun, must be roughly spherical, and must not be a satellite of a larger planet. The difference between a dwarf planet and a full-fledged planet is somewhat vague, and authorities are still working out their differences. The International Astronomical Union (IAU) has ruled that the difference between a planet and a dwarf planet is that a planet must have cleared its orbit of planetesimals, or be gravitationally dominant in its region of space. At any rate, the IAU's stamp of approval is required before a body can be recognized as a dwarf planet. At least three candidates are in line to receive this recognition: Quaoar, Sedna, and Orcus. More are being discovered all the time. As of this writing, two of the most recently discovered were two that have been evocatively nicknamed "Biden" and "The Goblin." Another, tentatively called "Farout," was discovered near the end of 2018. An impressive pink ice ball with a 300-mile diameter, it is the most distant object ever found orbiting the Sun. The radius of its orbit is calculated to be at least 11 billion miles. At that distance, Farout requires 1,000 years to complete one orbit.

Eris

Eris orbits the Sun once every 557 years and was found just a few weeks before the launch of the New Horizons spacecraft. Originally thought to be larger than Pluto, its very existence cast doubt on the definition of the word "planet." Later, more careful measurements showed it to be slightly smaller than Pluto, although more massive. Like Eris, the goddess of discord, whose machinations were responsible for the bloody Trojan War, the Pluto-like body Eris instigated doubt among IAU members who then voted to demote Pluto to dwarf status. This action sowed discord among members of the astronomical community, traumatized children who were forced to relearn the Solar System, and upset large numbers of partisan laypeople. For a time, the world of astronomy found itself in an atypical uproar. The battle over the status of Pluto has mostly subsided, but there are still those who hold steadfast opinions one way or the other.

Haumea (how-may-ah)

Haumea is named for a Hawaiian goddess of birth and fertility. It was discovered in 2003 and has an elongated shape like a football, meaning that it does not actually satisfy the official definition, which specifies that a dwarf planet must be round. Its dimensions are 1,220 × 940 × 620 miles. This odd shape may be at least partly due to Haumea's rapid rate of rotation, which is once every 3.9 hours. Haumea is part of a group of eight other Kuiper Belt objects that have similar orbits. Their strange orbits and Haumea's fast rotation may have resulted from a collision of Haumea and a large object in the far past, or there may be another explanation for this strange behavior, such as an undiscovered massive body regularly passing through their region of space.

Unlike most Kuiper Belt objects, Haumea has more rock than ice. Its ice is a thin coating over a rocky interior. Haumea's surface is marked by a large red spot, which is probably the scar from an energetic impact that has exposed the interior. Two moons of Haumea have been discovered and are named for daughters of Haumea in Hawaiian mythology. The larger moon, Hiʻiaka (hee-ee-ah-ka), is named after the goddess of the hula. The lesser moon, Namaka, takes its moniker from a Hawaiian water spirit.

Makemake (mah·keh·mah·keh)

Makemake is named after the creator god of the Polynesian inhabitants of Easter Island; the name alludes indirectly to its discovery near the date of Easter in 2005. Makemake is reddish in color, has a diameter of around 900 miles, and orbits the Sun every 306 years.

Sedna

Discovered November 14, 2003, Sedna is almost as red as Mars and has a long elliptical orbit. Its diameter of about 1,000 miles makes it more than two-thirds as large as Pluto. Sedna is named after a sea goddess of Inuit mythology who is known as the "Mistress of the Sea" and plays a central role in Inuit creation myths. In one legend, she is a giant, the daughter of a creator god, and becomes the ruler of the monsters of the deep. In another version, she is a beautiful maiden who rejects marriage offers from all until she is abducted by a bird spirit. In almost every myth, she loses her fingers, which then become sea creatures such as seals. Traditional Inuit hunters hoping to harvest sustenance from the sea still offer her ritual propitiation.

Sedna has a highly unusual elliptical orbit similar to several other KBOs, including Biden and the Goblin. After extensive analysis of these orbits with supercomputers, Cal Tech researchers Mike Brown and Konstantin Batygin came to the conclusion that the existence of a mysterious body five to ten times more massive than Earth could have caused these particular KBOs' anomalous orbits. This new body, which despite fitting into Brown and Batygin's computer models has yet to be found, is referred to as Planet Nine.

Quaoar (kwah-war)

Quaoar was named for the creator god of the Tongva people, a tribe indigenous to the area around Los Angeles. Quaoar is approximately one-half as big as Pluto, at 700 miles in diameter, and orbits the Sun in a comparatively circular path at an average distance of almost 43 AUs. It has a 100-mile-diameter moon named Weywot (way-wot) after the son of Quaoar.

Pluto

By far the most famous Kuiper Belt object is the former planet Pluto, named after the ancient Roman god of the afterlife, called Hades by the Greeks. Its diameter of 1476 miles definitely establishes Pluto as the largest known object in the Kuiper Belt. Its composition is 66% rock, with the remainder consisting mostly of water ice with traces of more exotic ices, such as frozen nitrogen, on its surface. There is likely a liquid ocean hundreds of miles below Pluto's surface.

Pluto possesses a large moon, named Charon after the ancient Greek mythological boatman who ferried the souls of the dead into Hades across the River Styx. Charon is so large, almost exactly half the size of Pluto, that scientists consider Pluto and Charon to be a binary system. The two are so close in size and distance that they have become tidally locked with each other, meaning that the same sides of Pluto and Charon face each other perpetually. Four other moons are also part of this system: Nix, Styx, Kerberos, and Hydra. These moons are all smaller than 70 miles in diameter and spin on their sides. The names selected for the smaller moons reflect a connection to the Roman god of the underworld. Nix, mother of Charon, was the Greek goddess of night and darkness. Styx, as already mentioned, was the river marking the boundary of the underworld. Kerberos (in other contexts sometimes spelled Cerberus) was a fierce three-headed dog who guarded the entrance to Hades. Hydra was a nine-headed, snakelike monster which Hercules was tasked to kill as the second of his twelve labors. The image at right, from the Hubble Space Telescope, shows Pluto, Charon, Nix, and Hydra. Hydra is farthest to the right.

Pluto from Hubble

Neither Voyager 1 nor Voyager 2 sailed by Pluto on their journeys through the Solar System, and from Earth, three billion miles away, the best image of Pluto was a fuzzy ball of about 36 square pixels. To record clearer, more detailed images, scientists proposed an ambitious and daring flyby, sending a spacecraft from Earth to less than 8,000 miles from Pluto's surface. This mission became known as New Horizons.

New Horizons

When the New Horizons spacecraft was launched toward Pluto in early 2006, Pluto was still considered a planet. However, while the probe was en route, the majority of attendees at a meeting of the IAU in August 2006, voted in favor of changing Pluto's status, and New Horizons became the first probe in history to be launched toward a planet only to have that planet reclassified as something else before the probe's arrival.

When New Horizons began its mission in 2006, it was called "the fastest spacecraft ever launched." Its launch speed was greater than the escape velocity of the Solar System, a first for any spacecraft. Its initial speed was greater than 37,000 miles per hour. Then, as it journeyed toward the outer Solar System, the relentless pull of the Sun's gravity continuously slowed its progress, but it was accelerated with a gravity assist from Jupiter. When it flew by Pluto, it was traveling at 30,800 miles per hour.

New Horizons

New Horizons (right) is a half-ton probe the size of a hot tub filled with electronic devices and topped with a radio dish. It also carries onboard some of the dust-like remains of Clyde Tombaugh, who discovered Pluto in 1930.

Its seven scientific instruments bear memorable names like PEPSSI and ALICE. One of these is REX, a radio-science package capable of measuring surface and atmospheric temperatures and determining how temperature varies with altitude. PEPSSI is the Pluto Energetic Particle Spectrometer Science Investigation. It analyzed the density of ions and plasma from Pluto's atmosphere. An instrument called SWAP measures the Solar Wind Around Pluto, as well as how fast the atmosphere escapes and the atmosphere's interaction with the solar wind. LORRI is the LOng-Range Reconnaissance Imager, a telescopic camera for high resolution, long-distance photographs. It was used to map and study Pluto's geology. SDC is the Student Dust Collector. This experiment, which analyzed space dust hitting New Horizons during its voyage to Pluto, was run by students. The instrument known as RALPH is a camera and spectrometer that measures the wavelengths of visible and infrared light to make color composition and thermal maps of Pluto's surface. ALICE makes spectroscopic measurements of ultraviolet light to study the atmosphere of Pluto and search for atmospheres around Charon and other Kuiper Belt objects. New Horizons also carries a radioisotope thermoelectric generator, a nuclear-powered device that creates heat from nuclear reactions and then converts some of that heat into electric current.

NASA's planetary exploration projects have experienced a high level of success in recent decades, often outperforming their planners' expectations. Probes sent to Mars, Jupiter, and Saturn have revealed many unanticipated features and phenomena of our planetary neighbors. The incredible amount of data they delivered has helped to answer many questions while raising even more, and New Horizons continued this grand tradition of success. Its cameras and instruments revealed much more about Pluto than anyone ever knew before, or even imagined. The public was entranced; more than 500 front-page newspaper stories were written about New Horizons, and its webpage received almost two billion views.

After nine years of flight, New Horizons flew by Pluto on July 14, 2015. The probe skirted just 7,750 miles above Pluto's surface and began radioing back detailed pictures. The pictures, when they finally reached investigators on Earth, revealed a complex and varied landscape. Some areas of Pluto appear heavily cratered, indicating extreme age. Their reddish coloration may come from frozen methane acted upon by sunlight and converted into organic compounds known as tholins. The whole globe evidently has been tinted by a thin layer of such chemicals; it is a planet of red, brown, and pink ice. Adjacent to Pluto's ancient cratered terrain are utterly different landforms, including rippled fields or dunes, and craggy, 15,000-foot mountains. The peaks are most likely made of water ice; at Pluto's surface temperature of minus 390 degrees Fahrenheit, frozen water is as hard as granite. Other materials, such as nitrogen ice, are not sturdy enough to form such massive mountains of material without collapsing under their own weight.

Among the surprising discoveries by New Horizons was the amount and variety of geological activity observed on Pluto. Previously, scientists were confident that Pluto, so small and distant from the Sun, was a geologically dead ice ball. However, the first photos released by the New Horizons team showed a huge volcano on Pluto, 100 miles wide or more and three miles high, making it probably the largest volcano in the outer Solar System (right, in the center of the photograph). Furthermore, the flanks of this giant were found to be free of impact craters, indicating that it must have erupted relatively recently. The "lava" this massive volcano spews out is molten ice—that is, water—which emerges from the caldera and then quickly freezes on the flanks of this remarkable volcanic mount.

Huge Volcano on Pluto

Polygonal Depressions on Pluto

Other signs of geological activity abound on Pluto. Relatively flat nitrogen glaciers rim extensive plains which are devoid of craters but are marked with polygonal depressions (left). Here, relatively warm material may be seeping up from below. Because of its smoothness, scientists estimate the age of this region at less than 20 million years, meaning that the region has been geologically active very recently. Mysterious features of this plain resemble slime trails like those left by snails. Perhaps these are due to phenomena similar to what propels the boulders across Racetrack Playa in Death Valley. There, the trails are due to the action of ice rafts on the smooth desert surface; on Pluto, the trails may be made by water ice boulders sliding on a thin film of liquid nitrogen. At -400°F, not much energy would be required to cause such events. The necessary energy may come from a property of water called latent heat of freezing. Water possesses a highly unusual property, almost unique among the substances of the universe. When it freezes, it expands to form a less dense crystalline lattice, in the process releasing a small amount of heat. Thus, although it may seem counterintuitive, as water freezes at the surface of a pool, liquids below may absorb some of this escaping heat and become warmer.

What is the energy source driving Pluto's geological activity? One idea is that oceans deep inside the planet hold heat from billions of years ago when Pluto first formed. This is plausible because Pluto's geographical features are consistent with the presence of a subsurface ocean. Another idea is that heat is generated in Pluto's interior by the decay of radioactive nuclei. Whatever is causing it must be a widespread phenomenon, because Pluto has a large number of Kuiper Belt cousins that are geologically active also.

Despite its alien appearance, Pluto has an earthlike feature: its blue atmosphere, which extends about 4000 miles, or halfway, towards Charon. The atmosphere consists primarily of nitrogen with some methane. When New Horizons looked back at Pluto after its flyby, the backlit planet appeared blue (below). Nitrogen appears blue in sunlight just as in Earth's sky.

Pluto's Blue Nitrogen Rim

Scientists wonder why Pluto's atmosphere did not dissipate ages ago. Because of Pluto's highly elliptical orbit, it is sometimes closer than Neptune is to the Sun, and its frozen gases can warm up enough to sublimate, creating a temporary atmosphere. When it is farther from the Sun, its temperature drops and the atmosphere freezes, delicately wafting to the surface as nitrogen snow. Every million years or so, Pluto's North Pole faces the Sun for 1,000 years in a so-called "super summer." Pluto's atmospheric pressure, though low by Earth standards—roughly equivalent to the pressure at an altitude of 50 miles above Earth's surface—is sufficient to permit the presence of liquids on Pluto's surface during such periods of heating. It is speculated that during Pluto's "super summer," liquid nitrogen on Pluto may form rivers and lakes.

New Horizons revealed surprises about Pluto's moon Charon. The probe gathered data that allowed scientists to create the first detailed color and topographical maps of Charon. The picture of Charon that emerged rivals our knowledge of large satellites of planets like Jupiter and Saturn that were studied by the Voyagers. The probe did not approach the four smaller moons as closely as it did Charon, but it was able to gather data on their sizes, rotations, and shapes, and snapped images from which scientists have been able to construct partial maps of each. There was enough data for scientists to be able to estimate the age of each moon's surface.

Details of images from the July 2015 flyby suggest the very real possibility that Charon could once have held a liquid ocean below its surface. The side of Charon photographed by New Horizons showed faults, ridges, scarps, and valleys, some of which are more than four miles deep. Also present are odd, isolated massifs. At Charon's North Pole is a region called Mordor after a fictional land in *The Lord of the Rings*. The fictional Mordor is an evil place, home of the dreaded Mount Doom, a fiery volcano. Charon's Mordor is red with tholins, and it is not clear where they came from. With no atmosphere, Charon does not have the ingredients to make tholins. Theories have been advanced suggesting that the tholin material somehow crossed the 10,000-mile distance from Pluto and settled on Charon's North Pole. Possibly during previous Plutonian "super summers," tholins or their precursor elements sublimated and drifted to Charon.

Charon (see next page) very likely has experienced expansion in the past, causing its landscape to buckle while ripping open faults and crevasses. Charon's outer layer is mostly water ice, and when the moon was young, it was warmed by radioactive decay and the moon's internal heat. Scientists believe that Charon may have once been warm enough to melt subsurface ice into an ocean. That

Charon

ocean would have later refrozen and caused the cracking and faulting that New Horizons photographed.

New Horizons is not done yet. In January 2019, after careful planning and aiming by NASA experts, it will reach another Kuiper Belt object, this one bearing the unimpressive name 2014 MU_{69}. 2014 MU_{69} was not even discovered when the New Horizons spacecraft was launched in 2006, and this 4.5 billion-year-old frozen object lies a billion miles beyond Pluto. Possibly as small as 20 miles across, 2014 MU_{69} may harbor pristine building blocks from the beginnings of the Solar System and could thus furnish new insights into how the Solar System first formed. After gathering as much data as it can from 2014 MU_{69}, New Horizons may be directed to rendezvous with other Kuiper Belt objects before speeding off in the direction of the constellation Sagittarius, all the while sending us electronic "postcards from the edge" of the Solar System.

The Disney Mystery

Walt Disney created the first sketches of his animal character Pluto in 1930, the same year Pluto the planet was found. Officially, the name for the planet came from an eleven-year-old English girl, Venetia Burney, whose grandfather forwarded her suggestion to Clyde Tombaugh at the Lowell Observatory. Tombaugh was delighted by the name, in part because the first two letters of "Pluto" are the initials of Percival Lowell. Coincidentally, Miss Burney's great-uncle had been responsible for the naming of Mars's moons Phobos and Deimos in 1878.

So the planet Pluto and the cartoon dog named Pluto entered the public lexicon almost simultaneously. For decades since then, third-grade students memorizing the nine planets were also watching cartoons of Mickey Mouse and his dog. Thus in many minds was forged a lifetime connection. Whether there was a connection between the name of the planet and the dog remains a mystery. Walt Disney never revealed to anyone how he selected the name. At any rate, it is known that the element plutonium was named after the dwarf planet, not the cartoon dog.

One of the very first photos of Pluto taken by the New Horizons spacecraft, on July 13, 2015, clearly shows the Texas-sized, heart-shaped icy plain surrounded by ices covered in red tholins. That picture appears at the head of this chapter.

'Oumuamua (Oh-oo-moo-ah-moo-ah)

A mysterious and unexpected interstellar visitor was first sighted on October 19, 2017, by the Pan-STARRS 1 telescope in Hawaii and initially classified as a comet. After several more days of observations, astronomers were able to calculate that this intruder came from the direction of the Lyra constellation and that its true origin was outside our Solar System. Dubbed "'Oumuamua" (Hawaiian for "messenger from afar, arriving first," or "scout"), this bizarrely shaped rock resembling a block-like spindle, streaked almost directly toward the Sun, and then, in December 2017, made a sharp turn under the power of the Sun's gravitation. Increasing in speed from 15 miles per second to 27 miles per second, it accelerated in a manner analogous to what happens when NASA probes receive a gravity assist from a massive planet. Then the odd interloper skirted Earth's orbit, passing within a mere 15 million miles before speeding off in the direction of the Pegasus constellation, eventually to enter the realms of other stars.

Fascinated researchers speculated that 'Oumuamua's dark red color, which may indicate the presence of tholins, was due to hundreds of millions of years of bombardment by stellar radiation. There is no satisfactory explanation yet for the body's unusual shape, ten times longer than it is wide (1300 feet by 130 feet). The spindle was observed to revolve on its axis once every 7.3 hours and seemed to exhibit characteristics of a stony, metallic object. No dust was observed in the object's vicinity, although its trajectory seemed to indicate that it was expelling some gas or vapor.

In 'Oumuamua's official designation, 1I/2017 U1, the "I" stands for interstellar. It is the first time the International Astronomical Union has used this designation. By the time it leaves us, 'Oumuamua will have crossed the entire Solar System, an achievement that is surely not unprecedented but has never before been observed. It will have passed through the Oort cloud not once, but twice. Although its journey has spanned millions of years, observers on Earth did not see it coming; once it was noticed, researchers only had two weeks for telescope viewing. (If it had been on course to strike the Earth, there would have been no way to prevent a devastating collision, at least not at our present level of preparedness.) Although 'Oumuamua came from the direction of Vega in the constellation Lyra, when 'Oumuamua was in that particular section of space approximately 300,000 years ago, Vega was nowhere near. This strongly implies that 'Oumuamua is a wandering body unattached to any solar system and doomed to wander the ocean of space bearing its "strange red light" like the Flying Dutchman of legend.

Its odd shape caused some starry-eyed observers to suspect that 'Oumuamua might be a manufactured object, possibly a piece of technology from some distant, space-faring civilization. Researchers, however, were unable to gather any evidence in support of this highly speculative supposition despite scanning the object for eight hours across multiple frequencies.

The passage of 'Oumuamua so unexpectedly and rapidly through our neighborhood could be viewed as a narrow escape, since the object barely missed striking us, coming within a whisker of Earth when considered on a galactic scale. But at the same time, it represents a missed opportunity. If only we had the technology and the power to hitch a ride on this galactic traveler! A human-made probe or transmitter on this wayfaring rock could have transmitted for millennia and eventually traversed the very nucleus of the galaxy and sent us first-hand intelligence as to what awaits humanity in that region of space. There will be more such fleeting opportunities in the future—scientists estimate that one or two such objects charge through our Solar System each year, unnoticed because of their small size and great distance—but we need to decide in advance that we will expend the resources necessary to take advantage of these opportunities, or else they will surely be missed.

Conclusion

The Kuiper Belt and the Oort Cloud are distant, mysterious, cold, dark, and virtually empty zones of interstellar space. A mark of their deep mysteriousness is that their existence was only very recently hypothesized, and the barren rocks populating them have just begun to be discovered, one by one. This region of space surely holds great importance and promise, as its matter has subsisted here from the epoch of our Sun's birth 4.5 billion years ago. To answer the most profound questions of the origins of the planets, moons, and asteroids in our Solar System, and of our own origins, this is where we must look. With the Voyager and New Horizons missions and others soon to come, we have begun that process. Here, at a boundary once considered the edge of the Solar System, we are enthusiastically entering upon an enterprising adventure of supreme importance.

When we pass this boundary—and it is to be fervently hoped that someday we will—we will truly become explorers of interstellar space. After centuries of viewing the stars, we will have finally begun to journey toward them. Sliding along the gravitational contours of space into the byways of the galaxy, we will have taken the first step toward becoming a galactic species.

Putting Pluto and Charon into perspective with the Earth and Moon

23
Planet Nine

Planet Nine (Artist's Conception)

Then felt I like some watcher of the skies
When a new planet swims into his ken;
John Keats

...a riddle wrapped in a mystery inside an enigma.
Winston Churchill

"Planet Nine" is the name provisionally assigned to an outsized icy planet proposed to exist in the outer Solar System. In January 2016, researchers in California found evidence that there is a giant planet tracing an eccentric, highly elongated orbit far from the Sun. The researchers, Konstantin Batygin and Mike Brown, made their discovery through mathematical modeling and computer simulations. The planet has not yet been observed directly; an artist's conception is shown above.

Gravity imposes order and balance, that is, a certain amount of mathematical predictability, on the Solar System. Using the inherent predictability of gravity, Batygin and Brown ran a large number of computer simulations of planetary motions. They found that including an additional massive body improved their models' accuracy. A number of Kuiper Belt objects orbiting beyond Neptune follow trajectories that tilt together and loop outward rather than lying flat in the plane of the Solar System. The simulations showed that these strange orbits could have been "sculpted" by a far-flung body approximately three times the size of Earth and from five to ten times as massive. It is as if the postulated planet, not yet seen by astronomers, was making itself known through a "gravitational silhouette."

Brown notes that the hypothetical ninth planet is sufficiently large—5000 times the mass of Pluto—that there is no question it would qualify as a true planet. If it exists, then it would dominate more space than any of the other planets do, which, according to Brown, would make it "the most planet-y of the planets in the whole solar system."

Batygin, an assistant professor of planetary science, states, "For the first time in over 150 years, there is solid evidence that the Solar System's planetary census is incomplete." Brown is well known for his actions in the dramatic controversy that led to Pluto's demotion from full planetary status, and he even wrote a book about it called *Why I Killed Pluto*. He seems to be anticipating that the discovery of a new planet will nudge him back into the good graces of the Pluto-loving public. He asserts, "All those people who are mad that Pluto is no longer a planet can be thrilled to know that there is a real planet out there still to be found. Now we can go and find this planet and make the Solar System have nine planets once again."

Brown has suggested naming Planet Nine after Nox, the Roman goddess of night and shadow—surely fitting for a planet that has remained hidden from human eyes for so long. However, even if he discovers the planet, its name will not be up to him. Names of planets are chosen by the International Astronomical Union, the same people who decide whether a given space rock is a planet or not. In addition to Nox, there has been no shortage of other suggestions: Terminus, the Roman god of borders; Fortuna, the Roman goddess of luck; Solo, as in Han Solo from *Star Wars*; Tartarus, a dungeon for sinners in Greek mythology; Olaf, a character from the Disney movie *Frozen*; and even George, a possibly tongue-in-cheek suggestion from the researchers.

Planet Nine (Artist's Conception)

Due to its extreme distance, at least 20 billion miles from the Sun, the planet would reflect very little light and thus evade detection by telescope. It is expected to have an apparent brightness at least 600 times fainter than Pluto's. Brown's opinion is that Planet 9 is likely to be an ejected ice giant, like Uranus and Neptune, with a rocky core sheathed in ice and enveloped in a comparatively shallow stratum of gases. Astronomers generally believe that the Jovian (giant) planets—Jupiter, Saturn, Uranus, and Neptune—were once close to the Sun but migrated outward due to collisions and gravitational effects. This mode of thinking is referred to as the Nice model, after the city in France where it was originally developed. If the existence of Planet Nine is confirmed, it will fit the Nice model perfectly. It will be a fifth giant planet and will have completed the lengthiest migration of any of them. Computer simulations carried out in 2011 support this idea, although at that time they were rejected as implausible. Complex programs intended to model the early development of the Solar System kept giving rise to scenarios in which a fifth gas giant took shape in addition to the familiar ones Jupiter, Saturn, Uranus, and Neptune. If Planet Nine came into existence in this way, a true sibling of the other planets, then it may have been flung outward by the gravitational power of the other planets, particularly Jupiter, and banished to outer darkness sometime in the distant past.

The Nice model is not the only possible explanation for the presence of an additional planet in our Solar System. An alternative theory is that Planet Nine is a rogue. A rogue planet is one that has been ejected from its home system and consigned to wander the barren emptiness of interstellar space. Occasionally, one of these itinerant nomads meanders into the gentle gravitational grasp of a passing star and is captured. Now, instead of roving aimlessly, this homeless orphan snuggles comfortably into a new path about its surrogate star. The existence of rogue planets has not been confirmed, but it is almost a certainty. It has been verified, for instance, that orbiting bodies can dramatically increase their velocity when they pass by a more massive object. If the velocity increases enough, it might exceed the escape velocity for that solar system. Likewise, pairs of stars that orbit each other as they simultaneously orbit a black hole are known to exist. As they approach the black hole, their rotational acceleration increases. Usually, the black hole eventually devours both. Occasionally, however, one of the stars is consumed as the other is tossed free. The doomed member of the pair bequeaths a portion of its rotational energy to its sibling, enabling the dramatic, speedy escape. The survivor becomes a vagrant destined to leave the galaxy altogether or possibly find a new orbit around the galactic center. Planet Nine may have departed its system of birth by a similar mechanism.

Another version of this origin story is that Planet Nine might have been stolen by our Sun from another star near the time of its birth some four billion years ago. Early in its history, the Sun, like most stars, formed in close proximity to others from the same cloud of dust. These newborn stars could have sailed close to each other, occasionally snatching the outermost children from their neighbors. Where the hand of binding gravity weakens to a gentle tug, wayward offspring are more easily stolen, since a slight nudge proves sufficient to displace them from their home. Purloined objects can include comets, asteroids, and planets—the ultimate theft in broad daylight. This means that Planet Nine could be almost any type of planet. It could be an enormous rocky planet of the Earth type, or an oversized icy planet of the Pluto type, or a small gas giant (a sort of mini-Neptune). Fanciful artists' conceptions of a rocky, magma-streaked Planet Nine are shown at the beginning and end of this chapter.

Whatever Planet Nine's origin and type, it still must follow an elliptical path around the Sun. Furthermore, to account for the observed orbits of the Kuiper-Belt objects, its orbit must be very elliptical, varying from 19 billion miles to 112 billion miles distant from the Sun. The average orbital distance of roughly 60 billion miles is about 20 times the distance of Neptune from the Sun and would put Planet Nine's orbital period somewhere between 10,000 and 20,000 years. Obviously, these are much greater than the orbital distance and period of any other planet.

Finding Planet Nine in a telescope will be like spotting a BB from 20 miles away, but researchers have a fair idea where to start looking. The observed orbits of the Kuiper-Belt objects limit the possibilities, and the search has been narrowed down to a relatively small area of sky encompassing the constellation Orion. Scientists are searching for minute differences between pictures taken several weeks apart hoping to see a moving spot against the background of fixed stars. This is similar to how Pluto was discovered, but astronomers will use infrared instead of visible light to make the determination. A planet as massive as ten Earths will retain residual heat from its creation, producing infrared radiation. Visible light, on the other hand, would be utterly insufficient for detecting the planet since so little visible light is available at that distance to be reflected.

Since the other giant planets in the Solar System possess moons, so too could Planet Nine. Such moons, if any, could also experience tidal heating and possess water in liquid form, making life a possibility despite their existence in the frigid darkness of the Kuiper Belt. Scientists find this tremendously thrilling. To paraphrase Shakespeare, it is yet another case of nature exceeding what is dreamed of in our philosophy.

If Planet Nine is ever found, it will be one of the most exciting discoveries in the history of humanity's exploration of the Solar System. Will the tally of planets climb back up to its long-accustomed total of nine? Superfast computers and adroit programming based on time-honored laws of physics may soon cause the Solar System to give up one of its longest-held and most deeply hidden secrets. The tantalizing clues suggest that humankind's perennial quest to explore space, discover new worlds, and fathom the nature of the cosmos is about to enter a new age.

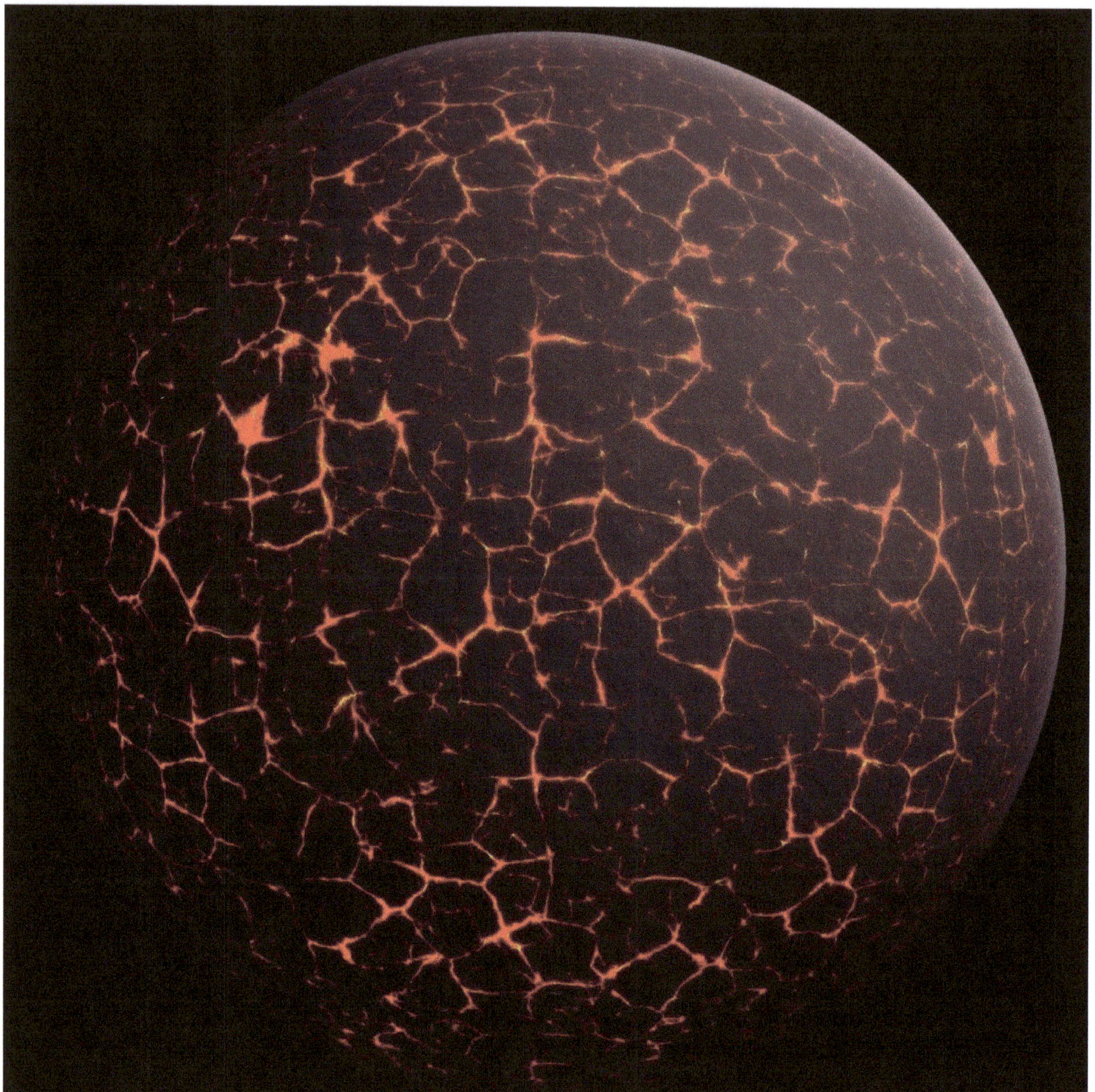
Planet Nine (Artist's Conception)

24
Constellations: Illustrating Humankind's First Primer

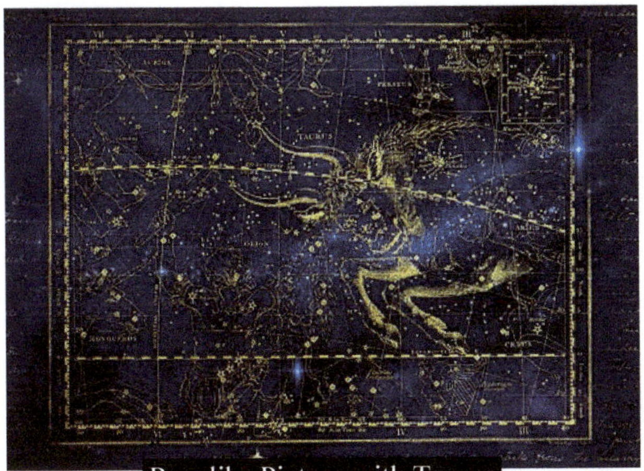
Rug-like Picture with Taurus

Canst thou bind the sweet influences of Pleiades, or loose the bands of Orion?
Job 38:31

The immense deserted night set up its formation of colossal figures that seeded light far and wide. Aldebaran trembled, throbbing far above, Cassiopeia hung her dress on heaven's doors, while the noiseless chariot of the Southern Cross rolled over the night sperm of the Milky Way.
Pablo Neruda

Humans, by nature, are wonderers, observers, and learners. In their migrations across the surface of the Earth, our ancestors nightly witnessed a subtle transformation as the dark azure sky became a gathering multitude of flickering lights against a backdrop of velvet blackness. Seemingly just beyond reach, this majestic dome was speckled with thousands of pinpoint fires forming pictures and patterns that came eventually to be recognized as wordless messages from a realm more elevated and ethereal than that of mere mortals.

Galaxies, stars, nebulas, and glowing clouds of dust display colors, images, and formations that inspire artist, astronomer, and commoner alike. The grand sweep and splendor of the stars break like thunder on the human mind, illuminating a cascade of images that produce an everlasting aura of power and beauty. As these glowing beacons nightly and annually traversed the sky in their stately and cyclical way, their beauty brought poetry to the lips of wordsmiths even before the dawn of the age of writing. Meanwhile, the power of the stars was their ability to serve as weather forecasters, compasses, clocks, and guides. Knowledge of their movements conferred clear and tangible survival advantages upon those who comprehended this lingua franca of the heavens. The stars steered ancient navigators upon the Mediterranean Sea and the South Pacific, foretold the rising of the Nile River, and sowed the early seeds of literature, mathematics, and science. Stars gave rise to calendars, which anticipated the changes of the seasons, thus facilitating the planning and management of agriculture. Since the earliest prehistoric times, stars have been an integral component of the development of humankind culturally, socially, and educationally.

To our ancestors, the night sky was one of the original, indispensable keys to survival. Early humans had to understand the sky's patterns to survive, and passing on this understanding from one generation to the next created culture. The first observations of the cosmos provided the initial impetus that thrust our forebears onto the path toward modern civilization.

Culture, science, and religion were all intricately intertwined. In early times, astronomy and astrology were one and the same. The pyramids of Egypt were constructed to align with certain stars so that the entombed pharaohs would have a clear path to follow to the afterlife. In China, conjunctions of stars were taken to indicate the rise or fall of a ruling dynasty.

The relationship of the heavens with all aspects of human culture contributed to some of humankind's most significant discoveries and creations. The stars have beckoned like beguiling goddesses, tempting the greatest of human minds to discover hidden knowledge, fathom arcane enigmas, and seek their inspiration among the starry skies. Myriads of stargazers yearning to explore the vast expanses of the universe have marveled at nighttime's exquisite, glowing panorama. From our earliest inklings of their greatness and importance, stars have captivated our imagination, guided our steps, and invigorated our will to transform our world and ourselves.

Constellations and Myths

The star formations we call constellations inspired the earliest humans to create wondrous fables and stories. These tales, often recounting the exploits of heroes and gods, sought to explain the patterns among the stars and served as aids to memory. The constellations, enshrining beasts, heroes, and royalty among them, became their own mnemonics with their individual stories. This is an example of humankind's innate and irresistible urge to explain the natural world and to see patterns everywhere.

The scattering of stars in the night sky is seemingly chaotic, but as one continues to look, one quickly learns to recognize groupings—patterns and pictures that seem to demand interpretation and narrative. Humans instinctively and automatically supply that narrative and, incidentally, take great enjoyment in doing so. In our modern world, we still recall and recount the tales that were the earliest fruit of humanity's impulse to create stories that describe and elucidate nature.

The myths associated with stars formed a major part of the cultural foundations of ancient civilizations. These earliest tales predate writing and history; they were surely invented and passed down for hundreds of generations. Each succeeding society remembered the myths of its predecessors; the Latin names of the constellations we use today were borrowed from the Greeks, the Greeks borrowed them from the Babylonians, and the Babylonians must have relied on earlier sources, now mostly lost in the sand, dust, and fog of time.

In recent times, the star-patterns in the sky have been organized into 88 constellations. These constellations tessellate the heavens; there are no empty spaces in the fabric of fable and imagery that blankets the heavens from Polaris to the Southern Cross.

Stars: Building Blocks of the Constellations

Although stars appear to us as mere glowing points of light, they are actually gargantuan, flaming, nuclear furnaces trillions of miles away. Many are larger than the diameter of the Earth's orbit or even mighty Jupiter's, and with great size come immense power and brilliance. According to some law of universal justice written into the structure of the cosmos, the brightest-burning giants are the quickest to die, their brilliance extinguished suddenly and explosively when their comparatively short lives come to an abrupt fiery end. Their matter is violently expelled and flung outward into space, leaving behind a stripped, cooling, super-dense heart at the core of an expanding bubble: the glowing remnants of the star's former glory.

Stars make up asterisms, constellations, and galaxies. It is estimated that the universe possesses more than 160 sextillion (1.6×10^{23}) stars since there are about 400 billion galaxies in the observable universe, each containing about 400 billion stars. One hundred sixty sextillion is approximately sixteen times the number of grains of sand on all the beaches of the Earth (10^{22} grains). Furthermore, it is a thousand times greater than the number of atoms in one of those grains of sand, which would be around 4×10^{20} atoms. In the famous Persian chessboard fable, a wily mathematician tricked the sultan into promising to place a grain of wheat on the first square of a chessboard and double the number of grains on each succeeding square. The sultan did not know that filling the chessboard in this way would require about 20 quintillion (2×10^{19}) grains of wheat by the time he got to the last square. Even this immense number is still ten thousand times smaller than the number of stars. Yet all of these pale in comparison with the veritably huge numbers invented by the Greek mathematician Archimedes and introduced in his classic treatise, the "Sand-Reckoner." In this famous work, which was addressed to King Gelon of Syracuse, Archimedes gave his estimate of how many sand grains would be required to fill the entire universe. His calculations were based, amazingly, on the heliocentric model of the universe promulgated by one of his peers, Aristarchus more than 1700 years before the birth of Copernicus. Archimedes calculated that 10^{63} grains of sand would fill the universe, a number which is much larger than recent estimates for the number of stars. Archimedes was making the first tentative steps toward developing an intuition for how big the universe really is. The numbers he came up with, though vast by comparison with the type of numbers that had been used before, fell short of expressing the universe's true size. His "sand-reckoning" assumed a spherical universe that in today's units would measure about two light-years across, whereas modern measurements place the radius of the observable universe at 13 billion light years.

The true intellectual giants, those who have such stunning insights into reality, seem always to wend their way through the labyrinth of ideas and arrive at a central truth. In the "Sand-Reckoner," the Greek sage Archimedes, who is recognized as one of the three greatest mathematicians of all time and whose accomplishments in mechanics were impressive, performs an amazing feat of imagination and calculation. His other achievements were varied and just as astonishing. He presented explanations about pulleys, levers, buoyancy, the area of a parabola, the ratio of the volumes of a sphere and the cylinder enclosing it (his personal favorite), an approximation for π, and (almost!) integral calculus. He also invented the Archimedean screw, a device still used today for moving water. His creations, while mystifying at first, become obvious after his explanations. Archimedes, who was called "The Master" by his fellow philosophers, laid the foundation for later generations in physics, mathematics, and engineering.

In the "Sand-Reckoner," Archimedes does two surprising and impressive things. First, he adopts Aristarchus's unpopular but ultimately correct idea about how the Solar System is arranged, with the Sun at the center. Second, he presents the largest numbers ever conceived up to that time.

Today, such large numbers are familiar and indeed ubiquitous, and the heliocentric theory has been confirmed beyond any doubt, like Einstein's relativity or the expansion of the universe. However, Archimedes was launching a foray into abstract thought that was a thousand years or more ahead of his time. The depth and extent of the stunning achievements of this unique and extraordinary mind inspire profound admiration. Archimedes will forever occupy a position near the pinnacle of human attainment.

Also inspiring deep appreciation are the stars, luminous lighthouses in the cosmic ocean of space, incandescent lamps whose beacons of visible light punctuate the night sky and seem to the human eye and mind to form recognizable shapes. Those shapes are called "asterisms." They are *not* constellations. The naming of constellations is under the jurisdiction of the International Astronomical Union (IAU), which is the governing body of professional astronomers. Asterisms, on the other hand, belong to the common people, the storytellers, artists, and dreamers. The IAU has divided the sky into a set of officially recognized constellations that together encompass all of the heavens, not leaving out a single star or nebula, nor, for that matter, a single square inch of empty sky. Absolutely everything in the sky is assigned to one of the 88 constellations.

All constellations have Latin names; at the time the names were chosen, Latin was the primary international language of scholarship. Stars within each constellation were identified with Greek letters, Greek being the secondary international language of scholarship. These letters were often (but not always) assigned in alphabetical order by brightness. Many stars, particularly the bright ones, also have common names. These names are often Arabic because for centuries, the leading astronomers of the world were Arabic-speaking Muslims.

The stars which constitute the constellations are all nearby. They are the *pharos* illuminating our local shores—the nearest, the biggest, and the most memorable. However, even these nearest stars are much, much farther away than our planetary neighbors. Even if we were to journey to a vantage point on Neptune or Pluto, our picture of the heavens would remain fundamentally unchanged. What we can see from Earth and what we could see from Pluto are almost exactly identical.

Orion

Orion is one of the most prominent of the constellations, boasting some of the brightest stars of the winter sky. It rises early on winter evenings in the Northern Hemisphere and is visible throughout the night, while in summer it rises late. On an August morning, you can see it come up over the horizon at around 5:00 AM PST. In winter early in the evening, Orion lies in the East at 8:00 PM PST.

The Orion Constellation

Orion's belt is distinctive and easy to spot. It consists of the three stars (from bottom to top in the picture at right): Alnitak, "the girdle"; Alnilam, "the string of pearls"; and

Mintaka, "belt." These names are Arabic. Alnitak is actually a triple star consisting of three blue giants, one of them a supergiant. Mintaka is a complex system containing no fewer than five stars, from a type O giant all the way down to a dim 14th-magnitude star that may not even be gravitationally bound to the others. The two biggest and brightest stars in the system orbit each other every 5.5 days. The stars of the belt are fairly distant from Earth at distances ranging from 800 to 2000 light years; that is, they are farther away than most of the stars we can see in the sky. Nonetheless, because they are giant stars and very luminous, even at these great distances they appear bright to us. The presence of these multiple star systems should not surprise us; astronomers have observed that the majority of bright "stars" are actually systems consisting of two or more stars, while dim stars tend to be singletons.

The bright blue star Rigel, in the lower right corner of the picture on the previous page, represents Orion's left foot or knee. "Rigel" comes from an Arabic word meaning "foot." Rigel is a triple star system. Rigel A is a blue supergiant of 17 solar masses that is 500 times brighter than Rigel B and 85,000 times brighter than our Sun. Rigel B is in actuality two stars that orbit a common center of gravity approximately every ten days.

The Hunter's right shoulder is Betelgeuse. (The pronunciation of "Betelgeuse" is uncertain. It is typical today to say "Beetlejuice," but this would not satisfy a purist. The dictionary gives "bet-el-jooz.") The name is Arabic, like the others so far, but has become a bit scrambled over the centuries. The first letter was originally a Y, and the Arabic name would translate as "hand of Jauza" (a giant goddess), but European scribes mistook the Arabic Y for a B.

Betelgeuse is very young for a star at only 10 million years of age; however, it is already nearing the end of its life. Within a million years, Betelgeuse is expected to explode in a supernova that will be as bright as the full Moon, but for the time being it is a variable red supergiant that fluctuates in size from as large as the orbit of Mars to larger than the orbit of Jupiter. Betelgeuse, one of the monsters of the nighttime sky, is among the largest and most luminous of known stars, occasionally emitting more light than 100,000 of our Suns.

Orion Nebula

On the lower left of the constellation as viewed from Earth is the star Saiph, which represents the right foot or leg of the Hunter depending upon how he is depicted. Saiph is an Arabic word meaning "sword," and refers to the sword of the goddess Jauza, mentioned above. Saiph appears dim by comparison with the other Orion megastars, but this is deceiving, since Saiph is intrinsically very bright (18,000 times more luminous than our Sun) and appears dim to us only because it is 1400 light years away. Also, much of its energy output is in the ultraviolet range, which is invisible to the casual viewer. Saiph is the type of star known as a blue supergiant. About twenty times larger than our Sun, it has a surface temperature five times hotter. For Saiph to be this powerful, it is almost certainly burning helium in its core instead of hydrogen, a process which happens late in a star's lifetime. Saiph is likely to become a supernova soon, perhaps in as little as a million years.

Hanging from Orion's belt is a group of celestial objects representing the Hunter's sword (see the picture on page 272). The tip of the sword is the Orion nebula (previous page), which is 1350 light-years from Earth and is visible to the naked eye. This diffuse nebula is a giant cloud of gas and dust that has already created thousands of young stars, making this fecund volume of space one of the Milky Way's most prolific stellar nurseries. The concentration of stars in this 30 light-year diameter nebula creates so much heat that the surrounding cloud of dust and gas glows, resulting in what scientists refer to as an emission nebula. The Hubble Space Telescope has taken pictures of over 150 stars in the nebula, each surrounded by its own ring of gas and dust. Astronomers believe these are proto-planetary systems that will eventually leave the nebula and form new solar systems that will orbit within the Milky Way for billions of years. If humankind ever develops the technology to travel enormous distances safely through space, then one of these nascent systems could become a future home for our descendants.

An exception to the rule that stars tend to have Arabic names is the star representing Orion's left shoulder, Bellatrix. This is a Latin name that means "female warrior" or "Amazon." Bellatrix, the third brightest star in the constellation, is a blue-white giant star of about nine solar masses. A very hot and bright star, if it were as close to us as the Sun, it would be 6,400 times brighter than the Sun. Such a source of intense, high-energy radiation that close to Earth would quickly fry our planet and turn it into an arid, airless, cratered, wasteland resembling the barren planet Mercury.

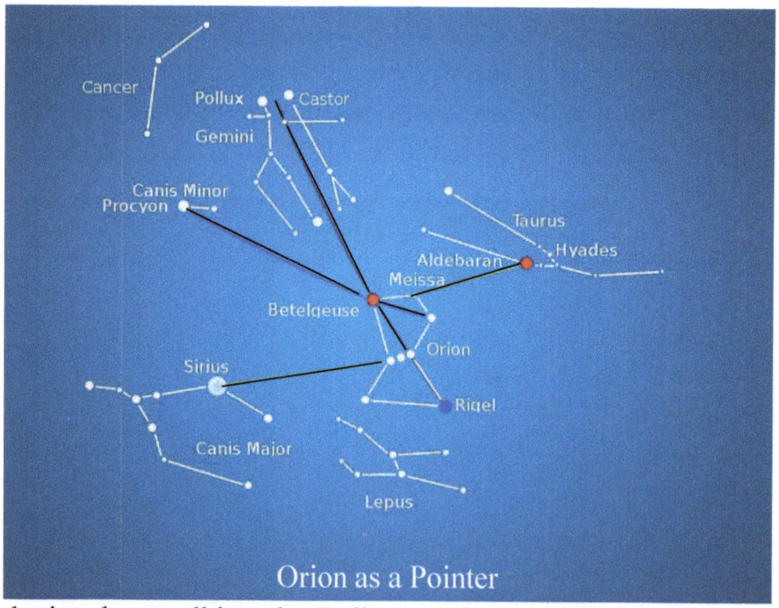

Orion as a Pointer

The bright stars of Orion can be used as pointers to find other constellations. Follow the slant of his belt downward to locate Sirius, the brightest star in the night sky. Sirius is the Dog Star and is the most prominent feature of Canis Venatici, the Hunting Dogs. Naturally they are at the feet of their master, the Hunter. The line from Betelgeuse to Meissa runs toward Aldebaran, which is blood-red in color and is the brightest star in Taurus, the Bull. The Hunter is usually depicted as stalking the Bull, sometimes shown as a specimen of the nimble, energetic, long-horned aurochs, a relative of domestic cattle. Beyond Taurus, along the same line, one can find the Pleiades, also known as the Seven Sisters.

Extending the imaginary line running from Rigel to Betelgeuse and continuing in the same direction, you will reach the bright stars Castor and Pollux in the constellation Gemini, the Twins.

Castor and Pollux are fraternal twins; however, Castor's father was Zeus, king of the gods, while Pollux's father was the Spartan king Tyndareus. Inseparable in life, the twins were reunited by Zeus upon their death and placed among the stars. Castor, whose heritage was immortal, is the brighter of the two stars, while Pollux, whose father was mortal, is dimmer.

Follow the line from Bellatrix to Betelgeuse and onward, and you will find Procyon, the brightest star in Canis Minor, the Little Dog. (Note that Orion is facing you, so his left shoulder is on your right as in the illustration below.)

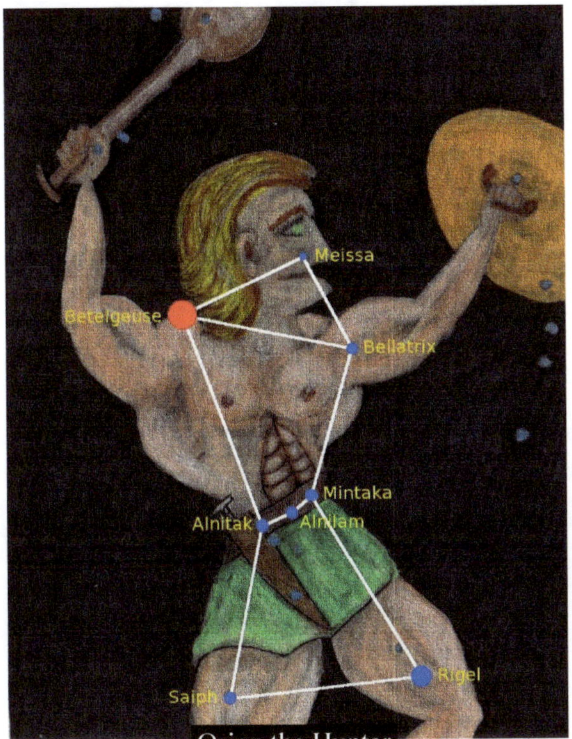

Orion the Hunter

Try this: Hold up three fingers of your left hand so that the tips of your fingers cover the three stars of Orion's belt. If your three fingers resemble a W, then you are facing west, while if your fingers resemble an E, you are facing east.

In Greek mythology, Orion was a son of the sea god Poseidon. He is described as handsome and exceptionally tall, and his club, fashioned of bronze, is unbreakable. Some ancient tales about him recount that Orion was slain as a punishment for his arrogance in the presence of the gods. In some versions, he was punished for sexual importunity. No matter the version, the cause of Orion's death is usually a fatal sting from Scorpio, the scorpion. For this reason, Orion dips below the horizon for two months in the summer. He is avoiding Scorpio, which rises during these months.

The constellation Orion was well known in many cultures. In Australia, Orion's Belt and sword are sometimes known as the Pot or the Saucepan. In South Africa, the three stars of the belt are known as *Drie Konings* (Three Kings) or *Drie Susters* (Three Sisters). In Spain and Latin America, these stars are called *Las Tres Marías* (The Three Marys).

The ancient Egyptians evidently aligned the Great Pyramids of Giza with the stars of Orion's belt. An aerial photograph of the three pyramids overlain with an image of the stars of the Belt shows a near-perfect match. Passageways inside the tomb of Khufu were constructed to point directly to Thuban, which at that time was the pole star, and to Alnitak in Orion. Because of the Earth's precession, those same passages now point to different stars, Polaris and Sirius, respectively.

Surprisingly, in ancient China, Orion was also identified as a hunter, although it is not clear how the Chinese arrived at the same interpretation of this constellation as the inhabitants of the Mediterranean basin many thousands of miles away. In modern China, the stars of the belt are referred to as The Balance Arm.

Ursa Major

You can probably spot the easily recognizable asterism of the Big Dipper, with its familiar shape of a pot with a bent handle, as in the image below. This asterism belongs to the constellation called Ursa Major, the Great Bear.

The two stars forming the front of the Big Dipper are the stars Alpha and Beta of Ursa Major, also known by their Arabic names Dubhe and Merak, respectively. Follow the line they form until you come to a smaller copy of the Big Dipper, the aptly named Little Dipper. The star you come to after having followed the line from the front of the Big Dipper to the tail of the Little Dipper is none other than Polaris, the North Star, friend of travelers and especially navigators from time immemorial.

Although the Big Dipper is not at the exact celestial North Pole, it never sets, and its motion throughout the night is a circle that surrounds the pole. Thus, anyone seeking to travel north could follow it confidently. In the United States in the mid-19th century, slaves attempting to flee to the North and find freedom were told to "follow the Drinking Gourd." Ancient Greek navigators traveling towards colonies in Sicily and Spain also knew to keep the Great Bear over their right shoulder.

The Greek myth to explain why the Great Bear circles the pole is the story of the beautiful nymph Callisto. Callisto was seduced by Zeus and bore him a son, Arcas. Zeus' wife Hera, who was continually enraged over her husband's peccadilloes, routinely chose to direct her malevolent rage toward the female objects of his lust, and thus in a fit of jealous pique she transformed Callisto into a bear. Callisto went about the rest of her life as a bear, facing all the difficulties and dangers inherent in such a lifestyle. One day she encountered her son Arcas in the forest, and, filled with joy, she rushed to embrace him. Arcas was horrified by what he perceived as a deadly attack from an angry she-bear and prepared to kill her with his spear. To avert this tragedy and save the nymph he still loved, Zeus at once intervened. He threw mother and son bodily into the heavens, where Callisto became the constellation of the Great Bear and Arcas became the star Arcturus.

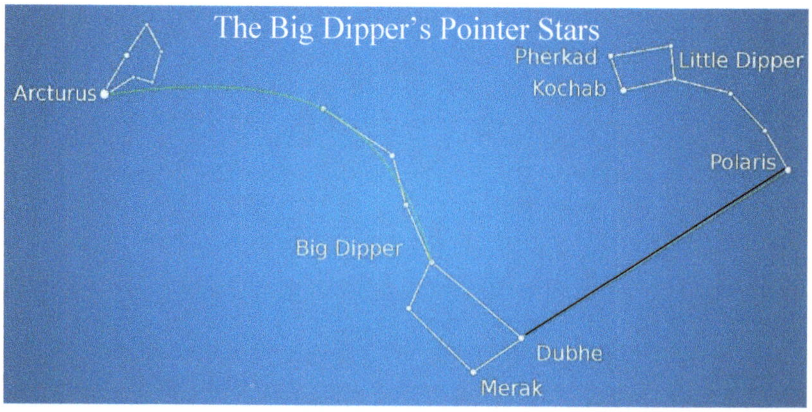

Hera could not abide that her rival had received the unmerited honor of being elevated to the status of a permanent dweller in the heavens. Thus, she conspired with the powers of the Ocean to ensure that the Bear would never be permitted to bathe. That is, its stars would never descend into the ocean as other stars do, but would continuously circle the North Pole. Arcas still follows his mother and can be found in the sky by tracing along the curve of the Big Dipper's handle and continuing until Arcturus is reached. This is known as "following the arc to Arcturus."

Other Constellations

There are innumerable stories about the mighty heroes, magical animals, and ancient gods that populate the constellations. Each ancient culture and civilization had its own set of myths, legends, and heroes. Many books have been filled with these stories without coming close to exhausting them. Different peoples in different times and places devised their own stories. There are some odd coincidences, however. For instance, ancient Mesopotamians saw a great bear where Ursa Major is, and so did the Native Americans.

The names of many of the constellations of the Southern Hemisphere are of recent origin. Unknown to the ancient Babylonians and Greeks, these constellations did not receive official names until after European explorers began to cross into the Southern Hemisphere on their voyages of discovery. Eventually, Southern Hemisphere constellations were given such non-mythological names as Tucana, the Toucan; Telescopium, the Telescope; or Indus, the Native American.

The Zodiac

Any discussion of the constellations must include the Zodiac. The word *Zodiac* comes from a Greek phrase meaning "circle of animal figures." The Zodiac constellations are special because they are arrayed along the ecliptic, the imaginary line which marks the path of the Sun on its daily journey across the sky. Dating back to the earliest written records of the first civilizations, this band of stars has been divided into 12 constellations, and we still identify them the same way today.

The "animals," some of which represent human figures (one, Libra, represents an inanimate object), are Aries the Ram, Taurus the Bull, Gemini the Twins, Cancer the Crab, Leo the Lion, Virgo the Virgin, Libra the Scales, Scorpio the Scorpion, Sagittarius the Archer, Capricorn the Goat, Aquarius the Water-Bearer, and Pisces the Fish. Since these constellations lie along the path traversed by the Sun over the course of a year, at any given moment the Sun is in one of these constellations. The zodiac constellation in which the Sun was located at the time you were born is your astrological sign, and many people believe that individuals born under the same sign share certain personality traits. However, today the majority recognize that our future is not determined by the stars overhead at the time of our birth, but by our innate abilities coupled with motivation and determination. The force of our will and the dynamism of our desires play greater roles in determining our destiny than do our native talents, much greater than any supernatural effects exerted by the shower of lights in the night sky. The power to succeed lies in us, not in our stars.

Unfortunately, much of the lore of the ancient mystics who developed the pseudoscience of astrology, which is the use of astronomy for predicting the future, has become anachronistic and is now recognized rightly as fictitious. In fact, because of the slow changes in the Earth's rotation over thousands of years, the original twelve constellations no longer lie exactly on the ecliptic, and the month with which each sign is associated is no longer its correct month according to where the Sun is. There is even a new member of the Zodiac club, the constellation, Ophiuchus the Serpent Bearer, which has slithered its way into the circle between Scorpio and Sagittarius while Aries now lies slightly outside. This, unsurprisingly, has not discouraged the writers, publishers, and sellers of horoscopes, who continue their arcane and non-evidence-based activities without apparent concern for what is actually happening in the skies above them.

Conclusion

Astronomers of these latter days have caught all the constellations and put them in their pens, cataloging and assigning every star to one or another of them. But the division of the night sky does nothing to mitigate the beauty of starlight, the awesome majesty of the grand and glorious sweep of stars across the curve of night, and the evocative and archetypal narratives that resonate in our beings from dozens or hundreds of centuries ago. These pictures in the sky have been with us humans since we first awoke and began to wonder. They will be with us still at the dread hour when human civilization draws to its end, or, alternatively, when we make the choice to survive and grow, to leap from this pale blue nursery across the steppingstones circling this white Sun and take our place out there among the stars, among the constellations; and then the teachers and storytellers of that distant age will not only speak of bears and hunters and dragons and lions but will point to the star shapes and poignantly intone: "Now, humans live there, too."

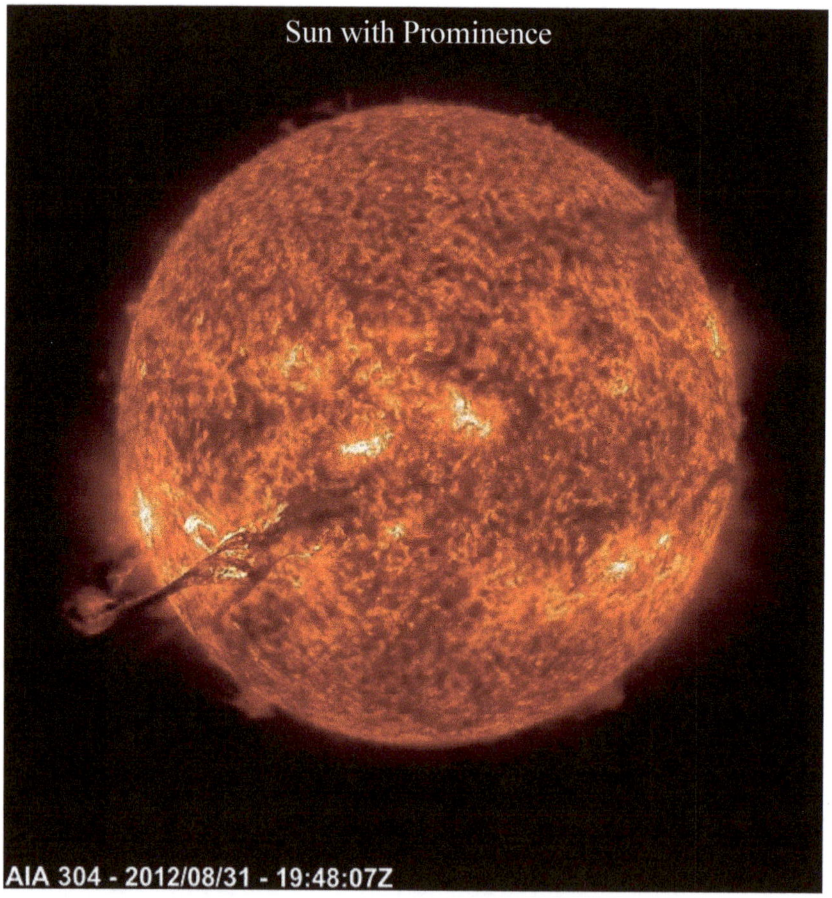

The Solar Dynamics Observatory, using the 304-angstrom ultraviolet wavelength, captured this image of a magnetic filament as it erupted from the surface of the Sun on August 31, 2012. Some of the particles from this event, after traveling at over one million miles per hour, reached the Earth on September 3, creating magnificent auroras.

25
The Existence of Extraterrestrial Life and Civilizations

Sometimes I think we are alone in the universe, and sometimes I think we are not. In either case, the idea is quite staggering.
Arthur C. Clarke

Through our eyes, the universe is perceiving itself. Through our ears, the universe is listening to its harmonies. We are the witnesses through which the universe becomes conscious of its glory, of its magnificence.
Alan Watts

The quest to discover extraterrestrial life has kindled the fires of human imagination to such an extent that it has entered the popular lexicon. In recent years, it has drawn the rapt attention of numerous astronomers so that every space probe, every radio telescope, and every new and elaborate device for scanning the far reaches of the electromagnetic spectrum seem to be aimed at enhancing the search. The human species is approaching the point where the discovery of other intelligent species is becoming possible. The search commences with an understanding of the precursors of life, those ingredients without which life, and therefore intelligent life, is impossible. Those vital substances originate within the very stars upon which we gaze wondering if they could possess planets like ours, with environments like ours, and homes to minds like ours.

The universe is vast, enduring, and filled with trillions of radiant stars, each in the midst of its own unique life cycle. Stars, like living organisms, pass from birth to death, and in between, they consume fuel. The fuel that powers living organisms comes from sunlight, food, or minerals. Stars, on the other hand, burn hydrogen as their fuel. During their immensely long lifetimes of millions or billions of years, they consume innumerable tons of hydrogen. One of the products of this process is energy in the form of electromagnetic radiation, which stars generously shower upon the universe. Electromagnetic radiation is one of the universe's gifts to life, for this energy invigorates living things by bathing them with warmth, nourishment, and stimulation as they come into existence, thrive, and evolve.

For stars to create this rain of radiation, they must continuously transform themselves by combining their lighter atoms into heavier ones in a process called nuclear fusion. This consumes a tiny portion of the atoms' mass, which is converted into surprisingly large amounts of energy that floods the cosmos. Average-sized stars like our Sun consume billions of tons of hydrogen per second, but they possess such vast reserves that they can continue burning at this rate for ten billion years or more before flickering out. To Earthly eyes, a Sun-like star displays a placid, benign appearance of serene stability and permanence, but internally it is a raging, volatile monster. As the consumption of hydrogen and its conversion into heavier elements proceeds, the drastic changes taking place in the star's core are virtually imperceptible from the outside. Its passage from blazing beast to shrunken cinder progresses slowly, subtly, and unnoticeably over eons.

The creation of the heaviest elements by nuclear fusion often requires multiple generations of stars as each succeeding supernova creates heavier elements that chemically enrich the cosmic seedbed. Stars are mega-factories for converting simple atoms into more complex ones. Those complex atoms with their large nuclei are another of the universe's gifts to life because complex atoms are the raw materials from which living organisms are formed. This process of combining atomic nuclei in the core of a star is also called heliosynthesis and is absolutely essential for the existence of carbon-based life.

Carbon, the basic building block of life on Earth, has the ability to join with itself and other elements to create chains, rings, and branches that can form the backbones of larger organic molecules. This modular construction process allows for an unlimited number of unique formations. Thus, the carbon chemistry of life is complex, versatile, dependable, and swift, ultimately culminating in the creation of the wondrous DNA molecule. Just as humanity likens the Solar System to an atom, we see the cosmos in a strand of DNA.

Through nuclear fusion, the universe in its guileless wisdom is performing the miraculous act so long sought by the ancient alchemists—the conversion of one element into another. The alchemists wished to change lead into gold, while stars change hydrogen into helium. Then, as the hydrogen is fused into successively larger nuclei, the process culminates in the monstrous supernova blasts that forge the heaviest elements. Humans, too, have learned how to perform this conversion by arranging for elements to be transformed in the hellish hearts of their nuclear reactors. The first successful practitioner of this arcane art was Enrico Fermi, who accomplished the feat in 1942.

In contrast to Fermi's splitting of heavy elements into lighter ones, heliosynthesis is the opposite process. The universe seems to operate under an overriding prerogative directed toward more complexity—and ultimately life. How the alchemists would have loved to wield the power that resides in a star! Those medieval wisdom-seekers also sought the elixir of life, and nothing could be more apropos than the elemental transmutations that transpire in the core of a star. Without the creation of those heavier elements, humanity would not be here today to gaze upon the universe's expansive beauty, nor would we be striving to fathom its enthralling mysteries or contemplate the sublime wonder of life itself. The universe's splendor, secrets, and sentience are but a few of the attributes that make the life within it so precious.

While the Earth is a fountain of life, it is also a monumental, mesmerizing mystery book. Secreted within and upon its majestic mountains, landscapes, and seashores lie obscure clues, many of which are deeply hidden. When pieced together they form a clear, linear, visual record of the critical moments in the 4.5-billion-year-long story of the transformation of this once lifeless, molten rock into a vibrant, verdant, breathing world. The tome that is Earth holds, for those who are willing to learn to read it, a fascinating and glorious revelation of the grand, sublime, and propagative artistry of the cosmos. It is an epochal narrative, which has culminated over the last 3.7 billion years with the rise of life.

According to these earthen records, life likely began in a gently sloshing tidal pool at the edge of a vast ocean or near an energetically gushing hot-water vent on the ocean floor. Perhaps the sporadic bubbling in the primordial soup of a warm little pond, or crystalizing lattices in the clays of drying mud contributed to the genesis of Earth-life. In any case, it is a triumphant story, occurring when myriads of atoms, after combining into astonishingly complex but still inanimate molecules, were spurred on by some as yet unknown stimulus and leaped in unison across the gulf separating the abiotic from the biotic to become a nascent cell.

The first cells resembled today's bacteria. Simple by comparison with later-arriving plants and animals, they were nevertheless extraordinarily complicated. They adapted to the conditions on the early Earth and thrived. Their phenomenal success continues up to the present day. Bacteria are so numerous and adaptable that experts cannot even agree on the number of bacterial species that exist. Some specialists estimate there may be as many as ten million species of bacteria spread throughout all levels of the Earth's biosphere. One recent NASA study found 314 different kinds of bacteria above the Atlantic Ocean in the upper troposphere alone. Of the various forms of bacteria, some are known to be able to survive even the severities of outer space. A sample of bacteria inadvertently left on the Moon inside a camera entered dormancy, survived three years, and then, after astronauts brought them back to Earth, were able to resume their previous activities and mode of existence.

Bacteria can survive in almost any environment. They thrive in water heated past the boiling point and in polar seas super-cooled below freezing. Some species flourish in the crushing pressures at the bottom of the deepest ocean trenches, while others do not die in the vacuum of outer space. There are bacteria that thrive in oxygen, while others can live anaerobically. Some bacteria can endure ultraviolet radiation, and others prefer infrared. They can live in acid or alkaline environments. A number of rare bacterial species do not even need carbon-based food for nourishment, but can directly metabolize iron, nitrogen, or sulfur.

Without bacteria, more complex forms of life could not exist on Earth. Bacteria facilitate vital activities, including digestion. Microorganisms enable the soil to produce plant life, form the basis for some of our most effective medicines, and have dictated the course of the development of life from Earth's early history through today. Microorganisms were responsible for altering Earth's anaerobic atmosphere and creating the life-giving, oxygen-rich atmosphere that we and other organisms find so essential, invigorating, and enjoyable.

Over eons, and because of their own self-assembling inclinations, cells eventually aggregated into multicellular organisms with functionally differentiated tissues, organs, and organ systems. The eventual result was a globe-spanning web of multihued flora and fauna with diverse lifestyles which allow them to occupy all available niches. The constituents of this vast web interact with each other as competitors and commensalists, predators and prey, disinterested observers, or, as humans, curious investigators of their shared world and the universe.

Amidst the discrete and mobile layers of Earth, there are biomes containing multifarious forms of life. On the ocean floor near volcanic chimneys, exotic tubeworms sway and blind yeti crabs feel their way along basaltic rocks; among human-made towers of steel, feral pigeons search for nesting sites while solitary bees search for flowers; close to the ocean's surface, greenback turtles bob, striped dolphins race, and ponderous, graceful humpbacks frolic with their calves; meanwhile, near the edge of space are found multitudes of microorganisms including living diatoms and bacteria.

The Earth's elegant bionetwork of the great and small pulses and throbs like a vast system of arteries flowing with life energy. Heat and light churn this rocky world and its ecosystems, pumping inorganic substances (air, water, and minerals) around and through the body of Earth, resulting in a remarkable and self-sustaining display of vitality, fertility, and adaptability. A panoramic, all-inclusive view of our world would reveal billions of creatures living eventful lives amongst a stunning backdrop of varied habitats. The earliest creatures developed simple sensory spots based on neurons to probe their local environments; simultaneously, they learned how to respond to those modest stimulations. Over time, sensory receptacles improved to become eyes, ears, noses, tongues, skin, and other less familiar sensory apparatuses such as the ampullae of Lorenzini found on the heads of sharks, strange electroreceptors for sensing the presence of hidden prey.

However, it is one thing to detect a stimulus, and it is quite another to interpret and act judiciously in response to those sensory signals. For this, there arose the need for neurons organized into networks and concentrated into functional and analytical modules—brains. The first brains were reactive. This kind of brain is called the reptilian brain, and its original architecture is preserved in the modern mammalian brain, but with additions and enhancements. From those earliest brains, the brains that developed were capable of deep cogitation and enabled creatures to make complex decisions such as whether to flee, pounce, or remain hidden.

The chronicle of the rise of the first biological kingdoms is captivating but is surpassed in significance by the recent six-million-year history of the rise of reflective primates and the more immediate 200,000-year history of the ascent of modern contemplative humans. Today, humankind realizes that our history is but a fleeting instant compared to the time that has elapsed from the beginning of the universe or even since the dawn of life. Furthermore, we recognize that adaptability guided by intelligence has catapulted our species remarkably quickly to its present position as the preeminent tool-wielding, problem-solving, story-telling life form on planet Earth.

All of Earth's creatures observe the same wonders and phenomena that transpire upon our planet, but only humans devise narratives to explain these natural events. We understand how a lightning strike shatters the atmospheric equilibrium to produce a deafening drumbeat of compressed air. We know that the precious precipitation upon which Earthly life depends is generated and replenished through the intricate interplay of clouds, rivers, and oceans. We are intimately acquainted with the mysterious attraction exerted upon the needle of a compass by the Earth's invisible magnetic field. We have come to appreciate the intricate choreography of the cosmic courtship dance that results in a solar eclipse. We have discovered and measured the accelerating expansion of the universe.

Many of Earth's creatures can see the stars at night, but only the human race can visualize itself roaming among those stars. Humans, with their unparalleled ability to solve problems, have been able to overcome the obstacles to achieve awe-inspiring goals such as walking on the moon, hurling probes to dart among the rings of Saturn, and sending out golden phonograph records to serve as heralds to the cosmos.

The underlying nature of the universe is not only comprehensible but has so far shown itself amenable to quantification through mathematical laws. It is quite remarkable that mathematics should be such an effective tool for describing the universe. As Galileo stated, and virtually every scientist since Galileo has agreed, "The great book of nature can be read only by those who know the language in which it was written, and this language is mathematics." Some of the most critical problems human beings have solved are mathematical problems involving complex geometry to visualize nature not as our senses perceive it, but as it actually can be measured. Master problem-solvers have made surprising discoveries such as: the Earth and all its kindred are spheroidal in shape, the speed limit of the universe is the speed of light, the cosmos sprang from a dimensionless singularity, and spacetime is a malleable fabric that permits minute and subtle waves to pass through it like ripples across an ocean. These findings are provisional, as are all scientific discoveries, but expertise in mathematics allows us to study them thoroughly, express our conclusions succinctly, and confidently make future predictions.

Further proof of humankind's complexity exists within the Earth's crust. Many of Earth's life forms have left fossilized evidence of themselves buried in deep strata of ancient rocks; however, only humans have left behind artifacts. *Homo sapiens* is the only organism on Earth capable of envisioning a complex tool in its mind and then actually building that tool in reality. These artifacts include a 20,000-year-old incised fibula of a baboon found near Ishango in the Congo, a 2100-year-old clockwork Solar System mechanism found near Antikythera in Greece, and a finely etched two-week-old microchip found discarded in a trash receptacle in central California.

These tools are snapshots that capture a piece of a particular human civilization's history, customs, and values. Tools reflect the adroitness and cleverness of their makers as well as represent the intertwined nature of cultural and biological evolution. Priceless artifacts like these not only convey how the members of a culture think and work, but are indicative of their fabricators' values. For instance, 15,000-year-old Altamira cave paintings graphically depict great hunts of a bygone era. Michelangelo's David strikingly manifests the sculptor's superb technique and reveals that his eyes and hands were the keen extensions of his kinesthetic mind as he chiseled away the excess stone from a twice-discarded block of Carrara marble until all that was left was the masterpiece. Exquisite tools, paintings, and statues eloquently proclaim, "Here is the mark of humanity."

The tools we continue to craft today have reached mind-boggling complexity. Knowing that the frontier of discovery exists at the limits of our ability to measure events, we have continuously refined our tools to be better at this task, especially at the extremes of nature. Therefore, we have created such wonders as the Hubble Space Telescope, the Scanning Tunneling Electron

Microscope, and the immense LIGO gravity-wave detector. Humanity has developed tools for listening to the roar of the Sun, for seeing beyond the red and blue wavelengths, and for sniffing the atmosphere for the presence of one molecule in a hundred billion.

Unfortunately, human civilization today has reached a point where it is capable of self-destructive behavior on a massive scale, but also, auspiciously, it remains capable of cooperation, kindness, and altruism. Here at the hour of our greatest peril and greatest achievements, balanced on the knife's edge of sanity, humanity's overwhelming desire to decipher the mysterious clues about the nature of space, time, and matter could lead to ever-greater accomplishments in the future. This desire to learn and know will continue unabated to drive us toward greater understanding of the physical universe to aid us in our search for life and extraterrestrial intelligence. In fact, the most vital and natural activity of the most advanced form of life on Earth may well be the search for life among the stars.

Physically, humans may be relatively weak and defenseless creatures lacking in size, strength, claws, armor, or fangs; nonetheless, humans find themselves robustly endowed with fortitude, resilience, stamina, and intelligence. We are inquisitive, introspective, creative, and expressive creatures who have invented writing, art, culture, and technology. By virtue of our ability to envision better futures for ourselves and our descendants, we have sculpted our environment and created a globe-girdling civilization comprising almost eight billion individuals. Our Earth is a sliding, shifting, encrusted, fragmented, layered sphere. We have plumbed the depths of the oceans, erected monolithic skyscrapers that sway when the ground shakes, traversed the oceans in submerged nuclear-powered vessels, built airplanes that can rise above the weather to skim the edge of space, and launched rockets capable of penetrating the void to safely carry astronauts to the Moon and back. We have also nurtured the soil and coaxed it into pouring forth abundant sustenance for us, developed life-saving medicines capable of extending life by decades, and tinkered with the genomes of creatures large and small, thus accelerating the process of natural selection. These impressive accomplishments have enhanced the quality of life of individuals and the human species as a whole. From Earth, *Homo sapiens* has now begun to venture into the realm of outer space in search of future abodes, life, and extraterrestrial intelligence.

Life in the Universe

The universe is big. It contains a very large number of planets that are virtually identical to Earth. We cannot know the exact number of such planets, but reasonable estimates, based on what is known about the size of the universe and the distribution of planets and galaxies within it can yield numbers so huge as to be almost impossible to grasp. As of July 2018, it is estimated that there are a trillion galaxies in the universe, and each galaxy, on average, may have 200 billion stars or more.

Such vastness is one reason for the widespread belief that extraterrestrial life must exist and that humans are not the only intelligent species in the universe. Guesswork, clairvoyance, and gut instinct, however, are neither satisfactory nor sufficient from the scientific point of view. To lay a more concrete foundation for our intuitions and estimates, a more rigorous, mathematical method is desired for addressing the question of whether there is life and, by extension, intelligent life, in the universe somewhere besides Earth. Humans ask the following vital questions: Are we alone?

Are there other planets with organisms similar to us and the other species with which we share the Earth? Are there other species, somewhere nearby or far away in space, with whom we can communicate? Are there undiscovered kindred spirits, monsters, or demigods roaming the celestial gravitational corridors? The answers to these questions have profound implications and should guide our planning and behavior as we spread our metaphorical wings and fly into the cosmos of an unknowable future.

Inquiries into the existence of extraterrestrial life and intelligence give rise to mixed feelings. While fraught with uncertainty and trepidation, they are also imbued with delight and anticipation. If alien life is discovered, and particularly if the alien life we discover proves to be technologically advanced, will pandemonium reign as the population of Earth descends into panic? Or will the populace experience a surge in enchantment and an outpouring of elation at the realization that we are not alone? This philosophical issue is suffused with theological, political, scientific, and technological ramifications, but if a technological civilization of extraterrestrials is ever detected, it will be *the* watershed moment in human history. It would not matter if they were far more advanced than us or only slightly superior—humanity would experience a new Renaissance…or stare extinction in the face.

The Drake Equation

In 1961, Frank Drake formulated a famous equation, which now bears his name. It is also known as The Green Bank Equation, after the location in West Virginia where the National Radio Astronomy Observatory is based. There, in 1960, Frank Drake used a radio telescope for six hours a day for four months to scan the sky for signals of intelligent origin, a project known as "Ozma." No promising signals were found, but less than a year later, Drake hosted a meeting in Green Bank on the topic of extraterrestrial intelligence. The meeting was attended by Carl Sagan and nine other intensely interested scientists. Sagan, the youngest of the attendees, was a postdoctoral fellow at the University of California at Berkeley. Perhaps inspired by the dolphin research of one of the scientists at the meeting, and sensing the simultaneous whimsy and grandeur of the topic of the meeting, the group nicknamed themselves "The Order of the Dolphin."

It was at this meeting that Drake presented the equation that would become his namesake. The equation was Drake's attempt to quantify the probability of the existence of intelligent extraterrestrials. It is based on seven factors, the values of which at the time were unknown or highly speculative.

Here is the equation: $N = R_* f_p n_e f_l f_i f_c L$

N represents the estimated total number of intelligent civilizations in our galaxy and is arrived at by multiplying together all the factors in the equation.

R_* is the average rate of star formation over the lifetime of the galaxy (in effect this quantity represents the number of stars in our galaxy today); f_p is the percentage of stars with planetary systems; n_e is the average number of planets per star that harbor suitable conditions for life to arise, survive, and flourish; f_l is the percentage of such planets on which life in fact arises; f_i is the percentage of such planets on which intelligent life evolves; f_c is the percentage of such planets on which a technical civilization develops that can transmit detectable signs of its existence; and L is the mean lifetime of that transmitting civilization.

Although we have been able to measure some of the terms in the equation more accurately since Drake's time, several remain wildly speculative. Some of the values could be off by ten orders of magnitude or more. Such wild guesses, inevitably influenced by preconceived notions and biases, must be examined with the utmost suspicion, especially since the use of a mathematical equation confers upon the answer a certain undeserved cachet of credibility.

Drake and his Green Bank group chose values for each of the factors in his equation to the best of their ability, according to contemporary astronomical knowledge. To calculate the value of R_*, they reasoned as follows: If the Milky Way is 10 billion years old (there is evidence for this estimate), and it contains 200 billion stars (today astronomers think there are perhaps twice as many), then the rate of star formation must be 20 stars per year. Reasoning further that the rate has probably slowed since the birth of the galaxy and that some of the stars would not be stable enough to support life, they finally arrived at an estimate of ten stars per year. This accords with more up-to-date estimates, which are on the order of about five stars per year. Experts at NASA and the European Space Agency calculate the number at seven.

The value of Drake's f_p, the percentage of stars with planets, can be estimated today much more accurately than in Drake's time. Using a variety of techniques, scientists have found thousands of exoplanets (planets in solar systems other than our own), an achievement undreamed of in the mid-twentieth century. Planets can be detected from the dimming effect as they pass in front of their star. They can also be spotted from the wobble caused by the gravitational force they exert on their star. It is also possible to find a planet by looking for the microlensing of light beams, another gravitational effect. All of these effects are very small and barely noticeable. However, super-sensitive instruments such as the Kepler Space Telescope and the Hubble Space Telescope have already discovered large numbers of new planets orbiting other stars. Their success at finding exoplanets has motivated the world's space agencies to accelerate the search and to capture and analyze the feeble reflected light of exoplanets for the telltale spectra that reveal the presence of biogenic gases such as oxygen, water, carbon monoxide, and ammonia. For a planet hunter, these are exciting times. Previously, it had been conjectured that these systems had either ejected or consumed most or all of their original planets. Now, astronomers are discovering planets orbiting about 90% of the stars they study, including binary star and multiple star systems, so our best estimate of f_p is 90% today.

New tools with extraordinary capabilities will soon be available to investigators hunting for those planets described by Drake's f_p. These will include NASA's Transitory Exoplanet Survey Satellite (TESS), which will cover 400 times the area covered by Kepler; the European Space Agency's PLATO, which will search specifically for Earth-like planets; the Giant Magellan Telescope in Chile, which will be ten times as powerful as the Hubble Space Telescope; and the mighty James Webb Space Telescope (JWST) due for launch in the third decade of the twenty-first century. The JWST has been designed for optimal sensitivity in the infrared portion of the spectrum, which will enhance its planet-hunting abilities. Among its most striking and unusual features is a set of mirrors coated in pure gold.

The true value of Drake's n_e is a mystery. It is impossible to know how many planets per star system are suitable for life because we do not know what life requires. It may well be the case that life cannot arise except under the white light of a G-type star. This is plausible because G-type stars are long-lived and stable. It may be that life can only arise in a binary planetary system such as our Earth-Moon system. In such a system, the companion stabilizes the primary and prevents it from wobbling wildly on its axis. Such wobbling could lead to erratic climate fluctuations, including million-year ice ages that could turn a life-friendly planet like Earth into a forbidding, frozen, ice ball sheathed in glaciers a mile thick. Our Moon maintains Earth's axial tilt, which is responsible for our yearly seasonal patterns and thus fosters various types of life cycles. Such life cycles of birth, death, and rebirth promote change and invoke the selection of the fittest. Our Moon also performs the crucial task of shielding us from the impact of wandering asteroids and other space debris. The pockmarked far side of the Moon bears mute and picturesque testimony to the thousands, perhaps millions, of asteroids that have been intercepted by our faithful companion. Any one of those asteroids, if it had struck Earth, could have caused mass extinctions or even the extermination of all life.

It may be that life can only arise at a certain distance from a galactic core. A planet too close to the center of the galaxy could be overwhelmed with deadly radiation, while if it was too far away, there likely would not be enough essential elements such as carbon, phosphorus, potassium, or sulfur due to lack of supernovas. It may also be that life can only arise on a planet with a strong magnetic field to deflect lethal solar and cosmic radiation. Furthermore, it may be that life can only evolve on inner planets in a star system with gas giants occupying the outer orbits so that the inner planets are protected from errant comets and asteroids. All these factors could profoundly influence developing life, and in certain cases they could provide protection from life-exterminating, catastrophic events. This would grant time for the developing organisms to become sufficiently advanced to be able to aim their telescopically augmented eyes and dish-shaped, amplified ears towards the cosmos.

If all of the requirements suggested above prove to be necessary, the value of n_e could be quite small. It seems apparent that liquid water is required for life, but even of this we cannot be certain. And while the Earth has oceans of liquid water that have existed for billions of years, how many other planets or moons do? Several moons in our Solar System apparently have underground oceans of water that may have existed as long as Earth's. Enceladus, Europa, and Titan exhibit all the raw materials thought necessary for life. Since gas giant planets seem to be plentiful throughout the galaxy, and since they are known to possess large numbers of moons, the value of n_e could actually be quite high.

Ultimately, we do not know what is required for life to arise. We do know that each of the conditions listed so far is present in the one case of intelligent life of which we are aware: our own. We do not know which of these circumstances and conditions are absolutely necessary; nor, for most of them, can we determine how frequently they are present throughout the galaxy.

Extrapolating from one known case to the entire universe could lead to greatly exaggerated estimates.

To gather verifiable data on what proportion of planets elsewhere in our galaxy are suitable for life, the careful study of other star systems will be necessary. With current advances in planet-hunting technology, the very near future not only promises numerous splendid research opportunities but also a hundredfold increase in the probability of discovering rocky, warm water planets in a star's habitable zone. Such planets could easily sustain an oxygen atmosphere and support a colony of human settlers. Furthermore, the time may be ripe for researchers finally to begin to find worlds teeming with extraterrestrial life and already occupied by intelligent little green beings.

If conditions on a planet are suitable for life, what is the probability that life will arise there? This value is described by f_l in the equation. Most scientists agree that if conditions are suitable, life is very likely to come into existence. The observation that life seems to have appeared on Earth rapidly after conditions became suitable seems to support this view. While this reasoning is not entirely fallacious, it is not completely convincing, either. There may turn out to be a wide gulf between "habitable" and "inhabited."

Just because life arises does not mean that intelligence will emerge, and the fraction of planets on which intelligence does arise is f_i in the equation. Intelligence could be viewed as the ability to contemplate one's own existence, but beyond that, the concept is imprecise. Perhaps the definition ought to include story-telling, problem-solving, and tool use. Dolphins and octopi are intelligent, but they would not, and arguably could not, make use of their intelligence to fabricate metal objects, elaborate dwellings, or delicate electronics. There was life on Earth for billions of years before a species meeting our definition of "intelligent" evolved. The process might have required a billion more years if life had followed a different fork in the labyrinth of evolutionary byways, or intelligence might not have evolved at all. Human intelligence evolved only very recently among the countless varieties of Earthly life. Intelligence, as some biologists are fond of saying, is not necessarily a survival advantage and, thus, there was no "need" for it to arise. On the other hand, intelligent life occurred on the only planet we know of with life, so from the observed relative frequency of occurrence, which is almost surely an overestimate, the value of f_i would be 100%.

Drake's f_c refers to the percentage of planets with intelligent life that come to host "technical civilizations." Drake and his group estimated the value of f_c at 100%, but the term "technical" is broad. Even a primitive civilization has technology, but its technology is not sufficiently advanced to be interesting or useful to us. For instance, ancient cave dwellers had technology: fire, pigments made of ocher and charcoal, hammerstones, and a selection of flint tools including scrapers, knives, and axes. However, as we investigate the issue of intelligent life in the universe, we are presumably hoping to find civilizations like our own that will be able to communicate with us or might even be able to leave their home planets on exploratory adventures. Then the word "technical" must refer, at least, to the ability to communicate over galactic distances. By this yardstick, to qualify as "technical," an alien civilization would need to have developed the ability to send and receive radio transmissions. However, radio is by no means the only way to communicate. An extraterrestrial civilization might use laser beams, gamma rays, or even gravity waves for communication. It is conceivable that they might be so advanced that they employ communication technology of a type we have never imagined. Even today, we observe that less and less human communication is being carried on radio waves as other modes, such as fiber optics, increasingly replace methods that are more antiquated.

Arriving at an estimate for Drake's L requires ruminating on the lifetime of technical civilizations, and this inevitably means thinking about if and when our own civilization will cease to exist. A million years is a long time for any species to endure; 99.9% of all the species that have ever lived on our planet have vanished. This is pithily encapsulated by the Sagan quote, "Extinction is the rule. Survival is the exception." And to paraphrase Neil deGrasse Tyson, while the human race may feel an obligation to survive, the universe is under no obligation to ensure humanity's survival.

Martian Canal Builders

As far as we know, typical radio-broadcasting civilizations may only last an average of 100 years before destroying themselves. This is based on our own culture, which began making radio broadcasts in 1906, and which, as many citizens are keenly aware, may well be on the verge of destroying itself. War, disease, famine, environmental degradation, or an asteroid strike could very quickly either destroy the human species entirely or impair our ability to send and receive radio signals. Most futurists are more optimistic. Humans have overcome formidable challenges in the past, and the goal of long-term survival is not unattainable. The original estimate by Drake and his colleagues at the Green Bank conference was that a technical civilization would last, on average, 10,000 years. An optimistically long estimate, or disturbingly short? We do not know.

Modern humans arose about 200,000 years ago ushering in the Sapiezoic era, the era of "wise life." In this era, the activities of human beings will significantly modify the environment. Our intelligence may retard or even prevent our eventual demise, especially if we advance our space program to the point that we are able to deflect threatening asteroids and avoid the perils of Malthusian overpopulation—with its inevitable accompaniments of famine, disease, and warfare—by colonizing other planets. On the other hand, intelligence may be the cause of our undoing, as would be the case if our society were hurled back into the dark ages by nuclear war, anthropogenic climate change, or a devastating, self-inflicted pandemic.

Another threat is the slow increase in the Sun's luminosity. This will not become a problem for hundreds of millions of years, but according to what we know about the life cycles of stars, it will definitely happen. The increased heat will boil away our oceans, and our species will either have to take flight to another planet or somehow, perhaps using unimaginably advanced technology, re-engineer the functioning of the Sun's nuclear furnace. In any case, the humans of such a distant future, having spent hundreds of millions of years adapting to changing conditions, would undoubtedly not seem human to us at all.

In the pursuit of excellence, *Homo superior* will eventually arise. What will this new incarnation of humanity be like? Will the rule of "tooth and claw" be rescinded? Will the Earth then be home to a docile species from which aggression, anger, and rapaciousness have been bred out or biochemically deleted from the genome? Where then will be the will to survive? Weakness often equals death, and the least fit are eliminated. There is no reason the human species should not expect to survive for thousands or even millions of years, but, to reiterate, we simply do not know. There is no basis for calculating an average until the estimation and guesswork have been hammered into shape on the anvil of experience. This is the case for most of the values in the Drake Equation. However, like Drake, we can do our best to generate a set of plausible values for the various factors in his equation:

$N = R_* f_p n_e f_l f_i f_c L$

$R_* \approx 5$	About 5 stars are born in the Milky Way each year.
$f_p \approx 90\%$	Most stars have planets.
$n_e \approx 1$	This assumes that each star has, on average, about one Earth-like planet.
$f_l \approx 100\%$	If conditions are right for life, life will certainly arise.
$f_i \approx 10\%$	Where life arises, intelligence will follow about one-tenth of the time.
$f_c \approx 10\%$	An intelligent species will develop a technical civilization in 10% of cases.
$L \approx 10{,}000$	A technical (broadcasting) civilization will exist for about 10,000 years before either being destroyed or descending into a previous state of technological backwardness. The value of this factor includes the possibility of new civilizations arising from the ashes of preceding ones and allows that civilizations may develop methods of communication that are undetectable by us, at which point they would, in effect, disappear from our view.

These very rough numbers, which reflect a sort of estimate of the informal consensus that exists among experts in the field, yield an approximate answer of N = 500 technical civilizations in our galaxy. This finding differs significantly from the original estimate formulated by Drake and his colleagues of 10,000. In light of the uncertainties outlined above, neither result can be considered reliable. The most uncertainty exists for n_e, for which the real value may be as small as 0.000001. Such a number would mean that there would only be approximately one technical civilization in every 2,000 galaxies. Then we would be very alone in the universe. On the other hand, if L is much larger than our estimate above—if technical civilizations last, on average, for 1,000,000 years—then the whole Milky Way would be teeming with diverse, multicellular, highly differentiated forms of life.

However, if 500 is the correct number of broadcasting species in our galaxy (which is estimated to be 100,000 light years across), then the nearest civilization is likely to be 300 light years away, and communicating even at the speed of light would be agonizingly slow. If there were 10,000 communicating species, then several species could lie within a twenty-light-year radius of Earth. Which raises the question: Why have we not heard from them?

Fermi's Paradox

Beyond the nebulous results generated by the Drake equation, what basis do we have for believing that there is extraterrestrial intelligence? The physicist Enrico Fermi brought his brilliant and incisive intellect to bear on the problem and in a flash of intuition produced the seemingly straightforward yet perplexing concept now known as the Fermi Paradox.

Fermi is famous for his phenomenal achievements in physics, including the first controlled nuclear chain reaction, which took place under his supervision in Chicago in 1942. He was actively involved in the Manhattan Project, the project undertaken by the United States government during World War II to construct a fission bomb. He was on duty as an observer at the explosion of the first working nuclear bomb, an event code-named Trinity, on July 16, 1945. At 5:30 a.m. on that day, Fermi stood behind a board containing a piece of dark welding glass ten miles from the explosion and measured the strength of the blast by releasing small pieces of paper as the air blast passed him, which occurred approximately 40 seconds after the explosion. The scraps of paper fell from a height of six feet and were carried about eight feet by the blast wave. From this, Fermi was able to estimate the power of the bomb with remarkable accuracy. He estimated that the bomb was as powerful as 10,000 tons of TNT, which differed from more exact later calculations by less than a factor of two. According to Fermi's wife, Laura, "He was so profoundly and totally absorbed in his bits of paper that he was not aware of the tremendous noise."

Another of Fermi's accomplishments was the naming of the neutrino, a mysterious subatomic particle with very small mass and no electric charge. "Neutrino" means "little neutral one" in Fermi's native Italian. He determined the neutrino's characteristics through the theory of beta (electron) decay. His 1934 paper reporting his findings was so far ahead of its time that it was not published in English until 1968.

The famous Fermi Paradox was proposed, according to an oft-recounted anecdote, very casually. On the way to lunch one day in 1950, Fermi and physicists Edward Teller, Emil Konopinski, and Herbert York were chatting about alien intelligence and interstellar space travel. A short time later, as the men were having lunch and the conversation had moved on to other topics, Fermi suddenly exclaimed, "Where is everybody?" The others instantly understood that Fermi was talking about the aliens. In between bites of lunch and snippets of small talk with his colleagues, he had calculated the probability of Earth-like planets and the existence of extraterrestrial intelligence. He had concluded that if intelligent life existed in our galaxy and space travel was possible, the entire galaxy should be filled with life, even if it took millions of years. In that case, the Earth should have been visited by extraterrestrials early and often. Why then is there no tangible evidence of such visits?

There are many possible solutions to the Fermi Paradox, the simplest of which is that there is simply no intelligent life in the galaxy other than humans. Another possible explanation is that perhaps the barriers to interstellar space travel are insurmountable. We humans have successfully traveled through space and returned from a distance of 240,000 miles. Journeying to another star would require traversing distances 100 million times farther. When we voyaged 240,000 miles to our own Moon, the trip took three days. At that rate, the trip to Alpha Centauri, the nearest star system, would require over 800,000 years. There are immense technological hurdles to be overcome before we can hope to become an interplanetary species of the kind whose absence perplexed Fermi.

Perhaps space travel can never be made practical, requiring too much time and too much energy. In addition space travelers would be pelted with deadly radiation, the vacuum of space has no life-sustaining oxygen, and microgravity leads to bone loss and muscle weakness. With the level of space technology we presently employ, not a single one of these obstacles can be overcome. Our present technology is absurdly inadequate for traversing the tremendous distances between star systems. The obstacles and dangers constitute insurmountable physical barriers to interstellar exploration, at least for the foreseeable future.

Humans who spent hundreds of thousands of years traveling through space would no longer be recognizably human. Cosmic rays would have mutated their DNA, and, in the absence of normal Earth gravity, bones and organs would come to function differently. It is not hard to imagine these former *Homo sapiens* as blobs of boneless protoplasm floating through the corridors of their generational arks on a flight to a distant planet on which they would never be able to set foot. Aliens attempting to visit Earth might face similar difficulties.

SETI

Since 1960, humankind has been listening to outer space in hopes of hearing the sounds of another civilization. The acronym SETI stands for "Search for Extraterrestrial Intelligence" and refers to any systematic program of investigation in this field by an Earth-bound researcher. Since near the beginning of the twentieth century, soon after the invention of radio, researchers have been listening to the heavens for signs of intelligent life. Both Guglielmo Marconi and Nikola Tesla reported in the 1920s that they had received radio transmissions from the planet Mars. (Evidently, they were mistaken, as their observations were never confirmed.) Today, radio receivers like the 1000-foot radio telescope at the Arecibo Observatory in Puerto Rico scan the skies for transmissions, and more sensitive facilities continue to come online. Although the Arecibo radio telescope is not officially part of any SETI program, the data it gathers while investigating normal astronomical phenomena are passed along to SETI researchers who pore over it for any signals that might indicate the presence of an intelligent civilization. NASA first funded a SETI program in 1971 while Frank Drake was a staff member. To date, we have not heard anything that remotely resembles the resonances, rhythms, or reverberations of intelligence—only the noise of static. The great silence is deafening. Its profundity is overwhelming and produces a numbness resulting from the growing awareness of the insufficiency of our understanding of the prerequisites for life.

Radio waves constitute only one band of the electromagnetic spectrum. They are easy to detect since they pass readily through our atmosphere. It makes sense that aliens might use them for communication, since we do. However, radio waves are not the only possibilities. If aliens were using gamma rays, for instance, we would need to put equipment in Earth orbit or beyond to detect them. Those short-wavelength rays are absorbed by air long before reaching detectors on the Earth's surface. Furthermore, if aliens are sending out gravitational-wave messages, we will need many more and better gravitational-wave detectors. A small number of such detectors are already working, such as the Laser Interferometer Gravitational-Wave Observatory (LIGO), and more are planned. Visible light is another possibility. Alien civilizations might communicate using powerful lasers. Such bursts are electromagnetic in nature just like radio waves, but with smaller wavelengths, and we should, in principle, be able to detect them. It might also be possible to detect alien emissions that are not strictly messages. For example, extraterrestrial civilizations might power their spacecraft with highly focused microwave beams, and these would probably be detectable on Earth.

SETI is not generously funded. Governments do not like to lavish money on programs like SETI which are perceived by many voters as irrelevant and inconsequential. (Carl Sagan personally lobbied Senator William Proxmire on the importance of SETI in the early 1970s and asked him to stop attacking NASA's SETI program.) Today, researchers still face mockery and indifference. Much of the money to fund SETI now comes from wealthy philanthropists who, like most humans, believe that alien intelligence exists, that its discovery is ardently to be wished for and diligently sought after, and that no expense should be spared in pursuing such a profoundly important goal.

The Final Analysis

Our civilization may survive for only a fleeting instant of cosmic time. If so, then let the cosmos sing, let it reverberate with a resounding fanfare, rich and resonant, which announces: "Here can be found the relics of humanity, which left an undying legacy of discovery and remained, until its demise, the consummate searcher." But if we survive, we will have crossed the dangerous straits through which we are now sailing and will become able, at last, to journey to distant worlds. Then the universe will record that our species, having sprung from its first home on a small rock orbiting an average star in an out of the way arm of an ordinary galaxy, was able to seed the cosmos and populate the galaxies while tirelessly searching for knowledge, beauty, and life.

We do not know if our unending search for life will ever bear fruit. Some suspect that there is an overarching principle of nature that compels life to evolve from the inanimate. Assertions that life develops easily are based upon conjectures that it is not a random formation of organic molecules, but is the likely result favored in an energy-rich cosmos. This would mean that there is a biological imperative that prevails in a universe of inanimate elements and compounds driving the inevitable development of life.

While the simpler forms of life may well be abundant, the development of complex, intelligent life might be a rare exception rather than the rule. If it transpires that the Earth is unique, that only *Homo sapiens* guides its future with value-laden plans and that humanity is the only intelligent species capable of transmitting messages in the cosmos, then humanity would represent all the hopes and dreams that could possibly exist in the universe. In this scenario, we would epitomize the only spark of sentient life and the only flame of culture anywhere. If that once-in-a-cosmic-lifetime flash of intelligence were to flicker out for good, then the tragedy would be more than monumental; it would be immeasurable and for all time, a bittersweet cry into infinite silence.

Therefore, if we are alone in the universe, it behooves us to lavish the utmost care upon our species. If we are alone, it is imperative for us to nurture and enhance our physical selves while we enrich and expand our minds and strive to increase our moral wisdom. If we are alone, then we are the only embodiment of sentient intelligence in the cosmos and we will need to accept the solemn responsibility of preserving ourselves. If we were to snuff out the candle of our own existence or, through inaction, let some cosmic event end the story of humanity, then it would be the end of consciousness, the end of the first and last species capable of contemplating its own ontology and its own destiny. We would be a universal anomaly that winked into existence and then just as suddenly departed. The event would be the deepest possible tragedy. There would be no intelligence left anywhere to mourn the loss or even to recognize it as a loss.

Such thoughts may represent the acme of hubris. History has shown that every time we believe we are the center of the universe and that humanity is somehow privileged or exalted, we soon learn that this is not the case at all. Instead, the exact opposite proves true. Earth is not the center of the universe, there is no privileged position in the cosmos, and the latest version of excellence will always be surpassed. While we are superbly adapted to life on Earth, there are millions of other planets like Earth. It is hard to believe, given the plethora of places and possibilities that could give rise to life, that we are alone in the universe—especially since there have already been 13.7 billion years of cosmic history and at least that much more to come. Therefore, it follows that other wellsprings of life may exist beyond Earth, and according to the maxim that *it is better to know, than not to know*, it is incumbent upon us to search for them.

Perhaps the critical consideration is that most humans genuinely wish and hope that extraterrestrial intelligence will be found someday. We deeply yearn to find kindred spirits. Humanity is an adolescent species. We are beginning to set aside the self-centeredness of childhood and seek connection with the universe beyond our limited domain. Adolescents can experience loneliness and a sense of otherness, and this impels them to reach out for reciprocation in relationships. Just as an adolescent grasps at opportunities to connect deeply with friends, so does the human race. The human race does not want to believe it is alone. Humanity's longings for connectedness with other beings and other worlds were once expressed in the garb of superstition and religion. Today, such feelings encompass the search for extraterrestrial life and the desire to communicate with such life. Many, like the late Stephen Hawking, voice grave existential concerns yet overwhelmingly still want to know if alien intelligence exists. Hawking worried that contact with extraterrestrial beings would be to our detriment, just as contact with European explorers was detrimental to Native American civilizations. However, he still supported Breakthrough Listen and Breakthrough Starshot, which proposes to send postage-stamp-sized probes to Alpha Centauri. "It's time…to search for life beyond Earth. The Breakthrough initiatives are making that commitment. We are alive. We are intelligent. We must know."

It is this primal impulse to know that compels us to send tough and ingenious probes to other worlds in our Solar System. It is this impulse that motivates us to accelerate our pace of discovery and to build ever more powerful telescopes capable of gathering more light and penetrating the dusty gloom of outer space to read the spectral signatures of distant worlds. It is this impulse that inspires us to send complex observatories into orbit and conceive of continent-spanning gravity-wave detectors. It is this impulse that led us to stand on the Moon and will someday lead us to stand on the surfaces of Mars and Titan and, someday, upon the surfaces of planets circling other stars. When we arrive there—and we will—we will be looking for one thing above all, the same thing we have been seeking up to now: life. And whether this magnificent, irresistible, and persistent search succeeds or fails, future generations of humankind will understand, appreciate, and value humanity's quest for life. This quest will become an enduring legacy that we joyfully bequeath to our posterity to nurture until (or if) the dream is brought to its successful culmination. The fanciful mathematical constructs of creative dreamers and the musings of illustrious physicists of bygone centuries will become irrelevant on that day when, at last, we stand face to face with living beings from another world.

Artist conception of a magnetar, a type of neutron star with an exceptionally strong magnetic field.

26
Stars: All that Glitters is Dust

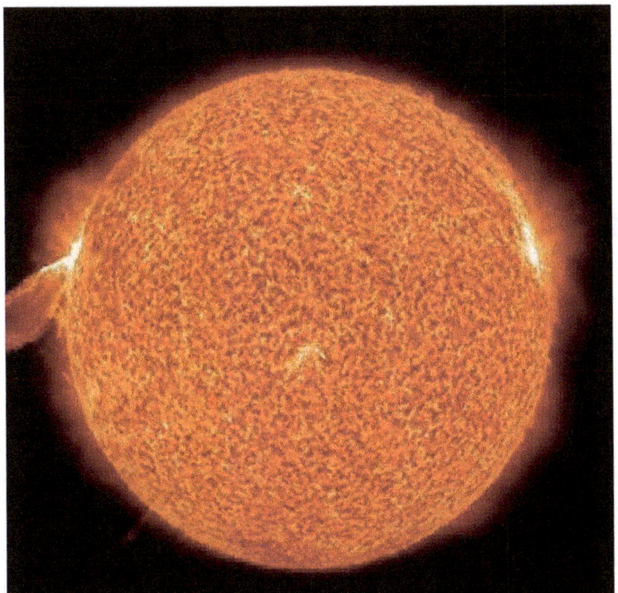
Sun with Prominence

If the stars should appear but one night every thousand years how man would marvel and stare.
Ralph Waldo Emerson

For my part I know nothing with certainty, but the sight of the stars makes me dream.
Vincent Van Gogh

We are stardust; we are golden; we are billion year old carbon.
Joni Mitchell

We are privileged to live in the time of stars. In the distant past there were no stars, and in the distant future the last stars will cease to glow. But at the present moment, we live in an age when stars fill the cosmos. Each evening as the stars glide into sight, they stimulate human eyes with the glowing radiance of a thousand gleaming lights. They inspire astronomers to ascend picturesque Hawaiian and South American mountaintops to peer at and photograph pink spiral galaxies, crab-shaped nebulae of emerald and sapphire, astonishing orange alien moons, red planets the color of flame, and stark solar eclipses. While some believe that "Euclid alone has looked on beauty bare," all the denizens of the sky, and stars in particular, possess their own exquisite splendor and project upon our emotions, our bodies, and our lives the aura of timeless, eternal beauty. It was no coincidence that the poet equated beauty with the night "of starry skies."

Evidence that the universe had a beginning is persuasive, and the scientific consensus is that the universe at first consisted of something called a singularity, which contained all the matter and energy of the cosmos. If this consensus belief is true, then there was a short period of time (280,000 years) when the early universe was opaque with a dense, impenetrable fog of hydrogen and helium and a tiny amount of lithium. Stars were born from this cloud and started to burn during this cosmic Dark Age, but their light would have been invisible. It was not until the universe had expanded after the Big Bang and cooled sufficiently that space turned transparent.

As the first stars caught fire, they blasted out immense quantities of heat and light, transforming the fledgling cosmos into a sunlit universe. Casting their beams through newly translucent space, stars ushered in the magnificent age of light. The instant of first ignition was the most extraordinary event in the history of the cosmos. From that moment on, there were sun kings, mighty lords of their own realms, sultans of light illuminating all creation and basking in the radiance of their neighbors.

Majestic entities highlighting the deep blackness of night, stars mark the essential contrast between matter and void, existence and non-existence. They energize the mind as much as they illuminate the space that surrounds them. The light they produce carries information about the mysteries of the cosmos and is the ideal messenger to traverse the ocean of space that encloses us. Here on Earth, light is the principal means, though certainly not the only one, by which we observe life and our surroundings. Light allows us to explore our own world, search for alien exoplanets, and investigate the universe.

The human intellect is able to draw from the subtlest clues important information about the past, the future, and the nature and destiny of the entire universe, space, time, and life. One reason we have the ability to do this is that we live in this wonderful age of stars, when we can lift our detectors to the sky and catch photons from stars of every size, color, and history.

Each shining light in our night sky tells a story that is different from all the others. The reading and deciphering of those cryptic tales has enthused Earthly observers for millennia, even as the stars guided our ancestors across the face of the planet and through the seasons of the year. Today, standing in awe of a universe that continuously reveals its grandeur, immensity, and diversity, we dare to unravel the mysterious narratives that express themselves in these marvelous packets of light, each of which is an emissary from a celestial giant, bringing news of the past from afar.

Light is incredibly nuanced. The photons of which it consists are tiny, massless particles that encapsulate energy while maintaining the nature of a wave. Once a photon comes into existence, it never ages or diminishes in any way. It flies resolutely and straightforwardly through the emptiness of space until it encounters something tangible, at which point it is either absorbed or reflected. From its inception, each individual photon manifests a single, unchanging wavelength that remains pristine, pure, and perfect throughout its existence. The wave characteristics stem from twin oscillating magnetic and electric fields, each perpendicular to the other and to the photon's direction of propagation.

Once emitted, a photon zooms through space in a timeless state. At the speed of light, time stops. This is a startling consequence of the law of relativity, according to which time and space harmoniously reshape each other to ensure that the speed of light remains constant for all observers. The discovery that this "universal speed limit" is a constant of nature was one of the great scientific breakthroughs of the twentieth century.

Light is the universe's primeval stylus that etches the cosmic tablets with clear, concise primordial language, providing cosmic documentation that details the recorded history of the universe and explicitly shows the history of the most distant galaxies as well as our own nearby Sun. Light is the fleeting heritage that the universe has bequeathed to humanity, for it alone has the absolute power to expose our eyes and minds to what transpires in places distant from us in time and space. With the transformative power of light, we can escape the confines of Earth to roam our galaxy and its environs. Light allows us to look back in time and view the first stars. Those objects, like a dim red dwarf star recently discovered in our galaxy, have ages in the vicinity of 13.5 billion years and are almost entirely lacking in elements heavier than helium. The information contained in the light of these objects, the first progeny of the Big Bang, helps explain the origin, expansion, and present state of the universe.

Light reveals the magnificent and the minuscule. It exposes the shapes of galaxies including our own, and it allows us to map our galaxy so that we know our own location within its immense borders. Light enables us to visualize the outlines of massive black holes at the center of almost every galaxy, and it shows us stars like Eta Carinae, an immense, exceptionally violent star that has already exploded once, creating a nebula, and will explode again soon. Light allows us to see speeding meteors like the one that shattered a peaceful Siberian morning over the city of Chelyabinsk, and light reveals the smiles of grandchildren as we tickle their ribs. Light from a display screen lets us picture the quivering atoms in the smallest of molecules. All visible things—the vibrant colors of a field of wildflowers, colorless droplets of water on a baby's cheek, black letters spelling out a literary masterpiece on a white page, yellowish grains of yam-flour dust by the margin of a Nigerian fire pit, blue skies, green pastures, bright red lava, and blue, white, and red stars—inform us of their nature by way of light.

The brilliant coruscations of stars flash packets of information across the emptiness of space and shower worlds with the gift of energy. Occasionally these energetic packets strike knurled ribbons of nucleic acid and slightly alter them, while more often these potent kinetic bundles encounter and are eagerly absorbed by the light-devouring rings of chlorophyll which ultimately power life. A photon's existence ends when it is absorbed by a piece of matter. The matter that absorbs the photon finds itself in an excited state and often gives up its excess energy by emitting a newly created photon of its own, or the absorbed energy may generate heat or power a chemical reaction, which is what happens in a plant's leaf and even in our eyes.

Many of the photons we encounter were born in fusion reactions in the core of a star. A photon thus created does not immediately streak out into space to illuminate the cosmos. A star's interior is so densely packed with atomic nuclei and electrons that the newborn photon is absorbed and reemitted countless times. When new, a photon is at its most energetic, but by the time it emerges from the star it has experienced many collisions, absorptions, and reemissions. The photon that eventually emerges is likely to be less energetic than when originally produced. Its meandering journey can require thousands (some scientists say millions) of years before it reaches the star's surface and at last flies free into the dark, cold emptiness of space.

We cannot see all manifestations of light, nor can we perceive every wavelength of this common yet enigmatic form of energy. Light propagates across the universe at a set speed; each photon is an individual packet of light that possesses a specific wavelength and vibrates at a particular frequency. Photons with smaller wavelengths have higher frequencies, and the most energetic are those with the shortest wavelengths and highest frequencies. A single light ray is a collection of photons of many different energy levels, and the different wavelengths constitute a rainbow of colors—red, orange, yellow, green, blue, indigo, and violet—which combine to produce the white light we see. However, there are many kinds of photons we cannot see.

What we see are the photons to which our light receptors are attuned. When the light is detectable by our eyes (rather than, say, an X-ray detector), that portion of the spectrum is called visible light. Our planet presents a blue face to the universe because its surface is 71% water, and the physical structure of a water molecule more readily absorbs low energy red photons and scatters or reflects more of the high-energy blue photons. Much of our planet's land surface reflects green light because the chlorophyll molecules in plants tend to absorb red and blue photons but reflect the green photons. Rainbows are colorful spectacles in which individual raindrops act as prisms to separate the colors. Auroras are sheets of shimmering colors that appear in the sky because of the impact of particles from the Sun on our atmosphere as they spiral down into the Earth's magnetic poles.

Our star, Sol, radiates more visible light than any other wavelengths. Some stars emit a preponderance of energy beyond the visible colors in the ultraviolet (UV) portion of the spectrum. While our eyes, which sense photons from only a narrow sliver of the electromagnetic spectrum, do not see UV energy, we do see the results of this energy upon our skin when it burns, blisters, and peels. These stars burn fiercely and emit tremendous amounts of UV but little visible light energy. While we believe these stars to be weak and dim, they are, in reality, fearsome beasts pummeling the surrounding space and matter with powerful invisible rays. Over half the stars in our night sky are small red dwarf stars that emit more infrared radiation than visible light and these stars also appear very dim to us.

If our Sun were to emit more UV energy than light energy, the Earth would become a roasted, burned-out cinder, and the habitable zone of our Solar System would be beyond the orbit of Jupiter. For a red dwarf (smaller than our Sun), this "Goldilocks zone" would be located where Venus is today.

A Cosmic Perspective

The existence, power, lifespan, composition, and movement of the stars are written in the silent language of light. Light is the Rosetta Stone of stars, enabling the knowledgeable to decipher their secrets, and we have come far in the last century in learning to read the light of stars. From the fact that visible light from a star's surface reaches our eyes, we know the star is there. From the color of the light, we can determine the temperature of the star's surface and how much power its nuclear engine is generating (per Einstein's wonderful equation $E = mc^2$). From the temperature, the star's lifespan can be predicted. This is because hot stars, which are invariably more massive than their cooler siblings, burn out more quickly.

If starlight is dispersed by a prism or a diffraction grating, the wavelengths of light are bent and spread to produce bands of colors in a spectrum. Within each star's spectrum are thin black lines indicating that some of the wavelengths of light are missing. These missing wavelengths have disappeared because particular elements present in the star are absorbing them. Since specific elements always absorb the same wavelengths, the lines, called "absorption lines," can be read like a fingerprint to discern which kinds of elements a star contains. This is how Cecilia Payne discovered in 1925 that our Sun is comprised mostly of hydrogen and helium.

The shift of the absorption lines from their expected locations (as observed on Earth) reveals a star's speed and direction. A wave-emitting object which is moving away from the observer appears to have longer wavelengths than the same object moving toward the observer. Once researchers realized that the absorption lines in starlight were repositioned as a result of this shift, known as the Doppler shift, it was a relatively simple matter to determine how fast and in what direction a galaxy's stars were moving. The capstone of this study was the discovery that the universe is expanding. All of the most distant stars show a Doppler shift in their absorption spectrum (the so-called "red shift") that indicates high velocity away from us.

The history of stellar spectroscopy is a fine example of how science advances. This phenomenally useful tool of modern-day astronomers did not burst onto the scene in a flash but required the work of numerous researchers in various countries over the course of many years. The spectral lines first had to be noticed, by William Hyde Wollaston in 1802. Then they had to be related to the elements; this was done by Robert Bunsen of "Bunsen Burner" fame and Gustav Kirchhoff, the formulator of Kirchhoff's laws of electrical circuits, in the 1860s. The shifts in the lines then had to be noticed, and this was duly accomplished by William and Margaret Huggins soon thereafter. Sensitive and delicate equipment was necessary and had to be invented—no small task, especially considering that for millennia no one even suspected the existence of such spectral lines. In the twentieth century, it was discovered that the absorption lines were due to the absorption of photons of specific wavelengths by the electrons in the atoms, yet the lines were known and utilized long before anyone suspected the existence of the electron. Centuries of steady work were required to reach the stage where this incredibly useful technique became a routine tool in every astronomer's toolkit. Today, spectrometers are ubiquitous, found in telescopes around the world, in orbit around the Earth, and circling other planets of the Solar System.

An expanding perspective

The energy of stars can herald the existence of life. We know that life has arisen near one hospitable, very common type of star in our neighborhood. As living organisms, we owe our existence to our star and ultimately draw everything we need for life from it. In addition, the Sun lets the universe know we are here, for our Sun's light reflects off our planet and all that our planet contains. This reflected light travels through the universe carrying the markers of oxygen and organic compounds, telltale signatures of the atmosphere through which it has passed: an atmosphere created by and capable of nurturing life. Starlight is thus carrying a message to the universe that we are here, and even if there is no one out there to see and read these reflected, starry communiqués, so that this priceless information secreted in rays of light exists solely for our benefit, then our sense of wonderment would not be diminished by one scintilla.

Dust in the Dark

The story of stars is not only a tale of light, but also of their component atoms of dust and gas. The ancient Greek philosopher Leucippus was the first to promulgate the notion of a smallest indivisible unit of matter, and his disciple, Democritus, developed the idea and called these indestructible units atomos, which means uncuttable. Although the modern concept of what an atom is and how it works is quite different from Democritus's, he was entirely correct in his belief

Horsehead Nebula

Eagle Nebula

North American Nebula

that atoms play a vital role in the existence of all we see. Today we know of many particles smaller than atoms, but these combine to form the atoms with which we are familiar. Electrons, protons, and neutrons unite to create a plethora of atoms, and atoms combine to form everything else, including dust.

To comprehend the universe, it is necessary to understand dust. Found throughout the universe, dust is the raw material of all the celestial bodies, including the gargantuan atomic furnaces that fuse ever more massive elements. Great nebular clouds of dust and gas await the nudge of some outside influence, whether the gravitational perturbations of a passing star or jets of energy and matter from the energized surroundings of a distant black hole, so that the dust and gas may start the slow and patient process of minting new stellar bodies.

These stupendous, light-absorbing clouds are conspicuous sailors of the sky and vital engines of stellar evolution. Here in the Milky Way galaxy, they form amorphous but occasionally very distinctive shapes—the Horsehead Nebula, the Eagle Nebula, and the North America Nebula, to give a few examples—due to chaotic variations in the accumulations of matter. These star-forming regions, or star-factories, amass gas and dust to form protostars, which, though quiescent in their infancy, will eventually transcend their lackluster present and transform into powerhouses of light and energy. Nebulae are the incubators of future celestial lords who eventually will be flung from their birth cloud into the wider galaxy, where they will follow their own destiny, burn brightly, and hold gravitational sway over immense dominions of space as newborn princes of light.

At the beginning of the universe, scientists say, the temperature was too hot for gas or dust to exist. There was nothing but superheated plasma, too hot for matter of any kind, too hot even for the fundamental forces such as gravity or electromagnetism to operate. As the universe cooled, conditions began to approach those that prevail today, and atoms of matter condensed into existence: the gases hydrogen and helium.

The gases in the early universe were not uniformly distributed. Clumping occurred, and the resulting unevenness meant that some areas of space contained more matter than others. The regions with higher concentrations of matter accumulated still more, until there was so much matter that the force of gravity became crushing. Eventually, the gravity became so intense in the centers of these spherical agglomerations of matter that massive amounts of heat and pressure were created. The result was a state of matter in which the superheated atoms were once again stripped of their electrons as in the early universe. Here the nuclei, dense cores of hydrogen atoms, were forced together until they began to undergo a surprising and dramatic process known as nuclear fusion. In nuclear fusion, two nuclei are united, resulting in the loss of a tiny amount of matter and the release of a relatively large amount of energy. In the giant, fusing gas-balls of the early universe, many fusion reactions took place, and the total amount of energy released was colossal as larger atomic nuclei were constructed. The first generation of stars was being born.

Eventually, the stars of this first generation would exhaust their hydrogen fuel and explode. In their death throes, they spewed forth all the heavy nuclei they had created in their fusion reactions, and the universe now contained something new: dust.

Young Star Cluster

Dust consists of tiny fragments of granular substances, whether carbon, lithium, iron, or other elements. In outer space, granules of dust attract each other by electrostatic force like cosmic dust bunnies. As they grow larger, approaching the size of pebbles, gravity becomes more important than the electrostatic attraction. It may seem surprising, but two small pebbles in outer space, separated by an inch or two, will float together and come gently into contact in a matter of a few minutes or an hour because of their mutual gravitational attraction.

In the dusty early universe, it was only a matter of time until additional rounds of coalescence began. New stars formed from the dust and remaining gas. These stars too came to the end of their allotted lifespans, and generation followed generation as stars continued flashing into existence, producing more heavy elements and spreading their substance throughout space with the power of massive supernova explosions.

Stars, then, begin as dust and gas, like the young, high-temperature stars in the image above, from the Hubble Space Telescope. Comets and asteroids also consist of dust, but unlike stars, they never accumulated enough mass to commence nuclear fusion. Planets are made of dust, and so is everything on them. This means that human beings are made of dust from the first and all the ensuing generations of stars.

Types of Stars

Stars are born in dust, exist for a time as fountains of gushing light, and then pass away in either explosive fury (blasting their outer layers into space, as shown in the image at right) or exhausted, smoldering cinders. Their mode of life during their period of stable light production and in their final moments of instability preceding death are determined by their size. The smallest of stars, those that gather to themselves a minimal amount of matter, barely achieve nuclear fusion. Known as brown dwarfs, these dim almost-stars are only slightly larger than a gas-giant planet, and the paucity of their light means that we see very few of them from Earth, although they may be very numerous throughout our galaxy and the universe. Slightly larger stars burn steadily for a long time. Their moderate mass means that they can burn their hydrogen fuel at only a sluggish pace, and since they consume their fuel so slowly they can endure for billions of years, glowing with a steady red or white light. It is estimated that red dwarfs are the most numerous of stars, representing some 80% of the stars in the universe. The white dwarf is a type of star of particular interest to humans, for under the beams of such a star the human race has arisen and thrived. Our Sun was once thought to be average, but now it is believed that the Sun shines more brightly than 95% of the stars in our galaxy.

Many scientists who hypothesize about extraterrestrial life suppose that only stars of a specific size can host planets with life. If a star is too big, it will burn out quickly, before life has time to evolve. If a star is too small, it will be giving out so little heat that a life-bearing planet would have to be very close, and would quickly become tidally locked to its star. This would not be conducive to life. One face of the planet would be sizzling hot, and the other face would be freezing cold, very likely with hurricane winds blowing at thousands of miles per hour in an unending, fruitless attempt to equalize the temperatures of the two sides.

When a dwarf star like ours dies, it expands over a few billion years, growing steadily hotter, and then gently blows off most of its matter until a tiny glowing remnant, about the size of Earth, is all that remains. Such stars, it might be said, never die—they just fade away.

Stars larger than our Sun behave differently. The heavier they are, the faster they burn off their fuel, and the more dramatic is their end. A supergiant star flames fiercely for a scant few million years before giving up its existence in a monstrous explosion known as a supernova. Afterward, the dense core of such a star, with all its protons and electrons blown away or super-compacted, may contain nothing but neutrons. Such a relic is referred to as a neutron star. Neutron stars are a mere 10 miles in diameter, contain the mass of several Suns, and are incredibly dense.

Stars with even more mass than the neutron stars become so dense after blowing off their excess matter that the gravitational force they exert is strong enough to prevent even light from escaping. Such stars are like revenants that reveal their ghostly presence only by how they warp space. They siphon light from the visible universe and clutch it so tightly within themselves that they are utterly, eternally black. They have been given the descriptive moniker "black hole."

Black holes are invisible, inscrutable, and seemingly inexplicable. Whereas astronomers can study the light that comes from a star, no information about black holes is forthcoming from this source, because no light emerges from the black hole's all-devouring maw. The perimeter of a black hole is the mysterious "event horizon," a point of no return from which no matter, light, or even information may emerge. (Stephen Hawking plausibly hypothesized that black holes give off a special type of radiation, now known as Hawking radiation, but this is not detectable with presently available techniques). Although black holes are invisible, it is strongly suspected that as matter streams into a black hole under the influence of its irresistible gravity, the flowing matter reaches such high velocities that it becomes super-heated and glows in a bright accretion disk. Structures fitting the description of an accretion disk have been detected by the Hubble Space Telescope. The artist's conception at right depicts a black hole with its accretion disk and powerful jets of matter shooting out in the direction of the black hole's axis of rotation.

Black Hole with Jets

Color

Stars range in color from red to white to blue. Color is an indication of surface temperature: red stars are relatively cool, while blue stars are the hottest. Our star is an average white star, neither as hot as the hottest stars nor as cool as the coolest.

Oddly, there are no green stars, and the fault does not lie with the stars but with our eyes. The human eye is an exquisitely adapted sensory apparatus for extracting photons of light from the environment and translating them into pulses of information in the brain. Cells in the retina called rods interpret the number of arriving photons as white if plentiful and black if only a few. Cells called cones are equipped with the biochemical ability to detect either red, blue, or green wavelengths of light. Although the light from many stars includes some wavelengths that if viewed alone would appear greenish, in all such stars the red or blue wavelengths combine with the green to make the light appear white to human eyes.

Groupings of Stars

A single star is a wonder on a grand scale, but when we scan the night sky, we see more than singletons. In fact, the majority of the visible lights in the sky are double stars at least, and stars are commonly found in groupings of three, four, or even more. Two stars that appear very close to each other are called a binary system although their apparent nearness does not mean that they are in fact close to one another. Sometimes one of the two is hundreds of light years more distant than the other, but the two just happen to lie along the same line of sight.

Within our galaxy, there are over 150 globular clusters, which typically contain about a million stars each. The stars in these huge ball-shaped clusters are all about the same age and are unusually old for stars in our galaxy. Since they are so old, the stars in these clusters are members of an older generation of stars and contain much smaller amounts of heavy elements than does, for example, our Sun.

Hubble Deep Field Survey

Mighty galaxies, of which our Milky Way is one, may contain billions of stars. Early astronomers who spied galaxies through their telescopes were captivated but had no way of knowing what they were looking at or how far away those strange, fuzzy patches were. Some hypothesized that these "island universes" were features of our own stellar neighborhood. More recently, using sophisticated methods such as variable stars and spectrometry for determining distances, it has been discovered that some galaxies are as much as 13.5 billion light years distant. Even stretches of space which appear empty to the naked eye are found to be filled with galaxies when examined by powerful telescopes, as shown by the famous Hubble Deep Field Survey, which revealed a large number of galaxies in a tiny region of "empty" space. The breathtaking image above shows the amazing variety of galaxies so detected.

Galaxies are awesome sights, impressive in their vastness, beauty, and geometry. They can form standard shapes such as ellipsoids and spirals, but many are irregularly shaped. Their unusual shapes are thought to result from the collision of two or more galaxies. When galaxies collide, each galaxy experiences deformations in its shape as a result of the gravitational power of the other. Such collisions are not rare, given the ages of galaxies and the age of the universe. Astronomers have evidence that both the Milky Way and Andromeda, a nearby galaxy in our local group, have been involved in collisions before. A German research team cited evidence that the Milky Way has been involved in 15 previous galactic mergers.

When it comes to galaxies that are colliding, size matters. Right now the Milky Way is cannibalizing the large and small Magellanic Clouds, two small galactic neighbors which have been orbiting it for millions of years and will soon be consumed. The Andromeda Galaxy, which is 2.5 million light-years away, and the Milky Way are on a collision course. They are the two heftiest galaxies in our local group and will combine in about four billion years to form a new supergalaxy which, although it has not yet come into existence, has already been christened with the euphonious appellation Milkomeda. Skywatchers 3.75 billion years in the future will have the privilege of viewing a spectacular sight: the now-nearby Andromeda Galaxy looming hugely and brilliantly as it approaches the Milky Way (artist's conception at right). In the collision, the Milky Way will lose its elegant spiral structure, and the two merged galaxies will become a single elliptical galaxy, according to computer simulations. Fortunately, there is no need to be concerned for the safety of our Sun in such a collision. Even when galaxies collide, the stars in each galaxy are so far apart that most likely none will strike any other. It is estimated that Andromeda and the Milky Way are of approximately the same size, but the supermassive black hole at Andromeda's center is thousands of times more massive than the Milky Way's (billions of solar masses compared to the Milky Way's millions). Once the two galaxies have combined, the two black holes will hunt each other, eventually uniting.

Andromeda and Milky Way

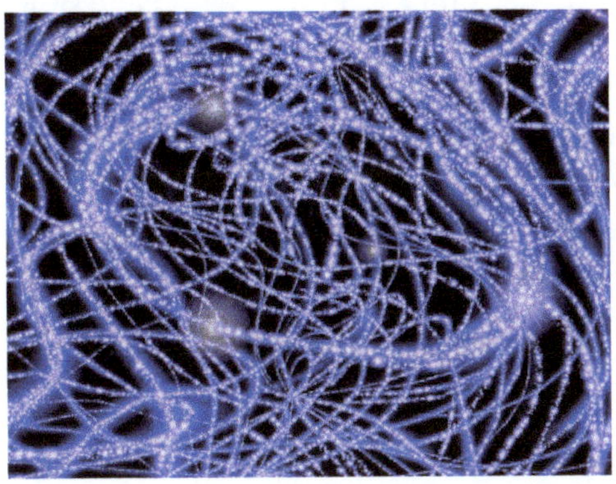

Astronomers in their quest to grasp and describe the incomprehensible vastness of space and the objects it contains do not stop at the galactic scale. They group galaxies together into clusters. For example, our galactic neighborhood contains 54 galaxies that are gravitationally bound to each other and are considered a cluster. The clusters, in turn, are grouped into superclusters, and ours is known as the Virgo Supercluster. Strings of superclusters are referred to as filaments (artist's conception at left), and these immense structures form a web throughout the length and breadth of space. The interstices of the web are the great voids, unimaginably broad expanses of empty blackness containing almost nothing: no visible galaxies, stars, planets, or dust.

The following are the brightest stars as seen from Earth, excluding the Sun.

Sirius—Brightest to Our Eyes

Sirius, although it appears to be a single star to the unaided eye, is actually a binary system. The two stars in the system, Sirius A and Sirius B, orbit each other every 50 years, maintaining an average distance of approximately 20 Astronomical Units (AU), which is 1.8 billion miles or about the distance from our Sun to the planet Uranus. However, the actual distance between the two stars varies much more than this average would indicate, ranging from 30 AU, the distance from our Sun to Neptune, to as little as 8 AU, the distance from our Sun to about halfway between Jupiter and Saturn. In 2019 Sirius A and Sirius B will reach their maximum separation.

The name Sirius means "sparkling" or "scorching," reflecting Sirius' brightness and its association with hot weather. It is the brightest star in the night sky, twice as bright as its nearest competitor, Canopus. Sirius A, a blue-white giant, is often called the Dog Star because it signaled the arrival of the hot, dry "dog days" of summer for the ancient Greeks. Sirius B is known as the "pup" and is a faint white dwarf about the size of Earth with one-fortieth the luminosity of our Sun. Surprisingly, despite its small size and relative dimness, Sirius B is just as massive as our Sun. Another surprising observation is that the pup's surface temperature is hotter than that of its larger and brighter companion. To experts, these facts suggest that the Sirius system originally consisted of two blue stars and has been in existence for only 250 million years. (For comparison, our Sun has existed for five billion years.) At one time, Sirius B would have been much more luminous than even the giant star Regulus (the 21st brightest star as seen from Earth) and would have been about five times as massive as our Sun until it blew off three-fourths of its mass in a relatively benign nova explosion. Before that explosion, Sirius B would have been the larger of the two stars, and astronomer-dinosaurs of the Jurassic Era, if any, would presumably have named it Sirius A.

Today, Sirius B is a mass of residual carbon and oxygen, the byproducts of various types of fusion reactions. When Sirius B approaches Sirius A at its closest in their orbital dance, it steals mass from its larger sibling because, despite its miniature size, Sirius B has more gravitational pull near its surface than its larger sibling does. Two stars as close to each other as Sirius A and Sirius B are referred to as contact binaries, meaning that their envelopes of gases make contact with each other at times. When this happens, the two stars can resemble a single star shaped like a giant, deformed peanut with one rounded lobe much larger than the other.

The Sirius star system is one of Earth's nearest neighbors at about only 8.6 light years. The Sirius system is on a trajectory which at present is bringing it closer to our Sun, and therefore for the next 60,000 years, Sirius will get brighter before moving away. Sirius will remain the brightest star in our heavens for the next 210,000 years.

Because of Sirius B's high temperature—a roasting 45,000°F at its surface—it produces most of its energy in the ultraviolet and x-ray regions of the light spectrum. If the human eye could see only ultraviolet light and X-rays, then Sirius B would appear to be the juggernaut of the system, completely overpowering Sirius A. Nonetheless, when all the wavelengths of light are considered together, the luminous energy from Sirius A surpasses Sirius B by a factor of a thousand.

Sirius B, while exceedingly hot and compact, no longer fuses atoms into heavier elements in its core and will steadily cool over the next two billion years, eventually becoming a cold, black relic of its former glory. Sirius A will continue to burn hydrogen for about another billion years, and then it will transform into a red giant as it starts to burn its helium. After burning its helium, Sirius A will shrink to become a white dwarf like its sibling and eventually become a massive, cold, bleak, planet-sized object. Thus, in about two billion years, the stars that form the Sirius system will disappear from view, while our Sun will grow approximately 20% more luminous and become a more prominent member of someone else's constellations. In four or five billion years our Sun will start to resemble Betelgeuse, in the process engulfing Earth and the other rocky planets of the Solar System as it enters the final phase of its existence.

It is unlikely that the Sirius system possesses planets. The gravitational chaos wrought by the two stars would likely have caused any planets to be either swallowed up or flung out of the system. If there are planets, it is unlikely that they would host life because of the relatively short time the two stars have been in existence—less than one twenty-fifth as long as the Sun, which has been providing dependable light, warmth, and gravity to Earth for almost five billion years.

Homer and Hesiod called Sirius *Cyon*, the Greek word for "dog." The name of the bright star Procyon (eighth brightest in the night sky), in the constellation Canis Minor, means "before the dog," since Procyon rises before Sirius. In Arabic, Procyon is called *Mirzan*, "the Roarer." Arabs imagined that Procyon announced the rising of Sirius with a roar.

Betelgeuse—Tenth Brightest

Betelgeuse is a striking red star in the constellation Orion and forms the right shoulder of the great hunter (the viewer's left). It is one of the largest visible stars, measuring between 130 million and 180 million miles in radius. (Our Sun has a radius of only about half a million miles.) If this monster were placed in the middle of our Solar System, then Mercury, Venus, Earth, and Mars would find themselves inside Betelgeuse's great flaming bulk.

Betelgeuse was first studied by John Herschel in the 1800s, who noted its brightening and dimming. This erratic behavior indicates that Betelgeuse has begun to burn unevenly, and most likely these fluctuations herald its imminent demise. Due to its immense size, mass, and relative closeness to us (640 light-years), when this star goes nova, it will appear as bright to us as the rest of the galaxy combined. It will temporarily be brighter than the full Moon and will probably be visible even in the daytime for several weeks.

Crab Nebula

When stars like Betelgeuse die, they blast immense expanding balls of gas and dust through space faster than any speeding bullet and create what is called a planetary nebula. (One of the most famous of these is the Crab Nebula (left), which was first noted by Chinese stargazers as a brilliant light on the morning of July 4, 1054. The Crab Nebula is 6,300 light years from Earth and is expanding at 1,000 miles per hour.) Betelgeuse is surrounded by several shells of material which have been expelled from the star over the last 100,000 years. The image at right, from the Herschel Space Observatory, shows Betelgeuse surrounded by this material. The material is spreading outward at a sluggish 50 miles per hour—speedy for a cheetah, but slow indeed for an astronomical phenomenon.

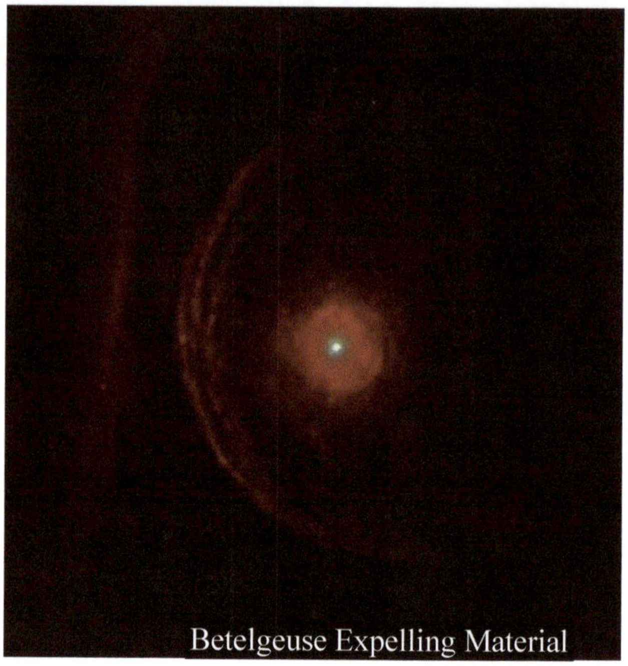

Betelgeuse Expelling Material

Alpha Centauri—Third Brightest

Alpha Centauri, like Sirius and many other lights of the night sky, is a multi-star system. Two moderately large stars, Alpha Centauri A and Alpha Centauri B, orbit each other every 80 years. Both are orbited every half million years by a smaller companion, Proxima Centauri, at a distance of a trillion miles.

This star system should occupy a special place in the minds of humanity, for Alpha Centauri A and our Sun are almost identical, with size, mass, and temperature within ten percent of each other. Yet Alpha Centauri, yellowish in color, is one to two billion years older than the Sun, and this accounts partly for the fact that it is 60% more luminous. Stars of this class get brighter as they age. Alpha Centauri B, yellowish-orange in color, is not as big as our Sun but still impressive, while the runt of the system, Proxima Centauri, is a red dwarf much smaller and cooler than the others, but with periodic flares that can double its brightness.

The stars of the Alpha Centauri system are the closest stars to Earth other than our Sun. The distance is only 4.3 light years, and this nearness has fired the imaginations of researchers and influenced popular culture. Many is the science-fiction tale that recounts a voyage to our nearest star-neighbor. In fact, the name bestowed upon the third member of the system reflects its nearness to Earth. “Proxima Centauri” means “the near one of the Centaur.”

A Russian billionaire plans to send a fleet of space probes to Alpha Centauri. The lightweight probes would be the thickness of a playing card and would be accelerated by laser beams to about one-fourth the speed of light. There are many technical hurdles to be overcome, as well as some financial and political ones, before this ambitious mission can become a reality.

Vega—Fifth Brightest

Vega with Dust Cloud

This bluish star is often called the "sapphire of summer." It is the fifth brightest star in the night sky, 40 times more luminous than the Sun, and is only about 25 light-years from Earth. Vega will be the pole star in a few thousand years as Earth's axis continues to precess. It will also be the brightest star in the sky one day. The Earth is speeding toward Vega, and in about 210,000 years Vega will surpass Sirius as the brightest star, hanging onto the title for about 270,000 years.

Vega is elliptical in shape; it is 2.78 solar diameters across its equator, while only 2.26 solar diameters from pole to pole. These figures strongly suggest that this star is spinning furiously, forcibly throwing its bulk outward. Vega has a donut-shaped disk of dust surrounding it, which may be a precursor to a solar system. Vega was the first star ever discovered to have anything resembling a solar system (in 1983), even though the orbiting material is dust instead of planets. Many hundreds of millions of years are required for a disk of dust to coalesce into planets. Unfortunately, this hot, fast-burning star will almost surely fail to endure long enough for its dust to become planets and the planets to give rise to life.

Polaris—48th Brightest

The present pole star, Polaris has been of immense practical value to human beings. For millennia, its position at the north celestial pole has enabled travelers, and particularly navigators, to find North with one glance at the night sky.

Polaris is a triple star system. The two brighter stars are a spectroscopic binary, which means that they are so close to each other that we can only tell them apart by examining their spectra and determining the differences in the elements of which they are composed. They orbit each other with a period of about 30 years.

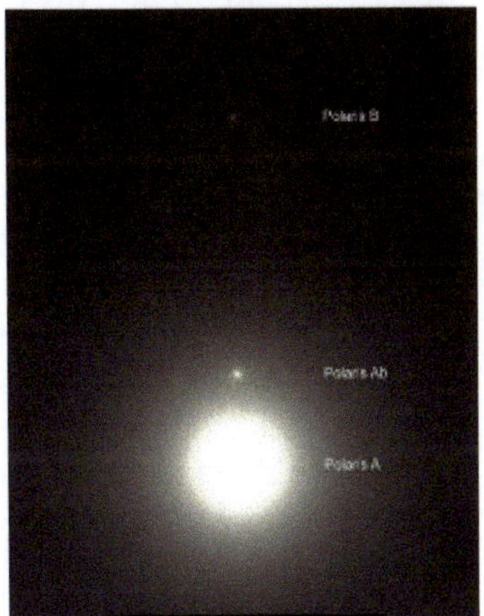

Polaris System

Polaris A is the type of star known as a Cepheid variable. Such stars occupy a place of supreme importance in the history of astronomy since their dependable variations in brightness enabled astronomers to make the first accurate estimates of stellar distances, including even the distances to other galaxies. Polaris A's fluctuations occur over four days. Polaris A is a huge star, some 50 times the diameter of our Sun. Even at a distance of 434 light years, farther than most other visible stars, it shines brightly in our sky.

Rigel—Seventh Brightest

Almost always the brightest star in Orion, Rigel is a blue giant, the sapphire to Betelgeuse's ruby, and is the most luminous star within a thousand light years of Earth. It is the seventh-brightest star in our sky and is as luminous as 40,000 Suns. Rigel has an extremely hot surface, at 20,000° F, making it four times hotter than our Sun. It is also 70 times wider and 17 times more massive than the Sun. The name Rigel is from Arabic and means "foot" because Arabic astronomers identified this star as the foot of Jauza, a goddess.

Rigel has a bluish-white companion, Rigel B, which is 140 times more luminous than our Sun. Rigel B resembles Regulus but is ten times farther away. There is also a Rigel C, which is very similar to Rigel B. Both Rigel B and Rigel C are subject to gravitational domination by the massive Rigel A, even though Rigel B and Rigel C are 50 times farther away from Rigel A than Pluto is from the Sun.

Antares—Fifteenth Brightest

The bright star Antares lies in the heart of the constellation Scorpius, "the Scorpion." Its name means "rival to Mars" because it is red in color like the red planet.

Like many of the stars we see in our sky, Antares has a small, dim companion, which was first seen in 1819. Over the centuries, controversy has arisen over the color of this companion. Many observers report seeing a greenish tint, but careful research into the matter has shown that the perceived green tint is an optical illusion caused by the proximity of bright-red Antares. The actual color of the companion is blue, indicating a high temperature. The two stars of the Antares system are 550 AU apart and orbit each other every 1,218 years. Within a few million years Antares will go supernova, but its companion will continue and probably never nova, gradually transforming instead into a white dwarf.

Both Antares and Betelgeuse are supergiant stars, and both are surrounded by a nebula. Their outer atmospheres pulsate, and both are distant compared with other visible stars. The distance to Betelgeuse is over 425 light years, while Antares is 600 light years away.

In the Realm of Superlatives

One of the most luminous stars ever detected is known as the Pistol Star after the Pistol Nebula in which it resides. Despite its intrinsic luminosity, it is too far away to be visible from Earth. We learn about such stars by training our greatest telescopes upon them and exposing light sensors for extended periods. The brightest star discovered to date bears the uninteresting name R1136a1. This monster also happens to be the heaviest, at 265 solar masses. This is close to the theoretical maximum mass for a star since stars even slightly more massive than this would quickly collapse and become black holes.

With regard to size, stars run the gamut from large to huge. The largest star ever discovered is a behemoth known as UY Scuti, which has a diameter more than 1,700 times greater than the diameter of our Sun, which is larger than the orbit of Jupiter. The hottest stars are the blue-white supergiants, such as one named Melnick 34, with a surface temperature in the vicinity of 100,000° F.

The smallest stars are about the size of a gas giant planet, such as Jupiter, but denser. One of these is the diminutive 2Mass JO5233822-1403022. Its brightness is 1/8000 that of our Sun. Another tiny star, named EBLM JO555-57Ab, has been discovered which is only slightly larger than Saturn.

The most distant star ever detected, nine billion light years from Earth, is nicknamed "Icarus" after the mythical young man who flew too close to the Sun. The official name of this star is MACS J1149+2223 Lensed Star 1, a name that refers to the method by which it was detected. Although a very bright blue giant, it would ordinarily be invisible even in the largest telescopes, but massive galaxies intervening in the billions of light years between us and the star magnify its light 600 times. This phenomenon is known as gravitational lensing.

The oldest star ever detected is the so-called Methuselah star, HD 140283, with an age comparable to the putative age of the universe. Scientists estimated its age by measuring its parallax, distance, and intrinsic brightness, and then applying the latest theories about burn rate, chemical abundances, and internal structure.

Despite their amazing and impressive characteristics, all of the stars listed in this section are, like the Pistol Star (which is 9,500 light years away), too far from Earth to be seen with the naked eye. As our telescopes see farther and farther, and as we improve our skills at interpreting what we see, the superlatives listed here will certainly be surpassed. Just as with the records associated with human achievement, when it comes to stars, records are made to be broken.

Conclusion

Stars, those massive conglomerations of ignited dust, burn brightly for an instant of cosmic time and then fade away or explode into fresh, ever-richer dust. In between, they find themselves surrounded by billions of smaller yet sometimes very consequential accumulations of dust. On one such planet arose a creature with a complex mind able and keen to gather knowledge. These creatures collectively came to comprise a gregarious species voraciously eager to revel in all that the experience of living has to offer, all the multifariousness of life, learning, and pleasure, including delight in the brilliance and mystery of stars.

There are many sources from which human happiness may spring. The German chess master Tarrasch, one of the first five grandmasters designated by the Tsar of Russia, wrote in the 1800s that chess had the power to make men happy. Van Gogh expressed similar sentiments when he stated that the stars made him dream. Just as the expressionists poured their creative energies into unmediated statements of their emotional responses to their subject matter, we feel our own emotions summoned by the nightly panorama. Euclid and the early Greek mathematicians laid a mathematical foundation 2,500 years ago when they began to appreciate the beauty inherent in a concise geometric proof. Our ancient ancestors discovered the beauty and the practical efficacy of the stars uncounted millennia ago. Today's astronomers not only enjoy studying light; they delight in the discovery of unforeseen truths and find great pleasure in the iron logic that generates a proof of some profound insight into nature.

Many are the cosmic mysteries that remain unexplained, awaiting new revelations and those with the insight to interpret them. Thoughts of the discoveries still to be made inspire exuberance about life and learning. New surprises are inevitable and will arrive speedily and steadily as we continue to reach outward and upward toward the beauty and majesty of those bright emperors of the night, the stars.

An example of a telescope used by William Herschel

27
Gravitational Waves

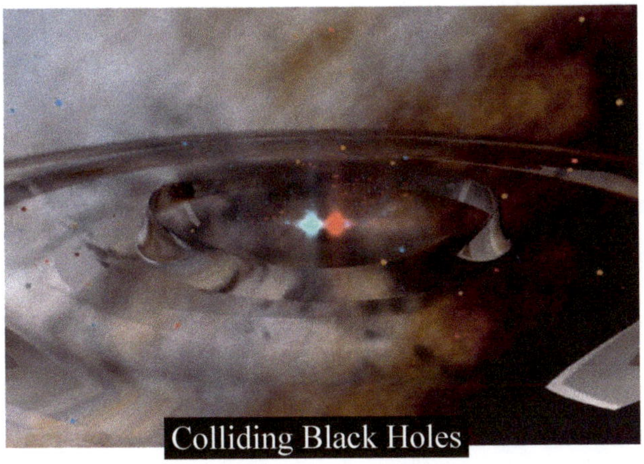

Colliding Black Holes

Lighter than a cork I danced on the waves.
Arthur Rimbaud

A long-sought and super-significant advance in astronomy occurred on September 14, 2015. On that date, two sophisticated and sensitive detectors, one in Livingston, Louisiana, and the other in Hanford, Washington, detected a fleeting and barely measurable phenomenon that until then had only been theorized about: gravitational waves.

The pair of detectors together constitute the Laser Interferometer Gravitational-Wave Observatory (LIGO). They are built on the mind-bending concept of the universe as an expanse of malleable fabric being unceasingly rippled and distorted by waves of gravitational energy. Ripples in spacetime, or gravitational waves, passing through the apparatus slightly change its length for an instant, with the result that perpendicular beams of laser light traveling within it shift ever so slightly out of phase, creating an event. The beams are set to cancel each other out perfectly until a gravitational wave is detected, at which point an audible chirp occurs, which is the epitome of beautiful music to the researchers. Each tiny peep marks the passing of a gravitational wave created by an energetic event exceeded only by the energy output of the Big Bang. That initial, minute peep in 2015 was the first piece of data ever collected in the embryonic science of gravitational-wave astronomy, as scientists made the invisible visible, and were able to view the indirect effects of the passing gravitational wave.

Every time a massive body accelerates, it transmits a frictionless, rolling shiver that silently, serenely, and shadow-like sails through nested spheres of space and time that receive the waves and pass them along. The violent collision of two black holes is such an enormous, catastrophic event that it literally shakes the fabric of the universe. The energy released in such an event is comparable to the combined energy output of every single star in the known universe, but not in the form of light—the released energy takes the form of invisible undulations that pleat the unseen fabric of spacetime.

Throughout its history, Earth has been massaged by this type of gravitational disturbance, but only now do we have the huge and phenomenally sensitive detectors capable of identifying them.

Located over 1800 miles apart, the two stupendously precise pieces of equipment are cousins of the famous Michelson-Morley interferometer that gave Einstein one of the key clues leading to relativity. Each detector boasts a pair of two-and-a-half-mile-long vacuum tunnels arranged in an L shape. The image at right gives an idea of the scale and perfect straightness of the interferometer arms. Inside each arm, laser light bounces back and forth between the world's smoothest mirrors. The laser light is 50,000 times more powerful than the laser beam in a typical pocket laser pointer. The device then measures any minuscule mismatch in the time it takes the beams in the different arms to complete the journey. The lasers act essentially as super-accurate rulers.

LIGO is so sensitive that it can be affected by vehicles on nearby roads, distant earthquakes, and even the tidal tugging of the Moon and Sun. Slight variations in temperature and even air currents can also throw off its readings. Countermeasures include seismic isolation, in which earthquake activity is identified as it happens and immediately compensated for. The mirrors are hung on four-stage pendulums to offset ground vibrations. The laser beams travel through extreme vacuum to minimize interference from molecules of gas. The two detectors acting simultaneously help in determining the direction from which the signal comes, and each helps verify the other's data.

History of the Study of Gravity

The 17th-century genius Sir Isaac Newton was the first to deduce that any two objects with mass exert gravitational forces on each other, but he was not able to shed any light on how a body can reach out across empty space and exert a force on a different body in a distant location.

Approximately 100 years after Newton, John Michell of England in 1783 and Pierre-Simon Laplace of France in 1796 independently extended Newton's concept of gravity, developing the idea of a "dark star." In Michell's case, he actually used the mass of our Sun to calculate the size of a black star and obtained a result that is remarkably modern. Laplace wrote that the "largest luminous bodies in the Universe may be invisible." Both men proved to be prophetic. The name "black hole" was not coined until 1967 by John Wheeler.

Albert Einstein spent a decade trying to determine how gravity is transmitted, and he finally concluded that space and time act in unison to create the effect we call gravity. To help understand how this happens, a frequently invoked simile for spacetime is to envision a flexible surface like a trampoline. Placing a boulder on a trampoline causes the trampoline to deform. A marble rolling on the now-bowed surface of the trampoline will travel a curved trajectory around the boulder. Einstein asserted that similarly, near an astronomical body like the Sun, the Earth and other orbiting objects appear to follow elliptical paths because the spacetime environment itself is warped. By 1919, astronomical observations had confirmed this remarkable claim and catapulted Einstein to international fame.

Einstein then generalized his discovery to include moving objects. Previously, he had dealt only with static cases, in which mass affects the curvature of spacetime but does not move. Now, he turned to dynamic cases, addressing the question of what would happen to the spacetime fabric in the presence of matter in motion. He realized that just as children jumping on a trampoline generate waves in the surface that ripple outward, matter in motion must create waves in the fabric of spacetime that ripple outward, too. Such waves, if they could be detected, would be gravitational waves.

The idea of gravitational waves represents general relativity's most significant departure from Newtonian gravity. In familiar situations, like the Sun exerting its gravitational pull on Earth, Einstein's predictions barely differ from those of Newton. However, Newton's theory contains nothing to suggest the existence of traveling gravitational disturbances.

Einstein made several approximate calculations about gravitational waves, and his calculations suggested something that he found disturbing. According to the calculations, the existence of gravitational waves would also imply the existence of a singularity, that is, a point in space where gravity and time are distorted so much that the usual laws of physics no longer apply. However, no singularities had been observed during his lifetime or any time before. Einstein co-authored a paper claiming that gravitational waves were impossible in the general theory of relativity, but the paper was withdrawn shortly before publication. At any rate, it was evident that if gravitational waves existed, they would be extremely weak, almost surely too weak to be detected.

Despite Einstein's misgivings, by the 1960s, scientists using more refined mathematical models convinced themselves beyond any doubt that gravitational waves were a feature of the general theory of relativity, and the race was on to detect them.

Detecting Gravitational Waves

In 1974, Joseph Taylor and Russell Hulse closely examined a binary pulsar (two co-orbiting neutron stars) and determined that these two tremendously massive objects were losing an amount of energy that matched the amount predicted by the gravitational-wave theory. Taylor and Hulse received the 1993 Nobel Prize for this work.

The direct observation of a gravitational wave is a daunting challenge, but the method is simple in theory. Traveling gravitational waves alternately stretch and squeeze spacetime like an accordion, but the effect is much smaller than a single atom. By precisely monitoring split-second variations in distance, scientists should be able to indirectly detect the passing wave.

Unfortunately, gravitational ripples lose strength as they travel from their source, like the diminishing circular waves expanding outward from the point on the surface of a pond into which a stone has been dropped. By the time gravitational waves reach Earth, the amount of stretching and squeezing they cause is so tiny that detecting it is like measuring the distance from Earth to Alpha Centauri with an accuracy better than the thickness of a dollar bill.

After decades of failures and controversy involving numerous researchers who sought to observe gravitational waves directly, physicists Kip Thorne and Rainer Weiss took up the challenge. They were later joined by the Scottish physicist Ronald Drever. Eventually, the team received the necessary funding to transform their vision into an immense, highly sensitive measuring device.

A grant of $250 million from the National Science Foundation provided the initial funding, and in 2002 the two interferometers comprising LIGO began operation, passively waiting in solemn silence for an incident signal to throw their exquisitely calibrated beams out of sync and change the world of science forever.

The wait lasted almost a decade without a hint of a signal. The decision was made in 2010 that the instruments needed to be seriously upgraded in sensitivity because, given the size of the universe and the number of galaxies, black holes should collide, on average, much more often than zero times in eight years—approximately two per year or even more often was the accepted estimate. After more than $200 million worth of upgrades, LIGO came back online in the fall of 2015, and it was ten times more sensitive than before. It was now capable of measuring changes in the lengths of its arms less than 1/10,000 the diameter of a proton. Less than two days later, a sudden, tiny alteration in the length of a laser tunnel produced a vibration that generated a sound akin to a drifting feather bouncing off a tuning fork. Seven one-thousandths of a second later, an identical note was heard 1,865 miles away at the other detector. The delicate vibrations matched what computer simulations predicted for gravitational waves produced by the final throes of orbiting black holes crashing together billions of light years away.

Scientists were able to recreate the details of the 2015 event detected by LIGO. The black holes which crashed together were each about 50 miles across. Scientists were able to infer that one of them had a mass of 36 Suns, while the other weighed in at 29 solar masses. As they spiraled into each other, at a distance of 500 miles they started to give off detectable gravitational waves, and then, following a cataclysmic collision, the two black holes coalesced into one. The resulting black hole had only 62 solar masses instead of 65. The remaining three solar masses were instantly converted into energy according to the equation $E = mc^2$. This tremendous amount of energy resulted in gravitational waves more powerful than the combined output of all the stars in the visible universe. Once emitted, the gravitational waves sped through the cosmos at the speed of light, until 1.3 billion years later, much attenuated but still detectable, they impinged upon the LIGO apparatus. By this time, the once-mighty wave had been reduced to a mere whisper of its former Herculean self.

More and better gravitational wave detectors will soon be built. The Virgo Facility, in Italy, is shown at right. In addition to colliding black holes, advanced detectors should in theory be able to observe gravitational waves from any sufficiently massive objects orbiting each other, such as neutron stars or pulsars. Even a sufficiently violent supernova

should give off detectable gravitational waves if the explosion is asymmetrical enough. Some researchers hope eventually to measure the gravitational waves created by the Big Bang or soon after.

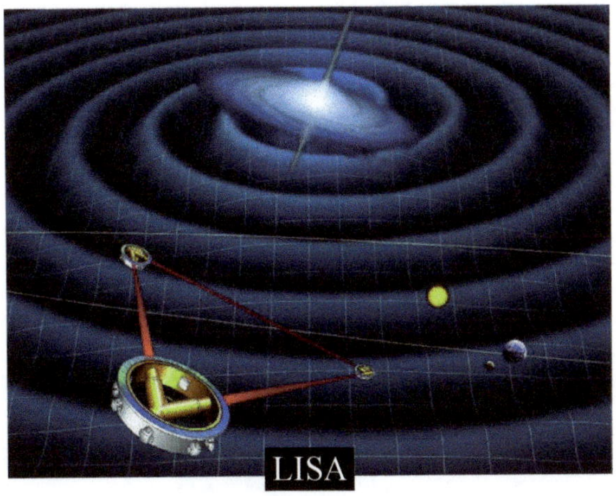

The European Space Agency plans to launch LISA (Laser Interferometer Space Antenna, at right) into solar orbit sometime in the fourth decade of this century. As it will consist of three separate spacecraft, its interferometer arms will not be constrained by the size of any physical building or terrain and will be almost two million miles long. This will allow sensitivity much greater than would be possible with any earthbound facility.

This research offers no short-term pecuniary gain. Quite the contrary—it falls squarely into the category of pure research, the type from which specific and concrete advantages (monetary or technological) come slowly, but history has shown that pure research has always proved beneficial in the long run. Patience, perseverance, and perspective will be required if we are to plumb the depths of reality using these immense interferometers. The instruments and data are far beyond our senses; we are attempting to decipher an unknown language without a gravitational Rosetta Stone to guide us.

History will look back on the detection of gravitational waves as one of those paradigm-shifting moments that altered the course of science. In the past, the science of astronomy has always depended upon an analysis of the electromagnetic spectrum of light to reveal the underlying structure of nature. Now, for the first time, science is starting to utilize an analysis of a different form of energy to interpret the natural world. When the telescope was invented, it substantially enhanced humans' ability to collect and analyze visible light, and with it, we experienced the splendor of vast new cosmic panoramas. During the 20th century, we broadened the array of light signals we could detect to include radio, microwave, infrared, ultraviolet, X-rays, and gamma rays—forms of light, but ones we cannot see. As we survey the heavens with ever more elaborate, ingenious, and sensitive instruments, the cosmic landscape grows more picturesque, varied, and richer each time.

A gravitational wave detector does not rely on electromagnetic radiation of any kind, and this makes it an entirely different kind of cosmic probe, with the potential of yielding even more dramatic discoveries. Light can be blocked. Any opaque material, like a window shade, can block visible light, and other wavelengths can be blocked as well. Gravitational waves, on the other hand, are neither impeded nor diminished by the presence of matter. It is as if this force is transmitted through a medium that is totally outside of our perception of physical reality.

Since gravitational waves are not diminished by matter, they should provide clarity heretofore never imagined. Thus, the astronomers of the world anticipate the construction of a network of detectors across the globe and out into space, which then will help triangulate sources of information and provide backup data in case one device is temporarily lost to mishap. With each improvement in astronomical detection, humankind's understanding of nature improves. Very likely, as we survey the distant reaches of the universe with ever more sophisticated tools, we will discover a dark landscape populated by objects and phenomena we have yet to even imagine.

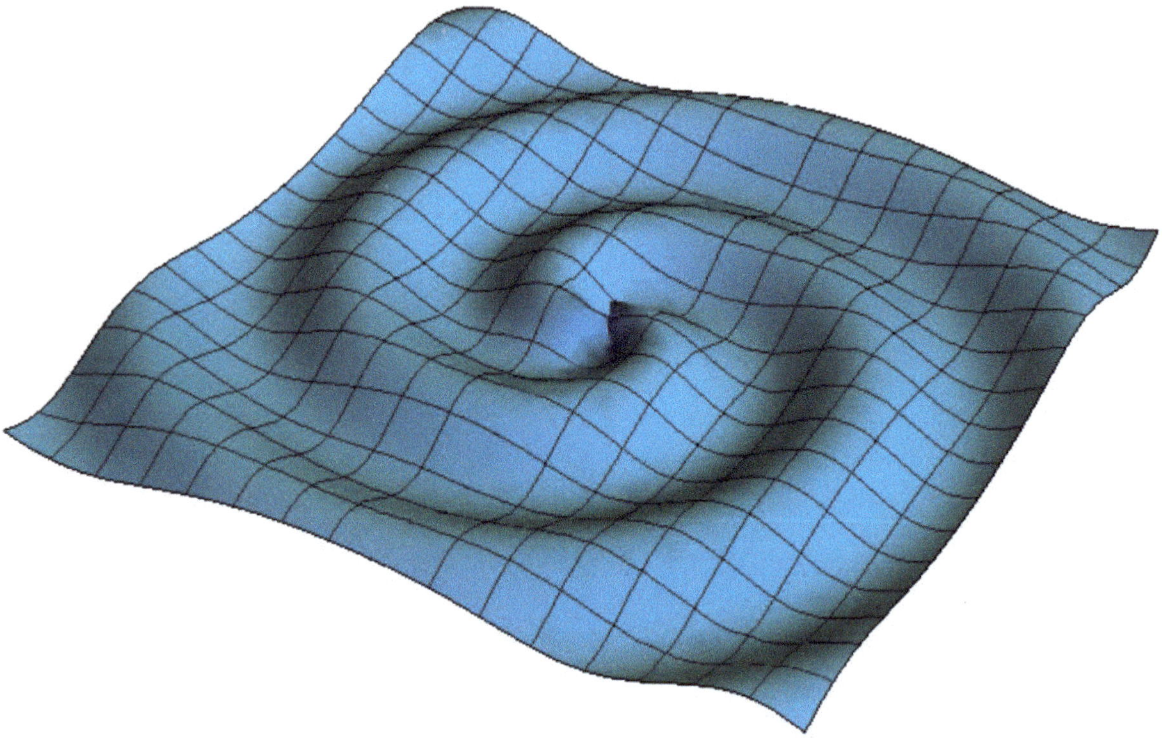

Artist's image of gravitational waves being created by a sizable gravitational event. Created using Wolfram Mathematica.

Afterword

Now, the fourth day evening, we descend,
make fast, set foot at last upon her beaches,
stand in her silence, lift our heads and see
above her, wanderer in her sky,
a wonder to us past the reach of wonder,
a light beyond our lights, our lives, the rising
earth...
Archibald MacLeish

Thus the sum of things is ever being renewed
...generations of living creatures are changed
and like runners pass on the torch of life.
Lucretius

Life finds a way.
Michael Crichton

The duration of a human life is but a fleeting instant compared to the vast sweep of cosmic time. Therefore, we must use our time creatively and continue to advance scientifically, technologically, and communally—at a walk, at a run, or inching forward on all fours, but always onward. On this planet, we fill our few allotted years with actions, ambitions, dreams, and memories. From birth, our bodies undergo an unending process of physical change, first transforming from tiny bundles of mewling joy to fledgling wanderers alternating from moment to moment between inquisitiveness and astonishment. Our senses become progressively more acute as we sprout to become miniature discoverers seeking new shapes, smells, sights, and sounds within our expanding realms. This varied sensory input is continuously stored in the growing warehouse of our minds. In our youth, we are initially focused upon learning, loving, and laughing. Later we shift our focus to becoming conscientious, contributing, and creative members of human society. As we reach the prime of life, our attention shifts to procreation, the raising of offspring, and our chosen careers. Then, if we are fortunate enough to reach an advanced age, we at last have the time to sit and ponder with deep appreciation the wider world in general and how stunning, singular, and superb our small spheroid truly is. The famous "Earthrise" photograph above, snapped by William A. Anders from the Apollo 8 spacecraft in 1968, captured the public's imagination and enabled many to appreciate the grand majesty of Earth in its entirety for the first time. Earth not only invigorates and brightens the surrounding space with its swirling clouds and azure oceans of water, but it also inspires curiosity. Examining this image, one cannot help but marvel at Earth's beauty and splendor and ask: Why is Earth so elegant, so magnificent—and so comfortable? It is only natural to pose the question: If the Earth is so right for humankind, then is there a profound reason for the existence of humanity? Does humanity have a purpose?

If we truly have the freedom to choose our own futures, then nothing, no one thing or entity, can assign a purpose to humanity. If we desire a purpose, then we must seize the initiative and forge our own futures. Purpose does not arrive on the wheels of inevitability but is fired and tempered in the consciousness of humankind. We are animals imbued with powerful survival instincts, intricate minds, and a gift for altruism. With these talents comes the ability to define our own roles and make our own choices. We can make our own decisions, develop our own enterprises, and follow our own particular path through life, a path which will depend on our strengths, needs, and proclivities. Guidance on this path can come from other humans we respect, but more importantly, our futures are shaped by our pasts. The hard-won information and understanding gleaned over generations intertwine with the memories created throughout a lifetime to drive away the shadows of the unknown and light the way to our futures and to happiness.

That future and that happiness will depend on the universe of which we are a part and into which we will expand. Our success as a species will depend on other planets, moons, stars, comets, and perhaps especially asteroids. We will find a way. So far, we have only discovered a tiny fraction of what we will need to know to thrive and survive over truly long time scales. For this reason knowledge of the Solar System, our galaxy, and the rest of the cosmos is of paramount importance to our species.

Although the book you are reading is complete for the time being, it will never be truly finished. The author has expressed tentative satisfaction and momentarily set the work aside, but the process of learning more about the universe cannot—it must not—come to an end. Like Olympian torch-bearers, all chroniclers of nature are part of an unending chain, striving to advance the light of knowledge as we surge forward along the lifeline of the universe towards space and eternity. Let those destined to tell the story of the exploration of space to future generations start now, lift the torch ever higher, and press resolutely onward.

At this moment, humanity is preparing to leave its cradle to embark upon a bold adventure that will propel us into the vast expanse that surrounds our diminutive planet. The attempt must be made. Humankind has always thrived upon courageous endeavors. If we can collectively muster the will, allocate the resources, and delegate the necessary tasks to ensure it will happen, then the greatest achievement of the present century will be the exploration of another planet by humans. Those who venture upon this audacious journey will be guided by reason, intuition, experience, and hope. We will send our intrepid explorers, brimming with energy and sky-high expectations on missions of discovery, assessment, and settlement—but only if we dare! The vanguard of humanity will carry with them into the cosmos all our hopes and dreams. Their voyages, heralding a new age of unique and unexpected challenges, will alter the very character of the universe as a sentient, organic, living species emerges to populate the far reaches of space. This is our destiny and our purpose if we choose to seize it.

When he presided over the Constitutional Convention of 1787, George Washington sat in an ornate chair emblazoned with a golden Sun. Benjamin Franklin reportedly quipped that he was unsure whether the Sun was rising or setting. In the photograph above, we see the Earth poised elegantly and poignantly in space. Is the Earth rising or setting? It is for us to decide.

Pythagoras with pyramid and the James Webb Telescope

Picture Credits

Cover Montage

Cover Voyager NASA/JPL-Caltech
https://images.nasa.gov/details-PIA17036.html

Cover Solar Prominences NASA/GSFC/Solar Dynamics Observatory
www.images-assets.nasa.gov/image/PIA18167/PIA18167~orig.jpg

Cover Black Hole
https://www.wpclipart.com/space/black_hole/black_hole_accretion_disk.jpg.html

Cover Earth
www.images-assets.nasa.gov/image/GSFC_20171208_Archive_e002131/GSFC_20171208_Archive_e002131~orig.jpg

Preface

xi Whirlpool Galaxy NASA Goddard
www.images-assets.nasa.gov/image/GSFC_20171208_Archive_e001925/GSFC_20171208_Archive_e001925~orig.jpg

Part I Quintessentially Human: The Search for Astronomical Knowledge

Ch. 1 Why Study Astronomy?

Pg.1 M81 NASA/JPL-Caltech/ESA/Harvard-Smithsonian CfA Jpl.nasa.gov
https://images.nasa.gov/details-PIA09579.html

1 Jet Plane and Moon pexels.com
https://www.pexels.com/photo/gray-airplane-illustration-1769394/

2 Crab Nebula NASA, ESA, J. Hester and A. Loll (Arizona State University)
www.images-assets.nasa.gov/image/GSFC_20171208_Archive_e002159/GSFC_20171208_Archive_e002159~medium.jpg

3 Pillars of Creation Pixabay 108004237
https://pixabay.com/go/?t=image-list-adobe&id=197798231

4 Two Galaxies NASA Goddard
www.images-assets.nasa.gov/image/GSFC_20171208_Archive_e000097/GSFC_20171208_Archive_e000097~medium.jpg

4 Galaxy NASA/JPL-Caltech/SSC
https://images.nasa.gov/details-PIA07901.html

5 Carina Nebula NASA/ESA/STScI NASA ID: PIA15985
www.images-assets.nasa.gov/image/PIA15985/PIA15985~orig.jpg

Ch. 2 Threads of Culture

7 Loom Pixabay1839920
https://pixabay.com/photos/craftsman-loom-craftsmanship-hands-1839920/

9 Bayeux Tapestry public domain 1924
https://upload.wikimedia.org/wikipedia/commons/thumb/2/27/Bayeux_Tapestry_scene32_Halley_comet.jpg/576px-Bayeux_Tapestry_scene32_Halley_comet.jpg

Ch. 3 Dogma versus Science

11 Crystal Spheres graphic by Joel Cardenas

12 Lithium Atom Pixabay2784853_960_720
https://cdn.pixabay.com/photo/2017/09/25/11/38/lithium-2784853_960_720.png

14 Interferometer Drawing
https://free-images.com/display/michelson1881c_png.html

Ch. 4 What is a Theory?

17 Nested Platonic solids By Johannes Kepler - Johannes Kepler: Mysterium Cosmographicum, Tübingen 1596, Tabula III: Orbium planetarum dimensiones, et distantias per quinque regularia corpora geometrica exhibens. Public Domain,
https://commons.wikimedia.org/w/index.php?curid=37300
https://en.wikipedia.org/wiki/File:Kepler-solar-system-1.png

18 Geocentric Theory Pixabay2026315
https://pixabay.com/vectors/earth-geocentric-jupiter-mars-2026315/

20 More Nested Platonic Solids By Johannes Kepler - Johannes Kepler: Mysterium Cosmographicum, Tübingen 1596, Tabula III: Orbium planetarum dimensiones, et distantias per quinque regularia corpora geometrica exhibens. Public Domain,
https://commons.wikimedia.org/w/index.php?curid=37300
https://en.wikipedia.org/wiki/File:Kepler-solar-system-2.png

20 Saint Mark's geocentric clock Image by Charles Safford

Ch. 5 Einstein

21 Einstein Pixabay1145030
https://pixabay.com/photos/albert-einstein-1921-portrait-1145030/

25 Curved Spacetime Pixabay102008713
https://pixabay.com/go/?t=image-list-adobe&id=102008713
Caption by Joel Cardenas

25 Gravitational Lensing Joel Cardenas 4-23-19

28 Mercury Precession of Perihelion Joel Cardenas 4-23-19

Ch. 6 Stonehenge

33 Sunrise Stonehenge Pixabay938998
https://pixabay.com/photos/sunrise-stonehenge-ancient-sky-938998/

34 Stonehenge Monument Stones Pixabay947348
https://pixabay.com/photos/stonehenge-monument-stones-947348/

34 Drawing of mortise/tenon
Drawing by Said Elmajzoub, Graphics by Andrew Hall and Christopher Myers

36 Trilithon artwork
Drawing by Said Elmajzoub, Graphics by Andrew Hall and Christopher Myers

36 Diagram of Final Stonehenge Setting drawing by Joel Cardenas

37 Stonehenge Pixabay741485
https://cdn.pixabay.com/photo/2015/04/27/06/15/stonehenge-741485_960_720.jpg

38 Stone Megalith Outdoors Grass Pixabay3186462
https://pixabay.com/photos/stonehenge-megalith-outdoors-grass-3186462/

42 Horsehead Nebula
www.images-assets.nasa.gov/image/GSFC_20171208_Archive_e001518/GSFC_20171208_Archive_e001518~orig.jpg

Ch. 7 History of Astronomy Part I

43 Astronomer in Alexandria
https://www.wpclipart.com/famous/science/Greek/Hipparchus_in_Observatory_of_Alexandria.png.html

47 Pythagoras Raphael, "School of Athens"
https://free-images.com/display/sanzio_01_pythagoras.html

49 Head of Kukulcán original artwork by Said Elmajzoub

49 El Castillo (Mayan temple) image by Joel Cardenas 4-23-19

50 The Caracol (Mayan observatory) free-images.com
https://free-images.com/display/el_caracol_in_chichen.html

51 Thales public domain
https://upload.wikimedia.org/wikipedia/commons/6/67/Thales-06.jpg

51 Anaximander Raphael, "School of Athens"
https://upload.wikimedia.org/wikipedia/commons/3/38/Anaximander.jpg

51 Eudoxus public domain photo of artwork
https://www.google.com/search?biw=1366&bih=651&tbm=isch&sa=1&ei=1GjLXM_0OH99AOS_5SoCQ&q=public+domian+image+of+eudoxus+cnidus&oq=public+domian+image+of+eudoxus+cnidus&gs_l=img.3...809065.812964..814191...0.0..0.218.1370.10j3j1......1....1..gws-wiz-img.WwY1EFs5Qpo#imgrc=dnhTKkYkPPNehM

52 Circling Stars Pixabay828656
https://pixabay.com/photos/star-trails-night-long-exposure-828656/

52 Aristarchus Public domain by author
https://en.wikipedia.org/wiki/File:Aristarchos_von_Samos_(Denkmal).jpeg

52 Eratosthenes public domain
https://upload.wikimedia.org/wikipedia/commons/a/a2/Portrait_of_Eratosthenes.png

53 Apollonius of Perga Published before 1924, author dead at least 100 years
https://en.wikisource.org/wiki/File:Apollonii_Pergei_Opera_1537_detail.jpg

53 Conic Sections graphic by Joel Cardenas

53 Hipparchus public domain
https://upload.wikimedia.org/wikipedia/commons/7/7e/Hipparchus_by_Raphael.jpg
54 Ptolemy free-images.com
https://free-images.com/display/ptolemy_1476_with_armillary.html
55 Alhazen free-images.com
https://free-images.com/display/alhazen_2_png.html
56 Omar Khayyam English Wikisource 58828
https://en.wikisource.org/wiki/File:Omar_Khayyam2.jpg

Ch. 8 History of Astronomy Part 2

57 Galileo Addressing Ecclesiastics public domain
https://www.google.com/url?sa=i&rct=j&q=&esrc=s&source=images&cd=&cad=rja&uact=8&ved=&url=https://storyminers.com/galileo-true-visionary/&psig=AOvVaw0oSoj5gHH38ClSVPQ5o6M7&ust=1554250181026009
58 Copernicus Pixabay67542
https://pixabay.com/photos/astronomer-man-nicholas-scratch-67542/
58 Brahe Public domain book from 1890
https://www.google.com/url?sa=i&rct=j&q=&esrc=s&source=images&cd=&cad=rja&uact=8&ved=2ahUKEwiYxoSxvqPhAhXJAYgKHYewDOkQjRx6BAgBEAU&url=https://en.wikipedia.org/wiki/Tycho_Brahe&psig=AOvVaw0XZP4y52-YLEp7OTGFGVwL&ust=1553816104325147
59 Kepler public domain
https://commons.wikimedia.org/wiki/File:Johannes_Kepler_1610.jpg
60 Galileo Pixabay public domain 1924
https://commons.wikimedia.org/wiki/Catergory:Galileo_Galilei#/media/File:Galileo.arp.300pix.jpg
61 Newton public domain 37337
https://commons.wikimedia.org/w/index.php?curid=37337
https://upload.wikimedia.org/wikipedia/commons/3/39/GodfreyKneller-IsaacNewton-1689.jpg
62 Halley's Comet free-images.com
https://free-images.com/display/halleys_comet_1910.html
62 Halley Public domain 1924
https://commons.wikimedia.org/wiki/File:Edmond_Halley 072.jpg
63 Herschel Pixabay11628
https://pixabay.com/photos/william-herschel-astronomy-11628/
63 Herschel's Giant Telescope free-images.com
https://free-images.com/display/short_history_astronomy_fig_59.html
63 Clyde Tombaugh public domain pub w/o ©
http://starchild.gsfc.nasa.gov/docs/Starchild/questions/question16.html
64 Henrietta Swan Leavitt public domain
https://upload.wikimedia.org/wikipedia/commons/3/3b/Leavitt_aavso.jpg

64 Fraunhofer public domain
https://upload.wikimedia.org/wikipedia/commons/9/9f/Joseph_von_Fraunhofer_Portrait.jpg

65 Annie Jump Cannon public domain
https://upload.wikimedia.org/wikipedia/commons/4/4c/Annie_Jump_Cannon_1922_Portrait_-_Original.tif

65 Cecilia Payne public domain wiki
https://upload.wikimedia.org/wikipedia/commons/3/3e/Cecilia_Helena_Payne_Gaposchkin_%2819001979%29_%283%29.jpg

65 Einstein public domain © not renewed
https://upload.wikimedia.org/wikipedia/commons/d/d3/Albert_Einstein_Head.jpg

66 Edwin Hubble public domain no © notice
https://upload.wikimedia.org/commons/f/f4/Studio_portrait_photograph_of_Edwin_Powell_Hubble.JPG

66 Penzias and Wilson Radio Telescope NASA GPN-2003-00013
https://upload.wikimedia.org/wikipedia/commons/f/f7/Horn_Antenna_in_Holmdel%2C_New_Jersey.jpeg

67 Gerard Kuiper NASA/JPL
https://upload.wikimedia.org/wikipedia/commons/0/0b/GerardKuiper.jpg

67 Jan Oort public domain, author has waived all rights
https://upload.wikimedia.org/wikipedia/commons/e/e4/JanOort2.png

69 Halley's Comet Woodcut 1690
https://www.wpclipart.com/space/comet/Halley/Halley_1690_engraving.jpg.html

Ch. 9 Curiosity: More Powerful than Rocket Fuel

71 Spacecraft above Earth
www.images-assets.nasa.gov/image/sts061-48-027/sts061-48-027~orig.jpg

Part II An Astronomical Tour
Ch. 10 Earth

79 Earth Pixabay11015
https://pixabay.com/photos/earth-blue-planet-globe-planet-11015/

82 Water in Three Phases Pixabay1149708
https://pixabay.com/photos/snow-mountain-water-steam-snow-1149708/

87 Aurora NASA Goddard
www.images-assets.nasa.gov/image/GSFC_20171208_Archive_e001871/GSFC_20171208_Archive_e001871~orig.jpg

Ch. 11 Moon

91 Near Side of Moon NASA/JPL/USGS
https://images-assets.nasa.gov/image/PIA00302/PIA00302~thumb.jpg

92 Phases of Moon spaceplace.nasa.gov
www.spaceplace.nasa.gov/oreo-moon/en/&psig=AOvVaw28z1lYQToipvva9ZGnMjNe&ust=1556151058058462
98 Schmitt with Flag NASA ID: AS17-134-20384
www.images-assets.nasa.gov/image/7036672/7036672~orig.jpg
99 Theia Striking Earth
https://www.wpclipart.com/space/solar_system/Earth/Earth_history/Theia_impacts_Earth_creating_moon.jpg.html
99 Footprint on Moon NASA
www.images-assets.nasa.gov/image/6901250/6901250~thumb.jpg

Ch. 12 Eclipses

101 Half Solar Eclipse NASA/Bill Ingalls
www.images-assets.nasa.gov/image/NHQ201708210302/NHQ201708210302~medium.jpg
102 Solar Eclipse Diagram
www.wpclipart.com/space/diagrams/solar_eclipse_diagram.png.html
102 Umbra from ISS NASA
https://www.images-assets.nasa.gov/image/iss052e056222/iss052e056222~medium.jpg
103 Diamond Ring Pixabay 152834
https://pixabay.com/vectors/solar-eclipse-eclipse-sun-flare-152834/
103 Baily's Beads NASA/Aubrey Gemignani
https://www.images-assets.nasa.gov/image/NHQ201708210100/NHQ201708210100~medium.jpg
103 Solar Eclipse Showing Corona NASA/Aubrey Gemignani
https://www.images-assets.nasa.gov/image/NHQ201708210101/NHQ201708210101~medium.jpg
105 Antikythera Device Actual public domain Therese Clutario
https://upload.wikimedia.org/wikipedia/commons/6/66/NAMA_Machine_d%27Anticyth%C3%A8re_1.jpg
105 Antikythera Device Reconstructed
https://free-images.com/lg/d39d/derekdesollaprice.jpg
106 Partial and Total Solar Eclipse NASA
https://www.imagesassets.nasa.gov/image/GSFC_20171208_Archive_e000234/GSFC_20171208_Archive_e000234~orig.jpg
107 Lunar Eclipse Diagram
https://www.wpclipart.com/space/moon/lunar_eclipse_illustration.png.html
111 Saros 136 map
https://eclipse.gsfc.nasa.gov/SEsaros/image/Saros145-big.jpg

Ch. 13 Mercury

113 Mercury
https://pixabay.com/photos/mercury-planet-surface-solar-system-11591/
114 Mercury Orbit Diagram graphic by Joel Cardenas

116 Mercury, Moon, Titan, Ganymede
Credits as already given for these four images, montage by Chris Myers

117 Mariner 10
https://free-images.com/display/trajectory_mariner_10.html

118 Messenger
NASA/Johns Hopkins University Applied Physics Laboratory/Carnegie Institution of Washington
https://www.images-assets.nasa.gov/image/PIA18145/PIA18145~orig.jpg

118 "Hollows" on Mercury
NASA/Johns Hopkins University Applied Physics Laboratory/Carnegie Institution of Washington
http://www.planetary.org/multimedia/space-images/mercury/hollows-and-pits-in-lermontov.html

119 Artwork of Mercury and Symbol original artwork by Joseph Cowart

120 Three Craters on Mercury
NASA/Johns Hopkins University Applied Physics Laboratory/Carnegie Institution of Washington
https://en.wikipedia.org/wiki/File:PIA19421-Mercury-Craters-MunchSanderPoe-20150416.jpg

121 Caloris Basin (false color)
NASA/Johns Hopkins University Applied Physics Laboratory/Carnegie Institution of Washington
https://images-assets.nasa.gov/image/PIA10359/PIA10359~small.jpg

121 Weird Terrain NASA/Johns Hopkins University Applied Physics Laboratory/Carnegie Institution of Washington
https://www.images-assets.nasa.gov/image/PIA12041/PIA12041~orig.jpg

121 Mercury Pole
NASA/Johns Hopkins University Applied Physics Laboratory/Carnegie Institution of Washington
https://www.google.com/url?sa=i&rct=j&q=&esrc=s&source=images&cd=&cad=rja&uact=8&ved=2ahUKEwiw463b6NfhAhVDTd8KHTTkA7AQjRx6BAgBEAU&url=https://en.wikipedia.org/wiki/Geology_of_Mercury&psig=AOvVaw0A8USDXbWKHcOHG0-LjHgr&ust=1555614111438215

122 Mercury, Venus, Earth, and Mars Pixabay 11060
https://pixabay.com/photos/planet-inner-planets-11060/

122 Unusual Mercury
https://pixabay.com/illustrations/mercury-stars-sun-solar-system-1226433/

Ch. 14 Venus

123 Bluish Venus nasa.gov
https://free-images.com/lg/8e80/venus_14.jpg

123 Venus and Earth Pixabay 11588
https://pixabay.com/photos/venus-earth-size-comparison-planet-11588/
126 Venera Lander free-images.com
https://free-images.com/display/venera_13.html
128 Hildr Lava River NASA/JPL
https://www.images-assets.nasa.gov/image/PIA00253/PIA00253~orig.jpg
128 7 Pancakes NASA/JPL
https://www.images-assets.nasa.gov/image/PIA00215/PIA00215~orig.jpg
129 The Tick NASA/JPL
https://www.images-assets.nasa.gov/image/PIA00089/PIA00089~medium.jpg
129 Tesserae NASA/JPL
http://photojournal.jpl.nasa.gov/catalog/PIA00085
129 Maat Mons NASA/JPL
https://pixabay.com/photos/maat-mons-venus-planet-surface-11584/
133 Whitish Venus
https://www.images-assets.nasa.gov/image/ARC-1980-A78-9090/ARC-1980-A78-9090~orig.jpg

Ch. 15 Challenges of Interplanetary Exploration

135 Spacecraft NASA/JPL-Caltech
https://www.images-assets.nasa.gov/image/PIA18017/PIA18017~orig.jpg
142 Rocket original artwork by Joseph Cowart
146 Martian Habitat
https://www.wpclipart.com/space/illustrations/space_illustrations_2/growing_vegatables_on_Mars.jpg.html

Ch. 16 Mars

149 Mars Pixabay 11013
https://pixabay.com/photos/mars-red-planet-planet-starry-sky-11013/
150 Phobos Pixabay 11017
https://pixabay.com/photos/phobos-moon-mars-i-11017/
152 Syrtis Major NASA/JPL/ASU NASA ID: PIA16437
https://www.images-assets.nasa.gov/image/PIA16437/PIA16437~orig.jpg
152 Polar region of Mars showing ice NASA/JPL/MSSS

https://www.wpclipart.com/space/solar_system/Mars/Mars_north_pole.jpg.html
153 Olympus Mons with New Mexico picture of Olympus Mons credit: NASA/JPL; graphic by Joel Cardenas
https://www.images-assets.nasa.gov/image/PIA02982/PIA02982~orig.jpg
155 Sojourner on Martian surface NASA/JPL
https://www.images-assets.nasa.gov/image/PIA01551/PIA01551~orig.jpg
155 Curiosity NASA/JPL-Caltech
https://www.images-assets.nasa.gov/image/PIA15960/PIA15960~small.jpg

156 Three Rovers in the Desert NASA/JPL-Caltech
https://www.google.com/url?sa=i&rct=j&q=&esrc=s&source=images&cd=&cad=rja&uact=8&ved=2ahUKEwiwgamI9oThAhUCAogKHVEFANcQjRx6BAgBEAU&url=https://www.slashgear.com/nasas-curiosity-is-biggest-mars-mission-yet-in-more-ways-than-one-06241771/&psig=AOvVaw2mFnvPpygvYkENbRgWvpm9&ust=1552765860634053

156 INSIGHT Lander NASA/JPL-Caltech
https://www.images-assets.nasa.gov/image/PIA19811/PIA19811~orig.jpg

157 Map of Mars free-images.com
https://free-images.com/display/marstopomap_pia02031_modest_cropped.html

Ch. 17 Asteroids

159 Earth with Approaching Asteroid
https://cdn.pixabay.com/photo/2018/05/03/13/31/climbing-3371341_960_720.jpg

160 Solar System with Asteroid Belt
https://www.wpclipart.com/space/asteroid/asteroid_belt_inner_solar_system.png.html

163 Space-Wheel Habitat
https://www.images-assets.nasa.gov/image/ARC-1975-AC75-1086-1/ARC-1975-AC75-1086-1~orig.jpg

164 Torino scale chart Graphic by Christopher Myers

165 PG boundary Kirk Johnson, Denver Museum of Nature & Science
https://nsf.gov/news/mmg/media/images/chicxulub2_f.jpg

167 Chelyabinsk Meteor Creative Commons Attribution 3.0 Unported license.
https://en.wikipedia.org/wiki/File:Взрыв_метеорита_над_Челябинском_15_02_2013_avi-iCawTYPtehk.ogv

167 Chunk of Chelyabinsk Meteor free-images.com
https://free-images.com/display/chelyabinsk_16x9.html

167 Chunk of Chelyabinsk Meteor Pixabay 184576
https://pixabay.com/photos/chelyabinsk-meteorite-shooting-star-184576/

168 Hoba Meteorite Pixabay 700836
https://pixabay.com/photos/meteorite-stone-hoba-namibia-700836/

170 Ceres NASA/JPL-Caltech/UCLA/MPS/DLR/IDA
https://www.images-assets.nasa.gov/image/PIA21079/PIA21079~orig.jpg

173 Asteroid streaking toward Earth
https://pixabay.com/illustrations/armageddon-apocalypse-earth-2104385/

Ch. 18 Comets

175 Comet Hyakutake free-images.com
https://free-images.com/display/comet_hyakutake_space_cosmos.html

176 String of pearls
https://www.images-assets.nasa.gov/image/ARC-1994-AC94-0353-2C/ARC-1994-AC94-0353-2C~orig.jpg

177 Tunguska trees
http://thenightsky.org/tunguska.html
179 Comet orbits graphic by Joel Cardenas
182 Artists' conception of Giotto
Creative Commons Attribution-Share Alike 3.0 Unported license.
https://upload.wikimedia.org/wikipedia/commons/8/88/Giotto_spacecraft.jpg
183 Coma of Halley's Comet NASA/ESA/Giotto Project

https://www.images-assets.nasa.gov/image/PIA17485/PIA17485~orig.jpg
183 Stardust collector (tennis racket)NASA/JPL
https://www.images-assets.nasa.gov/image/PIA03188/PIA03188~orig.jpg
183 Samples collected by Stardust NASA/JPL-Caltech/University of Washington
https://www.images-assets.nasa.gov/image/PIA02188/PIA02188~orig.jpg
184 Stardust probe NASA/JPL
https://www.images-assets.nasa.gov/image/PIA03183/PIA03183~orig.jpg
184 Workers with Stardust
https://www.images-assets.nasa.gov/image/KSC-98pc1870/KSC-98pc1870~orig.jpg
184 Hartley 2 NASA/JPL-Caltech/UMD
https://www.images-assets.nasa.gov/image/PIA13579/PIA13579~orig.jpg
185 Rosetta
https://www.wpclipart.com/space/ships/Rosetta/Rosetta_comet_chaser.jpg.html
185 Comet 67/P ESA/Rosetta/NAVCAM
https://www.images-assets.nasa.gov/image/PIA18899/PIA18899~orig.jpg
187 Mark Twain Pixabay1602117
https://pixabay.com/illustrations/mark-twain-american-author-writer-1602117/
187 Halley's Comet in 1910
https://free-images.com/display/halleys_comet_1910.html

Ch. 19 Jupiter

189 Jupiter compared to the Earth Pixabay11617
https://pixabay.com/photos/jupiter-planet-earth-11617/
192 Juno being built NASA/Kim Shiflett
https://www.images-assets.nasa.gov/image/KSC-2011-4633/KSC-2011-4633~orig.jpg
192 Juno probe artwork Pixabay1700974
https://cdn.pixabay.com/photo/2016/09/28/17/16/jupiter1700974_960_720.jpg
192 Jupiter with Juno in foreground
NASA/JPL-Caltech/SwRI/MSSS/Christopher Go Pixabay1700974
https://pixabay.com/illustrations/jupiter-planet-space-probe-1700974/
193 Galileo plaque on Juno NASA/JPL-Caltech/KSC
https://www.nasa.gov/sites/default/files/images/577242main_galileo-full_full.jpg
193 Lego figures NASA/JPL-Caltech/LEGO
https://www.images-assets.nasa.gov/image/PIA14414/PIA14414~orig.jpg
195 Cutaway view of Jupiter
https://www.wpclipart.com/space/solar_system/Jupiter/Jupiter_inside.png.html

196 Galileo orbiter NASA
https://images-assets.nasa.gov/image/PIA18176/PIA18176~medium.jpg
198 Jupiter resembling Starry Night NASA/JPL-Caltech/SwRI/MSSS/Gabriel Fiset
https://www.images-assets.nasa.gov/image/PIA21390/PIA21390~orig.jpg
198 Jupiter's belts and zones graphic by Joel Cardenas
200 Great Red Spot NASA/JPL-Caltech/SwRI/MSSS/Jason Major
https://www.images-assets.nasa.gov/image/PIA21772/PIA21772~orig.jpg
201 Jupiter's magnetosphere JAXA Inset Credit: NASA, ESA, Chandra, Hubble
https://apod.nasa.gov/apod/ap160406.html
202 Diagram of Jupiter's rings NASA/JPL/Cornell University
https://www.images-assets.nasa.gov/image/PIA01627/PIA01627~orig.jpg
204 The four Galilean moons NASA/JPL/DLR

https://www.images-assets.nasa.gov/image/PIA00600/PIA00600~orig.jpg
205 Io NASA/JPL/University of Arizona
https://www.images-assets.nasa.gov/image/PIA01223/PIA01223~orig.jpg
206 Loki erupting on Io NASA/JPL/USGS
https://www.images-assets.nasa.gov/image/PIA00010/PIA00010~orig.jpg
207 Io flux tube GNU Free Documentation License
https://upload.wikimedia.org/wikipedia/commons/thumb/2/2d/Jupiter_magnetosphere_schematic.jpg/800px-Jupiter_magnetosphere_schematic.jpg
208 Europa NASA/JPL/DLR
https://www.images-assets.nasa.gov/image/PIA00502/PIA00502~orig.jpg
208 Ganymede NASA/JPL/DLR
https://www.images-assets.nasa.gov/image/PIA01666/PIA01666~orig.jpg
209 Calisto NASA/JPL/DLR
https://www.images-assets.nasa.gov/image/PIA03456/PIA03456~orig.jpg
211 Descending umbrella of lava
NASA/Johns Hopkins University Applied Physics Laboratory/Southwest Research Institute
https://images-assets.nasa.gov/image/PIA09360/PIA09360~orig.jpg
212 Io and Europa in front of Jupiter NASA/JPL
https://www.images-assets.nasa.gov/image/PIA00144/PIA00144~orig.jpg
212 Jupiter with Great Red Spot NASA/JPL-Caltech/SwRI/MSSS/Kevin M. Gill
https://www.images-assets.nasa.gov/image/PIA22946/PIA22946~orig.jpg

Ch. 20 Saturn

213 Saturn Pixabay (cropped)
https://cdn.pixabay.com/photo/2012/08/25/22/22/space-54999_960_720.jpg
216 Rings NASA/JPL/Space Science Institute
https://www.images-assets.nasa.gov/image/PIA08275/PIA08275~orig.jpg

217 Waviness in rings near Daphnis NASA/JPL-Caltech/Space Science Institute
solarsystem.nasa.gov
https://www.google.com/url?sa=i&rct=j&q=&esrc=s&source=images&cd=&cad=rja&uact=8&ved=2ahUKEwjrzcyJq9jhAhXkYt8KHe2QDNgQjRx6BAgBEAU&url=https://solarsystem.nasa.gov/resources/17604/the-realm-of-daphnis/&psig=AOvVaw2xDuenaEB5RMnGsikw353T&ust=1555632000545538

218 Spikes in rings NASA/JPL/Space Science Institute
https://www.google.com/url?sa=i&rct=j&q=&esrc=s&source=images&cd=&cad=rja&uact=8&ved=2ahUKEwjBqoS9qdjhAhXMmeAKHSFXCU0QjRx6BAgBEAU&url=https://solarsystem.nasa.gov/resources/15115/the-tallest-peaks/&psig=AOvVaw3fBBujEfZRFuNi5NS-gXbc&ust=1555631537740513

219 Titan NASA/JPL/University of Arizona/University of Idaho
https://www.images-assets.nasa.gov/image/PIA20016/PIA20016~orig.jpg

220 Landscape of Titan jpl.nasa.gov
https://www.google.com/url?sa=i&rct=j&q=&esrc=s&source=images&cd=&cad=rja&uact=8&ved=2ahUKEwjwjMnjq9jhAhUkh-AKHetGCpoQjRx6BAgBEAU&url=https://www.jpl.nasa.gov/news/news.php?feature=6718&psig=AOvVaw3ROPYlAUl1OyDMEN8yfqbk&ust=1555632171088525

221 Titan Rocks with haze
https://www.wpclipart.com/space/solar_system/Venus/Titan_by_Huygens_probe.jpg.html

222 Enceladus, showing tiger stripes solarsystem.nasa.gov
https://www.wpclipart.com/space/solar_system/moons/Saturn_moons/Saturn_moon_Enceladus.jpg.html

222 Enceladus with geyser NASA/JPL/Space Science Institute
https://www.images-assets.nasa.gov/image/PIA11688/PIA11688~orig.jpg

223 Enceladus mission
https://www.wpclipart.com/space/solar_system/moons/Jupiter_moons/Europa_cryobot_concept.jpg.html

224 Iapetus free-images.com
https://free-images.com/display/moon_saturn_iapetus_japetus.html

225 Cassini spacecraft NASA/JPL
https://www.images-assets.nasa.gov/image/PIA04603/PIA04603~orig.jpg

228 Planet-circling storm, six images NASA/JPL-Caltech/Space Science Institute
https://www.images-assets.nasa.gov/image/PIA14905/PIA14905~orig.jpg

230 The hexagon storm NASA/JPL-Caltech/Space Science Institute
https://www.images-assets.nasa.gov/image/PIA20513/PIA20513~orig.jpg

232 Saturn in infrared NASA/JPL/STScI
https://www.images-assets.nasa.gov/image/PIA01268/PIA01268~orig.jpg

232 Shadows of the rings NASA/JPL/Space Science Institute
https://www.images-assets.nasa.gov/image/PIA20500/PIA20500~orig.jpg

233 Saturn from above NASA/JPL-Caltech/Space Science Institute
https://www.images-assets.nasa.gov/image/PIA21047/PIA21047~orig.jpg

233 Inside-out view of the rings NASA/JPL-Caltech/Space Science Institute
https://www.images-assets.nasa.gov/image/PIA21627/PIA21627~orig.jpg

234 Propellers NASA/JPL-Caltech/Space Science Institute nasa.gov
https://www.google.com/url?sa=i&rct=j&q=&esrc=s&source=images&cd=&cad=rja&uact=8&ved=2ahUKEwiw8rmXqtjhAhWkiOAKHTmcAagQjRx6BAgBEAU&url=https://www.nasa.gov/image-feature/jpl/pia21433/cassini-targets-a-propeller-in-saturns-a-ring&psig=AOvVaw1gTDm6sM_mTGddrWfU9vKp&ust=1555631766958461

234 Aldebaran through the rings NASA/JPL/Space Science Institute
https://www.images-assets.nasa.gov/image/PIA08283/PIA08283~orig.jpg

235 Prometheus, Pandora, Epimetheus NASA/JPL/Space Science Institute
https://www.images-assets.nasa.gov/image/PIA12717/PIA12717~orig.jpg

236 Atlas, Daphnis, Pan NASA/JPL-Caltech/Space Science Institute
https://www.images-assets.nasa.gov/image/PIA21449/PIA21449~orig.jpg

237 Pandora NASA/JPL-Caltech/Space Science Institute
https://www.images-assets.nasa.gov/image/PIA21055/PIA21055~orig.jpg

237 Prometheus, Rhea, Janus NASA/JPL/Space Science Institute
https://www.images-assets.nasa.gov/image/PIA12652/PIA12652~orig.jpg

238 Odysseus Crater on Tethys NASA/JPL/Space Science Institute
https://www.images-assets.nasa.gov/image/PIA11572/PIA11572~orig.jpg

239 Dione from Cassini NASA/JPL-Caltech/Space Science Institute
https://www.images-assets.nasa.gov/image/PIA18327/PIA18327~orig.jpg

240 Mimas as the Death Star NASA/JPL/Space Science Institute
https://www.images-assets.nasa.gov/image/PIA06258/PIA06258~orig.jpg

241 Titan in front of Saturn NASA Goddard
https://www.images-assets.nasa.gov/image/GSFC_20171208_Archive_e000212/GSFC_20171208_Archive_e000212~orig.jpg

242 Dunes on Titan nasa.gov
https://www.google.com/url?sa=i&rct=j&q=&esrc=s&source=images&cd=&cad=rja&uact=8&ved=2ahUKEwi98qSFrdjhAhUEneAKHWOPCeIQjRx6BAgBEAU&url=https://www.nasa.gov/mission_pages/cassini/whycassini/cassinif-20080502.html&psig=AOvVaw1OKEf-WXgTEtvBh-mjFQ9m&ust=1555632510450004

Ch. 21 The Ice Giants

243 Neptune NASA/JPL
https://www.images-assets.nasa.gov/image/PIA00046/PIA00046~orig.jpg

243 Uranus NASA/JPL-Caltech
https://free-images.com/display/uranus_planet_gas_giant_0.html

244 Delicate Arch Pixabay 896885
https://pixabay.com/photos/delicate-arch-sunset-rock-formation-896885/

246 Voyager II's golden record NASA/JPL
https://free-images.com/display/space_travel_voyager_golden.html
248 Uranus showing rings
https://www.wpclipart.com/space/solar_system/Uranus/Uranus_in_visible_light.jpg.html
249 Miranda NASA/JPL/USGS
https://www.images-assets.nasa.gov/image/PIA01490/PIA01490~orig.jpg
249 Neptune and Triton NASA/JPL
https://www.images-assets.nasa.gov/image/PIA02215/PIA02215~orig.jpg
250 Statue of the god Neptune free-images.com
https://free-images.com/display/god_myth_sea_neptune.html
252 astronomer and telescope
https://upload.wikimedia.org/wikipedia/commons/1/11/Observatory_in_Alexandria_at_the_Time_of_Hipparchus.jpg

Ch. 22 Trans-Neptunian Objects

253 Pluto solarsystem.nasa.gov
https://pixabay.com/photos/pluto-dwarf-planet-kuiper-belt-nasa-1315109/
256 Comparison of Dwarf Planets NASA/JPL-Caltech
https://www.images-assets.nasa.gov/image/PIA05567/PIA05567~orig.jpg
257 Pluto from Hubble
NASA, ESA, H. Weaver JHUAPL, A. Stern SwRI, and the HST Pluto Companion Search Team
https://www.images-assets.nasa.gov/image/PIA12453/PIA12453~orig.jpg
258 New Horizons NASA, Applied Physics Laboratory
https://upload.wikimedia.org/wikipedia/commons/4/4f/New_Horizons_Transparent.png
259 Huge Volcano on Pluto
NASA/Johns Hopkins University Applied Physics Laboratory/Southwest Research Institute
https://images-assets.nasa.gov/image/PIA20155/PIA20155~orig.jpg
259 polygonal depressions on Pluto
NASA/Johns Hopkins University Applied Physics Laboratory/Southwest Research Institute
https://images-assets.nasa.gov/image/PIA20201/PIA20201~orig.jpg
260 Pluto with blue nitrogen rim
NASA/Johns Hopkins University Applied Physics Laboratory/Southwest Research Institute
https://images-assets.nasa.gov/image/PIA19964/PIA19964~orig.jpg
261 Charon
NASA/Johns Hopkins University Applied Physics Laboratory/Southwest Research Institute
https://www.images-assets.nasa.gov/image/PIA19968/PIA19968~orig.jpg
263 Pluto and Charon with Earth and Moon NASA 20150713
https://upload.wikimedia.org/wikipedia/commons/f/f2/Pluto_Charon_Moon_Earth_Comparison.png

Ch. 23 Planet Nine

265 Planet Nine artist's conception graphic by Joel Cardenas
266 Planet Nine artist's conception graphic by Joel Cardenas
268 Planet Nine artist's conception
https://www.publicdomainpictures.net/pictures/40000/velka/unknown-planet.jpg

Ch. 24 Constellations

269 Rug-like picture with Taurus Pixabay3596307
https://pixabay.com/illustrations/constellation-bull-orion-aries-3596307/
272 Orion as seen in sky Pixabay2942261
https://pixabay.com/photos/orion-night-star-constellation-2942261/
273 Orion Nebula NASA/ESA/JPL-Caltech/IRAM
https://www.images-assets.nasa.gov/image/PIA13959/PIA13959~orig.jpg
274 Orion as pointergraphic by Joel Cardenas
275 Orion as hunter artwork by Manny Cazares, graphic overlay by Joel Cardenas
276 The Big Dipper pointer stars graphic by Joel Cardenas
278 sun with prominence
https://www.images-assets.nasa.gov/image/PIA18167/PIA18167~orig.jpg

Ch. 25 Extraterrestrial Life

279 Little Green Man free-images.com
https://en.wikipedia.org/wiki/File:Marciano_Gen%C3%A9rico.JPG
289 Martian Canal Builders
https://www.wpclipart.com/space/solar_system/Mars/Martian_canal_builders.png.html
295 Magnetar
https://www.wpclipart.com/space/stars_universe/magnetar_cloudball.jpg.html

Ch. 26 Stars

297 Star with prominence Pixabay11120
https://pixabay.com/photos/sun-solar-flare-sunlight-eruption-11120/
302 Horsehead nebula free-images.com
https://free-images.com/display/horsehead_nebula_dark_nebula.html
302 Eagle Nebula free-images.com
https://free-images.com/display/eagle_nebula_ic_4703_1.html
302 North America Nebula free-images.com
https://free-images.com/display/north_america_nebula_ngc.html
303 Young Star Cluster NASA Goddard
https://www.images-assets.nasa.gov/image/GSFC_20171208_Archive_e000123/GSFC_20171208_Archive_e000123~orig.jpg

303 exploded star
NASA, ESA, and K. Noll (STScI), Acknowledgment: The Hubble Heritage Team (STScI/AURA)
https://www.images-assets.nasa.gov/image/GSFC_20171208_Archive_e000210/GSFC_20171208_Archive_e000210~orig.jpg

305 Black hole with jets
https://www.wpclipart.com/space/black_hole/black_hole_accretion_disk.jpg.html

306 Hubble galaxies
https://www.wpclipart.com/space/stars_universe/galaxy/galaxies.jpg.html

307 Andromeda and Milky Way
https://www.wpclipart.com/space/stars_universe/galaxy/Andromeda_in_sky_in_3.75_billion_years.jpg.html

307 Filaments graphic by Joel Cardenas

309 Infrared Image of Betelgeuse By: Giuseppe Donatiello using WISE & Akari data
https://www.upload.wikimedia.org/wikipedia/commons/4/44/Betelgeuse_%2830036871174%29.jpg

310 Crab Nebula NASA.gov
https://pixabay.com/go/?t=image-list-adobe&id=66848507

310 Betelgeuse with material
https://www.images-assets.nasa.gov/image/PIA16680/PIA16680~orig.jpg

311 Vega with Dust cloud images-assets NASA ID: ARC-1983-AC83-0768-4
https://images-assets.nasa.gov/image/ARC-1983-AC83-0768-4/ARC-1983-AC83-0768-4~orig.jpg

311 Polaris System NASA/ESA/HST, G. Bacon (STScI)
http://hubblesite.org/newscenter/newsdesk/archive/releases/2006/02/image/fTransferred from en.wikipedia to Commons., Neaizsargāts darbs, https://commons.wikimedia.org/w/index.php?curid=3433727

314 a Herschel telescope
https://pixabay.com/photos/telescope-herschel-telescope-11624/

Ch. 27 Gravitational Waves

315 Colliding Black Holes NASA Goddard
https://www.images-assets.nasa.gov/image/GSFC_20171208_Archive_e000415/GSFC_20171208_Archive_e000415~orig.jpg

316 LIGO arm
https://free-images.com/sm/7e1e/ligo_x_interferometer_arm.jpg

318 Virgo free-images.com
https://upload.wikimedia.org/wikipedia/commons/thumb/5/55/Virgo_aerial_view_01.jpg/800px-Virgo_aerial_view_01.jpg

319 LISA
http://en.wikipedia.org/wiki/Special:FilePath/LISA-waves.jpg
http://lisa.jpl.nasa.gov/gallery/lisa-waves.html

320 gravity waves artwork Craig Blodgett on Wolfram Mathematica and Joel Cardenas

Afterword

321 Earthrise Pixabay11014
http://commons.wikimedia.org/wiki/File:NASA-Apollo8-Dec24-Earthrise.jpg

323 Pythagoras with pyramid
https://www.wpclipart.com/famous/religion/Pythagoras/Pythagoras_by_Knapp.jpg.html

323 James Webb Telescope
https://www.images-assets.nasa.gov/image/GSFC_2017120_Archive_e000162/GSFC_20171208_Archive_e000162~orig.jpg

Back Cover Montage

Back Cover Earth
www.images-assets.nasa.gov/image/GSFC_20171208_Archive_e002131/GSFC_20171208_Archive_e002131~orig.jpg

Back Cover Crater Eratosthenes on the Moon
www.images-assets.nasa.gov/image/as17-145-22285/as17-145-22285~orig.jpg

Index

G

H

I

J

K

L

M

N

About the Author

CHARLES MICHAEL SAFFORD

Charles Michael Safford is a Nevada educator. He has been teaching science and math classes to Nevada students since 1991. For many years, he also served as principal of Pershing County's Middle School and of its combined Middle School and High School. Under his leadership, Pershing County schools received a series of grants and awards, including recognition of Pershing County's middle school as an Exemplary School by the State of Nevada's Department of Education. In 2003, the *Reno-Gazette Journal* acknowledged his contribution to the state's educational institutions, by designating him as the Outstanding K-12 Principal for Northern Nevada. For the last decade, he has been teaching science and math at Coal Canyon High School, and he is also teaching advanced mathematics for Great Basin College as an Adjunct Professor. Presently, he is a trustee of the Pershing General Hospital Board and a member of the Pershing Healthcare Foundation. *Pebbles on the Cosmic Seashore: A Lifetimes of Astronomical Discoveries*, is based upon lectures he developed for his students at Coal Canyon High School.

Elizabeth Harvey, Ph.D.
Professor of History, Merced College

www.ingramcontent.com/pod-product-compliance
Lightning Source LLC
LaVergne TN
LVHW071345210926
848486LV00003B/232
9781938814716